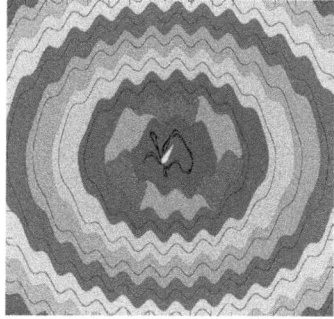

Elegant Automation

Robotic Analysis of Chaotic Systems

Elegant *Automation*

Robotic Analysis of Chaotic Systems

Julien Clinton Sprott

University of Wisconsin-Madison, USA

World Scientific

NEW JERSEY · LONDON · SINGAPORE · BEIJING · SHANGHAI · HONG KONG · TAIPEI · CHENNAI · TOKYO

Published by

World Scientific Publishing Co. Pte. Ltd.

5 Toh Tuck Link, Singapore 596224

USA office: 27 Warren Street, Suite 401-402, Hackensack, NJ 07601

UK office: 57 Shelton Street, Covent Garden, London WC2H 9HE

Library of Congress Control Number: 2023014120

British Library Cataloguing-in-Publication Data
A catalogue record for this book is available from the British Library.

ELEGANT AUTOMATION
Robotic Analysis of Chaotic Systems

ISBN 978-981-127-751-1 (hardcover)
ISBN 978-981-127-752-8 (ebook for institutions)
ISBN 978-981-127-753-5 (ebook for individuals)

For any available supplementary material, please visit
https://www.worldscientific.com/worldscibooks/10.1142/13445#t=suppl

Dedicated to the memory of John McCarthy, considered by many as the father of artificial intelligence.

John McCarthy, 1927–2011

Preface

This book was written by a machine. Well, most of it anyway. A human is writing this preface to explain how the book came about, and I also wrote Chapter 1 to lay the groundwork and explain more fully what is in the subsequent chapters, which were written almost entirely by a computer program that I have been slowly developing for the past thirty years.

I didn't plan to write another book. My last four books, *Elegant Chaos* [Sprott (2010)], *Elegant Fractals* [Sprott (2019)], *Elegant Circuits* [Sprott and Thio (2022)], and *Elegant Simulations* [Sprott *et al.* (2023)], the last two coauthored with others, were supposed to conclude my efforts to bring elegance to the scientific community. At age eighty, I hoped that others would continue the tradition with additional books in the series.

However, I had noticed for many years that some of the papers I was reviewing and reading in the literature had a certain sameness about them, and they were not always carefully written or complete. The author would say something along the lines of "here is a new chaotic system that has never been studied, and here are its properties...". I once said to my colleague Chunbiao Li, that this paper could have been written by a machine.

Some years and many such papers later, I realized that the programs I had been developing for finding and analyzing new chaotic systems could be easily extended to include the final step of writing the paper, ready for submission to the journal. And so, as a joke, I wrote such a paper, or rather I wrote a program that wrote the paper and submitted it to the *International Journal of Bifurcation and Chaos* to see if I could get it published.

To my surprise and delight, two days later, the editor informed me that he and the reviewer liked it, and it was published the following month [Sprott (2022b)]. I assumed that would be the last of it, but then Lakshmi

Narayanan, an editor at World Scientific Publishing, asked if I could make it into a short book. My first reaction was oh, no, not another book! I had been writing books continuously for most of the past decade and was still finishing *Elegant Simulations*.

With a bit more thought, I realized that it would be relatively easy to produce such a book, with each chapter written by the computer, analyzing a different chaotic system, some old and some new, with an emphasis on simple systems that I had earlier discovered or studied. In addition to the novelty of a computer-written book, there might be a market for a book that analyzes many chaotic systems in an identical fashion to facilitate comparison, perhaps as a first step in cataloging all known chaotic systems. In the process, I would have the computer simplify the systems to the extent possible, making them maximally 'elegant'.

This book is the result of that effort. It will be of interest to students and researchers for identifying and exploring new chaotic systems. The book assumes only an elementary knowledge of calculus. The systems are initial-value ordinary differential equations, but they must be solved numerically, and so a formal course in differential equations is of limited use.

You will get the most out of this book if you can write simple computer programs in the language of your choice or have access to software that allows you to solve systems of coupled ordinary differential equations and display the results graphically. The programs used in this book were written using the PowerBASIC Console Compiler on a simple Windows desktop personal computer.

There is no substitute for the thrill and insight of seeing the solution of a simple equation unfold as the trajectory wanders in real time across your computer screen using a program of your own making. A goal of this book is to inspire and delight as well as to teach. I hope you will enjoy reading and studying it as much as I did developing the program that wrote it.

I thank my many colleagues and collaborators who over the years have used my programs to find new chaotic systems and coauthor papers describing their properties. You have inspired and motivated me to continue my own studies and to produce this book.

J. C. Sprott
Madison, Wisconsin
March, 2023

Contents

Chapter 1

Introduction

This chapter will describe the computer program and methods used to write the subsequent chapters. The program identifies systems of coupled non-linear ordinary differential equations whose solutions are chaotic, simplifies them to the extent possible, analyzes them using conventional methods, produces the figures, and writes the accompanying text. Collecting some common information here minimizes repetition.

1.1 Background

The term 'artificial intelligence' was coined by John McCarthy and was first used in the title of a 1956 Dartmouth College summer workshop that many consider the birth of the field [Moor (2006)]. Despite the lofty goals and aspirations of early proponents, subsequent progress was slow although punctuated with occasional advances, and now finally seems on the brink of wide acceptance and use. Machines are increasingly assuming many intellectual tasks formerly performed by humans.

One such simple task, developed thirty years ago at a time when personal computers were still something of a novelty, was to search a class of dynamical systems to find equations whose solutions are chaotic [Sprott (1993a)]. Hundreds of new examples of chaotic systems were found in that way [Sprott (1993b)].

A slight modification of the program allowed it to search for algebraically simple examples of chaotic systems, and nineteen cases were found that were simpler than any that were previously known [Sprott (1994)] plus one case [Sprott (1997)] that is as simple as possible [Zhang and Heidel (1997)]. Hundreds of additional such minimal systems were subsequently identified [Sprott (2010)].

It was once the case that every new chaotic system that was discovered warranted a publication, and such papers are still occasionally submitted to the scientific journals. However, it is now generally agreed that new chaotic systems are publishable only if they exhibit some new phenomenon or are a simplification of the standard examples of such phenomena [Sprott (2011)]. Furthermore, such papers should include a careful and thorough analysis of the system.

The following chapters catalog fifty three-dimensional autonomous chaotic flows that have been proposed over the last half century and provide a minimal but consistent analysis of their properties in a format that facilitates easy comparison. Some of these systems are old and have been extensively studied, but others are relatively new and unfamiliar. Collectively, they exhibit most of the phenomena that have been discovered in low-dimensional chaotic flows.

The identification, simplification, and analysis of these systems is conventional and repetitious so as to be easily automated. The remainder of this chapter describes the methods used so that you can replicate them yourself.

1.2 Search Method

The search for an autonomous chaotic flow starts with a system of coupled nonlinear *ordinary differential equations* with at least three variables (x, y, z) that change in time t according to

$$\dot{x} = f(x, y, z)$$
$$\dot{y} = g(x, y, z) \tag{1.1}$$
$$\dot{z} = h(x, y, z),$$

where the overdot denotes a time derivative, $\dot{x} = dx/dt, \dots$. The system is *autonomous* because the time does not appear on the right-hand side. *Nonautonomous* systems can be made autonomous by adding a variable u with a time derivative $\dot{u} = 1.0$ and replacing the t in the equations by u.

The three functions f, g, and h are usually polynomial functions of the variables that include a number of fixed parameters (a_1, a_2, ...) that are generally but not necessarily coefficients of the various terms, at least one of which must be nonlinear. The equations have initial conditions (x_0, y_0, z_0) at $t = 0$. The search task is to find values of the parameters and initial conditions for which the solution of the system is chaotic, perhaps with some imposed constraints such as the sign or number of parameters.

The search method is relatively primitive and computationally intensive. It consists of choosing parameters and initial conditions randomly from a Gaussian distribution with mean zero and unit variance and holding their values constant while solving the equations numerically using the fourth-order *Runge–Kutta method* with an adaptive time step size [Press *et al.* (2007)] and calculating the value of the largest Lyapunov exponent (described later). The signature of chaos is taken as a decidedly positive Lyapunov exponent, generally a value greater than 0.001. It may take several to as many as thousands or even millions of trials to find a chaotic solution if one exists, and the orbit is calculated for a time of at least 1×10^4 to reduce false positives such as chaotic transients that eventually decay to a periodic or static value or escape to infinity.

1.3 Simplified System

Having found a chaotic solution, the program then centers the Gaussian distributions on the corresponding values of the parameters and initial conditions and continues searching. However, new chaotic cases are accepted only if the parameters are at least as 'simple' in the sense of having the same or fewer digits, preferably integers, and even better with values of zero or ± 1, with $+1$ preferred over -1.

In that way, perhaps after an hour or two of computation depending on the number of parameters, the size of the basin of attraction, and the robustness of the system (described later), the search will usually converge to a single or perhaps a small number of cases that are maximally simple. The resulting system is said to be 'elegant'.

Generally, the most elegant system will have four fewer independent parameters than the number of terms since the three variables plus time (x, y, z, t) can be linearly rescaled without altering the dynamics. However, in many cases, one or more of the parameters have values of ± 1. If any of the terms are not simple powers of the variables, there will be additional parameters. For example, a term such as $A \sin(kx)$ has two parameters since it has an amplitude A and wave number k. Some parameters may affect only the amplitude of one or more of the variables and thus can be set to ± 1 without loss of generality and are called *amplitude parameters*.

If there are several equally elegant cases, the one with the largest value of the Kaplan–Yorke dimension (described later) is chosen. Furthermore, the initial condition is readjusted to have simple values (usually a set of small integers) that are as close as possible to some point on the calculated orbit, while avoiding values that are at an equilibrium point or outside the basin of attraction (described later). Sometimes the search is too efficient, finding an elegant set of parameters in a tiny island of chaos remote from the chaotic mainland.

1.4   Equilibria

The analysis of a dynamical system usually begins with a calculation of the *equilibrium points* given by the values of (x, y, z) that simultaneously satisfy the equations

$$f(x, y, z) = 0$$
$$g(x, y, z) = 0 \qquad\qquad (1.2)$$
$$h(x, y, z) = 0.$$

Since the equations are nonlinear as required for chaos, the solution usually involves some trial and error and is not guaranteed to exist or that it can be found even if it does. Furthermore, it is common for a system to have multiple equilibrium points, sometimes even infinitely many, examples of which you will shortly see.

The method here takes one hundred trial values of (x, y, z) chosen randomly from a Gaussian distribution with a standard deviation of 10 and then uses the multidimensional *Newton–Raphson method* as described by Press *et al.* (2007) to home in on the equilibrium points. For the systems described in this book, it is likely but not certain that all the equilibria have been identified, except when there are infinitely many of them.

Having determined the values of (x, y, z) at the equilibrium points, it is customary to *linearize* the equations in their vicinity and calculate the *eigenvalues* from which their stability can be determined. A three-dimensional dynamical system will have three (not necessarily distinct) eigenvalues at each equilibrium point, and their values can be real, imaginary, or a complex conjugate pair. The method used here to calculate them is the *QR algorithm* from Press *et al.* (2007).

If any of the eigenvalues have a positive real part, then the equilibrium is *unstable* (small perturbations away from it will grow), and that will be the usual case for a chaotic system. However, if all the eigenvalues have a negative real part, the equilibrium is *stable*, and it is an *attractor* for initial conditions in its vicinity. If all the eigenvalues are real, the equilibrium is a *node*, and if there is a complex conjugate pair, the equilibrium is a *focus*.

If the real part of some eigenvalues is positive and others are negative, it is a *saddle point*. Saddle points are especially common in chaotic systems,

especially *saddle foci* in which the orbit approaches the equilibrium along
a unique line (the *stable manifold*) and then spirals outward from it in
a unique surface (the *unstable manifold*). If the eigenvalues are purely
imaginary, then the equilibrium is a *center*.

If the real part of all the eigenvalues is non-zero, the point is said to
be *hyperbolic*; otherwise, it is *non-hyperbolic*. If the eigenvalue with the
largest real part is zero, the linear system is *neutrally stable*, and its *nonlinear stability* requires a more complicated calculation, most easily done
by numerically following an orbit starting in its vicinity to see whether it
is attracted to or repelled by the equilibrium.

It is also instructive to calculate the *Poincaré index* (also called the
Poincaré–Hopf index) of the equilibrium point [Aleksandrov (2022)]. Imag-
ine that you walk in a closed loop (a circle for convenience) counter-
clockwise around the equilibrium point while holding a weather vane that
points in the direction of the local flow vector. The number of counter-
clockwise rotations that the weather vane makes after one circuit around
the point is the Poincaré index, which is necessarily an integer. In a two-
dimensional flow, the Poincaré index is $+1$ for a node or focus and -1 for
a simple saddle point, while larger values indicate a more complicated flow
in its vicinity. In three dimensions, the Poincaré index will depend on the
chosen plane, for consistency taken here as a surface of constant z that
intersects the equilibrium point.

A reasonable choice for initial conditions to find a chaotic attractor is the
vicinity of one of the equilibrium points (but not at the point!), especially on
its unstable manifold. If such an orbit goes to an attractor, that attractor
is said to be *self-excited*. If there are no equilibrium points for which this
occurs, the attractor is *hidden* [Leonov and Kuznetsov (2013)]. If there
are no equilibrium points or if all the equilibrium points are stable, any
coexisting attractors are necessarily hidden. Hidden attractors were once
thought to be rare and unusual, but many such examples are now known
[Jafari *et al.* (2015); Wang *et al.* (2021)], and there are several such cases
in the chapters that follow.

It is also useful to identify any symmetries that the system may have.
A system is symmetric if the sign of one or more of the variables can be
changed and the resulting equations are equivalent to the original ones. A
three-dimensional system has three types of such *involutional symmetries*.
Invariance with respect to a single variable is called *reflection symmetry*, invariance with respect to two variables is called *rotational symmetry*,
and invariance with respect to all three variables is called *inversion symmetry*. Many such examples will be shown.

A system can also be *time-reversal invariant* if the transformation $t \to -t$ leaves the system unchanged. Typically this occurs in the presence of one of the involutional symmetries and implies that the system has an attractor in both forward and reverse time, or, equivalently, that there exists a symmetric *attractor–repellor pair*.

Other types of *conditional symmetries* can also occur. For example, a system can have a *translational symmetry* if the equations are unchanged when a constant is added to one or more of the variables. Systems can also be symmetric with respect to a 90° rotation about one or more axes, and several such examples will be shown.

If the equations have a symmetry, any attractors must either share that symmetry or there must be a symmetric pair of attractors. Thus the identification of symmetries helps ensure that all the attractors for a system are found. It also portends the possibility of *spontaneous symmetry breaking* and *attractor merging* bifurcations.

1.5 Attractor

The attractor for a system is the set of points visited by the orbit in the limit as time goes to infinity. However, in the literature, it is customary to plot a finite segment of the orbit after the initial transient has decayed and call it 'the attractor'. For a three-dimensional system, the attractor can be a stable equilibrium (zero dimensions), a periodic *limit cycle* (one dimension), a quasiperiodic *torus* (two dimensions), or a chaotic *strange attractor* (a fractional dimension between two and three). Our interest here is in strange attractors, and the chapters that follow show many such examples.

Attractors only occur in *dissipative systems* in which energy is supplied usually in the form of positive feedback while it is dissipated in some form of friction. *Conservative systems* can also be chaotic in three dimensions, but they do not have attractors. Rather they have a three-dimensional *chaotic sea*, every point of which is eventually approached arbitrarily closely by the orbit. The chaotic sea will typically have embedded 'islands' of *quasiperiodicity* that lie on nested tori. Some such examples will be shown.

Since the strange attractor for a chaotic flow has a dimension greater than two, it must be viewed from several angles to get a complete visualization, and the plots show projections onto each of the three coordinate planes, along with a 'three-dimensional view', which is just a 45° rotation about the z axis. The plots also show a projection of the equilibrium points onto the various planes.

To make the plots more elegant, the orbit is shown in a rainbow of colors indicating the value of the largest (or least negative) *local Lyapunov exponent*. Red indicates the most positive values, blue the most negative values, and green indicates values close to zero. You can think of red as representing regions of poor short-term predictability and blue as regions of good predictability.

The local Lyapunov exponent is the exponential rate at which two nearby orbits are separating (or converging if negative). Its value is not usually a smooth function of position since it is determined by the past history of the orbit which can be rather different for two nearby points. It is necessary to follow the orbit for some time to allow the vector representing the direction separating the two orbits to become oriented in the direction of most rapid divergence (or least rapid convergence).

1.6   Time Series

Whereas the attractor shows the dynamics in the (x, y, z) *state space*, for many purposes, it is important to know how the variables change in time, $x(t)$, $y(t)$, $z(t)$. Such a plot is shown for each system in the chapters that follow. Each variable is displaced vertically for ease of viewing, but the horizontal dotted lines show the zero values. These plots represent a short portion of the infinitely long aperiodic orbit after any initial transient has decayed.

The time series plots also show the value of the largest local Lyapunov exponent at each instant of time. You will note that the values rapidly and repeatedly change sign, and it is seldom obvious that the average value is positive as it must be for a chaotic system. Furthermore, the plot of the local Lyapunov exponent is colored according to whether the orbits are separating fastest (or converging slowest) in a direction parallel to the orbit (red) or perpendicular to the orbit (blue), with intermediate cases shown in a rainbow of colors.

The vector direction usually changes as often and as rapidly as does its magnitude. Contrary to the way the Lyapunov exponent is usually visualized, most of the expansion and contraction is mostly in the direction parallel to the flow, and that component necessarily averages to zero since two orbits perturbed exactly in the parallel direction must arrive at their destination with nearly the same separation with which they started since they followed the same path in state space.

1.7   Lyapunov Exponents

The local Lyapunov exponent is determined by following two orbits initially separated by a small distance, typically $r_0 = 2^{-26} \approx 1.5 \times 10^{-8}$, and calculating the new separation r after one time step, typically $\Delta t = 0.01$. The Lyapunov exponent (base-e) at that time step is $\ln(r/r_0)/\Delta t$. After each time step, the separation is reset to r_0 but without altering the direction of the separation. It is necessary to follow the orbit for a while to allow the perturbation to orient itself in the direction of maximum growth, which usually happens rather quickly for a chaotic flow (in a time the order of the inverse of the positive Lyapunov exponent). Thus the local Lyapunov exponent depends on the past history of the orbit, and this fact accounts for much of the spatial structure in its value.

The *global Lyapunov exponent* is determined from a time average of the local Lyapunov exponent along the orbit. Because of the slow convergence, it is necessary to follow the orbit for a long time to get an accurate value of the global Lyapunov exponent. You can think of the positive and negative values of the local Lyapunov exponent as the head and tail of a coin that is repeatedly flipped to determine if there is a bias, and if so how large it is. Roughly speaking, to get an estimate of the bias accurate to n digits requires flipping the coin 10^{2n} times.

A three-dimensional autonomous flow actually has three Lyapunov exponents, only the largest of which is determined as just described. However, if the largest Lyapunov exponent is positive, it is easy to determine the other two. One exponent must be zero corresponding to the direction parallel to the flow since a perturbation in that direction will neither grow nor shrink on average. The third exponent must be negative for a bounded system and is easily determined from the value of the positive exponent and the sum of the three, which is given by the *trace* of the *Jacobian matrix* of partial derivatives $\partial f/\partial x + \partial g/\partial y + \partial h/\partial z$, time-averaged along the orbit if it is not constant.

The sum of the three exponents is the rate at which a small volume of initial conditions grows in time (or decays if negative). For a bounded orbit, this value cannot be positive, although it is zero for a conservative system according to *Liouville's theorem.*[Liouville (1838)] A negative sum implies the existence of an attractor of zero volume in the state space since any finite initial volume of initial conditions decays exponentially to zero. Any attractor with a dimension less than 3.0 must have zero volume in its three-dimensional state space.

Convergence of the Lyapunov exponents is tested by observing the fluctuations and drift in the running average of the local largest Lyapunov exponent, typically over a time window of 10^3. When the value is changing less than 5×10^{-5}, it is assumed to have converged to approximately four digits. Typically, this requires following the orbit for a time between 10^6 and 10^7, and it takes an hour or two of computation. The Lyapunov exponents are quoted to four digits, but the least significant digit is only an estimate.

From the three Lyapunov exponents, the *Kaplan–Yorke dimension* [Kaplan and Yorke (1979)] can be determined. Since the exponents are $(+, 0, -)$ for a three-dimensional chaotic flow, the dimension must be greater than 2 but less than or equal to 3. The fractional part is simply the value of the positive exponent divided by the magnitude of the negative exponent. This dimension is one measure of the *complexity* of the attractor (how space-filling it is).

All the strange attractors for the three-dimensional systems in this book have a Kaplan–Yorke dimension greater than 2.0 and less than 3.0, and thus they are *fractals* [Sprott (2019)] by definition. By contrast, a chaotic sea has paired positive and negative exponents that sum to zero and thus an integer dimension. Typically, the chaotic sea is a *fat fractal* [Umberger and Farmer (1985); Grebobi *et al.* (1985)] with a dimension of 3.0, but with an infinite number of ever smaller islands of quasiperiodicity. Sometimes, a system is *ergodic*, which means that the chaotic sea fills the entirety of the state space without any islands, and the search for low-dimensional chaotic systems that are fully ergodic is a topic of current research.

1.8 Basin of Attraction

The *basin of attraction* for an attractor is the set of all initial conditions that approach the attractor in the limit of $t \to \infty$. For a three-dimensional system, the basin is three-dimensional, and its volume can be finite or infinite, even the whole of the state space (*globally attracting*), and the basin must be *connected* (not consisting of multiple disjoint sets). The basin boundary may be either smooth or fractal.

Because such basins are three-dimensional and often complicated, it is hard to visualize them. Consequently, it is customary to show a two-dimensional slice through the basin. Infinitely many such slices are possible, but the cases that follow typically show the basin in the $z = 0$ plane if that plane intersects the attractor, although it is often shown in a constant-z

plane that passes through one or more of the equilibrium points, especially if they are stable and hence themselves attractors in which case they are plotted as dots in the figure.

The basin of the strange attractor or the whole of a chaotic sea is plotted in red. If there is more than one attractor, their basins are shown in contrasting colors. Initial conditions that give unbounded orbits or that lie on invariant tori are shown in white. The intersections of the attractors with the chosen plane are plotted in black. Thus you can see how close the attractor comes to the basin boundary in this typical plane and how carefully the initial conditions must be chosen to find the attractor. It is common for an attractor to come very close to its basin boundary in places because chaos often occurs just before the attractor becomes unstable with an unbounded orbit.

The scale of the plot is chosen according to the projection of the attractor onto the plane. Thus the basin typically includes regions not plotted, sometimes extending all the way to infinity, and it is seldom centered on the plot. It would be possible to replot the basin after its size and position have been determined, but this was not done in the interest of time.

Calculation of the basin is computationally intensive, since every point in the plane must be used as an initial condition and the orbit followed until it converges onto the attractor or clearly fails to do so. This is done graphically by testing each pixel in the plot and following the orbit until it overlaps one of the black pixels representing the attractor. Initial conditions near the basin boundary necessarily converge slowly. Such plots usually take several hours of computation and are accurate to the precision of the plot (typically 800 × 800 pixels).

1.9   Bifurcations

All the previous calculations involved a particular elegant choice of the parameters. It is also of interest to see how the dynamical behavior changes as the parameters are varied. A thorough study would involve testing the entire parameter space, which is computationally impractical if there are more than about two parameters. Consequently, the following chapters show only a variation of the single parameter a over the range of zero to twice its nominal value, which is thus at the center of the plot.

The most revealing dynamical signature is the spectrum of Lyapunov exponents. Unfortunately, the simple method described above for calculating all three exponents only works when the largest exponent is greater than zero. Consequently, the method used here follows Wolf *et al.* (1985) except that the Jacobian matrix of partial derivatives is calculated numerically rather than analytically since it is easier to automate and less prone to errors. Note that the Wolf *et al.* (1985) paper also contains a program for calculating the Lyapunov exponents from a time series, but that method is not appropriate or recommended for cases such as these where the dynamical equations are known.

The Lyapunov exponents show clearly the regions of chaos $(+, 0, -)$, quasiperiodicity $(0, 0, -)$, periodicity $(0, -, -)$, and stable equilibria $(-, -, -)$, as well as many of the *bifurcation points* where the dynamics change abruptly. Also shown is the Kaplan–Yorke dimension as determined from the Lyapunov exponent spectrum, and the local maxima of the x variable. The common *period-doubling route* to chaos is evident in some of the plots, but there are many other unusual bifurcations that beg for further examination.

For some of the cases, the initial conditions are taken at the maximum value of a (twice its nominal value) and not changed as a is gradually reduced to zero. For other cases, that method does not lead to the attractor, and the initial condition is reset at each value of a. Whenever the orbit is unbounded, nothing is plotted, and the value of a is decremented with the initial condition reset. Typically the orbits are calculated for a time of at least 1×10^4 for each of the 500 values of a, which requires several hours of computation.

1.10   Robustness

In a strict mathematical sense, a dynamical system is *robust* if all small changes in the parameters do not cause a qualitative change in the dynamical behavior. Otherwise, it is *fragile*. Thus robustness is a binary distinction and difficult to prove. However, for practical purposes, it is useful to have a relative measure of the robustness. One such definition is the amount (in percent) by which the parameters can be changed before the probability that the chaos is lost rises to 50%. The program used to calculate this quantity is given by Sprott (2022).

The calculation does not give a unique value since it depends on which parameters are varied and what initial conditions are used. The cases in the following chapters vary all the parameters that are not ± 1 and use the default fixed initial conditions. Nevertheless, the values are useful for comparison with what is shown in the bifurcation plots and with the other systems analyzed in the same way. Typically, the calculation considers a few thousand cases and requires about an hour of computation.

1.11 Conclusions

The program described here falls far short of anything that might be considered 'artificial intelligence'. All the program does is to fill in the blanks of some pre-written text with automatically calculated numerical values and blindly produces the figures. A seasoned researcher would surely notice many curious and unusual features in the plots that require further examination and discussion as would be expected in a research publication.

Some of these features may be numerical artifacts from the inevitable approximations and the trade-off between accuracy and speed. Most chapters required about a day of computation. Other features may be new and noteworthy. They are left here 'as is' to preserve the purity of the machine writing and to leave you the reader with a challenge and inspiration for your own investigation and publications until the day arrives that machines are sufficiently intelligent to do this too.

Chapter 2

JCS-08-13-2022 System

The system with the odd name JCS-08-13-2022 is the one that motivated this book. The entire research was done, the paper was written, and it was submitted to the journal in one 24-hour period on August 13, 2022. This system is of no particular interest except that it serves as a simple illustrative example of the method to be applied throughout the book.

2.1 Introduction

To illustrate the method, consider the dynamical system [Sprott (2022b)] given by

$$
\begin{aligned}
\dot{x} &= a_1 x + a_2 y + a_3 z \\
\dot{y} &= a_4 xy + a_5 xz + a_6 yz \\
\dot{z} &= a_7 x^3 + a_8 y^3 + a_9 z^3.
\end{aligned}
\tag{2.1}
$$

This system was chosen to be sufficiently complicated as not to have been previously studied but simple enough to allow easy analysis. All the linear terms are in the $\dot{x}$ equation, all the quadratic terms are in the $\dot{y}$ equation, and all the cubic terms are in the $\dot{z}$ equation to avoid degeneracies when interchanging variables.

Each term has a coefficient, labeled a_1 through a_9, and these are the parameters that are to be simplified by setting as many to zero as possible and as many of those that remain to ± 1, with a preference for $+1$, while keeping the largest Lyapunov exponent greater than zero and the sum of the Lyapunov exponents less than zero. This ensures that the system is maximally elegant with chaotic dynamics on a strange attractor.

The following sections were written by the computer program that performed the optimization, carried out the analysis of the resulting system, and produced the corresponding figures, all without human intervention.

2.2 Simplified System

After about 5×10^6 trials, of which 48 were chaotic, simplified parameters for Eq. (2.1) that give chaotic solutions are $a_1 = 0$, $a_2 = 0$, $a_3 = 1$, $a_4 = 1$, $a_5 = 1$, $a_6 = 0$, $a_7 = -0.1$, $a_8 = 1$, $a_9 = 0$. Thus Eq. (2.1) can be written more compactly as

$$\begin{aligned}
\dot{x} &= z \\
\dot{y} &= xy + xz \\
\dot{z} &= ax^3 + y^3
\end{aligned} \tag{2.2}$$

where $a = -0.1$.

Note that with five terms, Eq. (2.2) should have one independent parameter through a linear rescaling of the three variables plus time, and so the dynamics is completely captured by the single parameter a, which could be put in any of the five terms, albeit with a different numerical value.

2.3 Equilibria

The system in Eq. (2.2) with $a = -0.1$ has a neutrally stable non-hyperbolic equilibrium at $(0, 0, 0)$ with eigenvalues $(0, 0, 0)$ and a Poincaré index of 0 in the $z = 0$ plane, and the attractor is self-excited. Despite the neutral linear stability, the equilibrium is nonlinearly unstable. The system has no symmetry.

2.4 Attractor

Figure 2.1 shows various views of the attractor for Eq. (2.2) with $a = -0.1$ and initial conditions $(-1, 0, 0)$. The rainbow of colors shows the local value of the largest Lyapunov exponent with red indicating the most positive values (regions of worst predictability) and blue indicating the most negative values (regions of best predictability).

2.5 Time Series

Figure 2.2 shows the time series for the three variables along with the local value of the largest Lyapunov exponent (LL) for Eq. (2.2) with $a = -0.1$. Red color in the Lyapunov exponent indicates that the error is growing parallel to the orbit, while blue indicates growth perpendicular to the orbit.

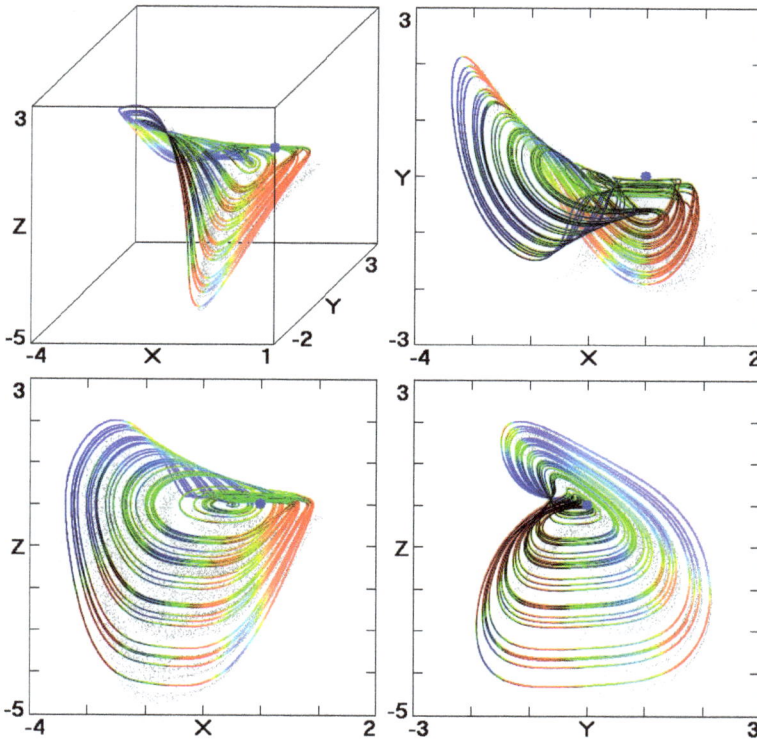

Fig. 2.1   Views of the attractor for Eq. (2.2) with $a = -0.1$ and initial conditions $(-1, 0, 0)$.

Note that the orbit passes through regions where the local Lyapunov exponent is strongly positive and other regions where it is strongly negative as is typical for a chaotic system and is also reflected by the colors in Fig. 2.1.

2.6   Lyapunov Exponents

The global Lyapunov exponents are determined by averaging the local Lyapunov exponents along the orbit. The values typically converge very slowly because of the large variation along the orbit, and an integration time of order 10^8 is required to obtain 4-digit accuracy.

The results of such a calculation for the system in Eq. (2.2) with $a = -0.1$ after a time of 2×10^6 are LE = (0.0249 , 0, -0.4182) with a Kaplan–Yorke dimension of 2.0595, where the last digit in the quoted values is only

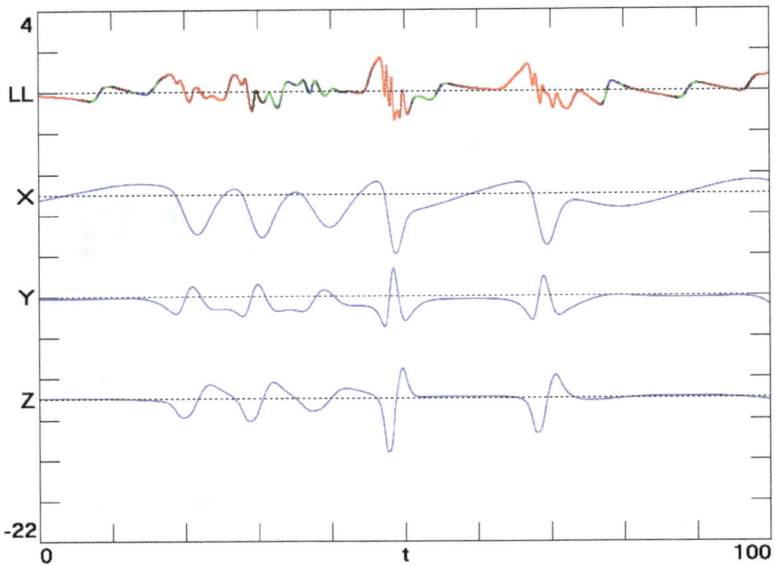

Fig. 2.2 Time series for the variables in Eq. (2.2) with $a = -0.1$ along with the local Lyapunov exponent (LL).

an approximation. The positive value of the largest Lyapunov exponent indicates that the system is chaotic, and the negative sum of the exponents (-0.3933) indicates that the system is dissipative with a strange attractor.

2.7 Basin of Attraction

Figure 2.3 shows (in red) the basin of attraction for Eq. (2.2) with $a = -0.1$ in the $z = 0$ plane. Also shown (in black) is the cross-section of the attractor in the same plane.

2.8 Bifurcations

Figure 2.4 shows the bifurcation diagram for Eq. (2.2) as a function of the parameter a from $a = -0.2$ to $a = 0$. The initial condition was taken as $(-1, 0, 0)$ at $a = -0.2$ and was not changed as a slowly varied toward $a = 0$. Each of the 500 values of a was calculated for a time of 1×10^4.

The upper plot shows the three Lyapunov exponents. The middle plot shows the Kaplan–Yorke dimension, and the lower plot shows the local

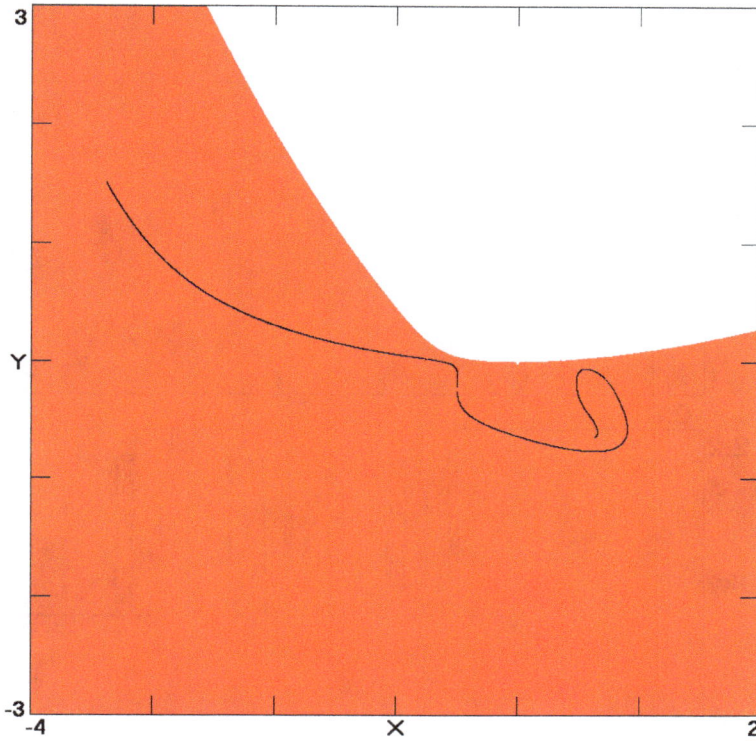

Fig. 2.3 Basin of attraction for Eq. (2.2) with $a = -0.1$ in the $z = 0$ plane.

maxima of x. The chaotic region is in the vicinity of $a = -0.1$, and the route to chaos is clearly shown.

2.9 Robustness

One measure of the robustness of a chaotic system is the amount by which the parameters can be changed from their nominal values before the probability of chaos decreases to 50% [Sprott (2022)]. For the system in Eq. (2.2) with $a = -0.1$ and initial conditions $(-1, 0, 0)$, after 5001 trials, it is estimated that the parameter a can be changed by 48% before the chaos is more likely to be lost than not. Thus the system is somewhat robust. This result is consistent with the data in Fig. 2.4.

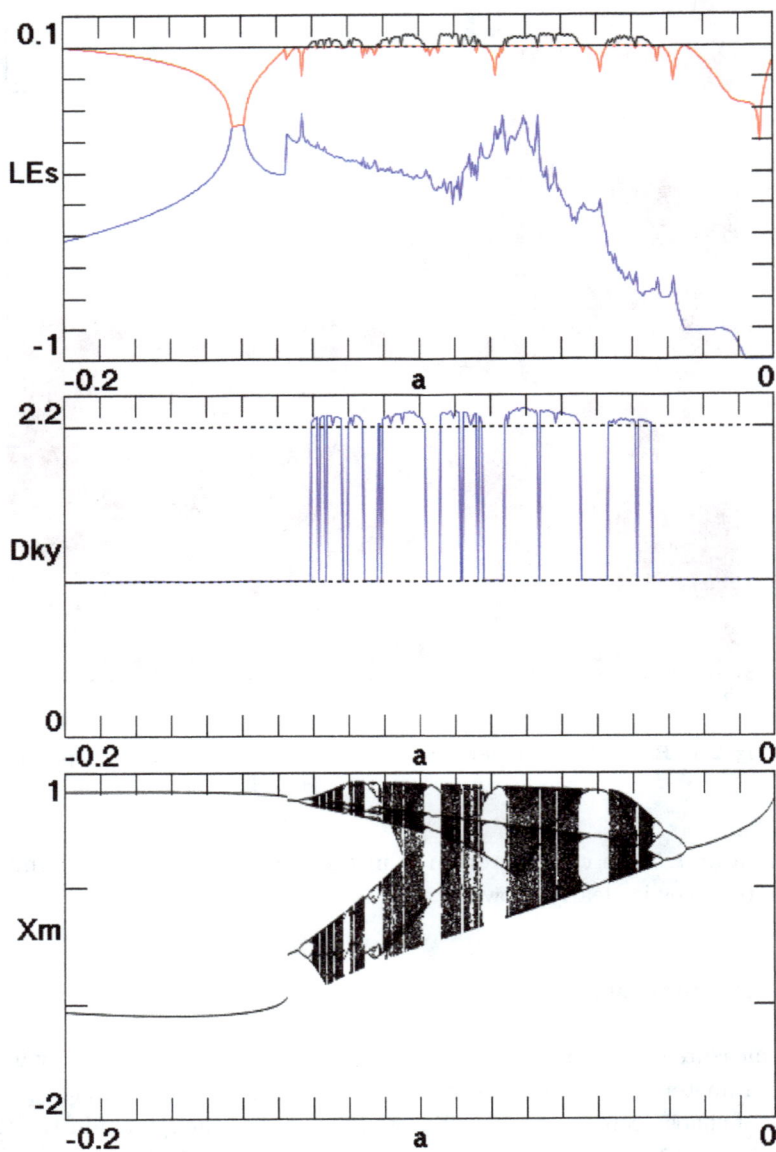

Fig. 2.4   Bifurcation diagram for Eq. (2.2) as a function of the parameter a.

Chapter 3

Lorenz System

The Lorenz system is the oldest and most famous example of an autonomous chaotic system. Despite being extensively studied, apparently no one had attempted to simplify the parameters to the extent possible. The resulting simplification gives a strange attractor coexisting with two stable equilibrium points. It is symmetric with respect to a 180° rotation about the z axis.

3.1 Introduction

The Lorenz (1963) system written in its most general form with an adjustable coefficient in each of the seven terms is given by

$$\dot{x} = a_1 y - a_2 x$$
$$\dot{y} = -a_3 xz + a_4 x - a_5 y \qquad (3.1)$$
$$\dot{z} = a_6 xy - a_7 z.$$

The usual parameters are $a_1 = a_2 = 10$, $a_3 = a_5 = a_6 = 1$, $a_4 = 28$, and $a_7 = 8/3$. It is known that chaotic solutions exist for $a_5 = 0$ [Zhou et al. (2008)], but other simplifications are possible [Sprott (2009)].

The following sections were written by the computer program that performed the optimization, carried out the analysis of the resulting system, and produced the corresponding figures, all without human intervention.

3.2 Simplified System

After about 4×10^5 trials, of which 309 were chaotic, simplified parameters for Eq. (3.1) that give chaotic solutions are $a_1 = 1$, $a_2 = 1$, $a_3 = 1$, $a_4 = 1$, $a_5 = 0$, $a_6 = 1$, $a_7 = 0.08$. Thus Eq. (3.1) can be written more compactly

as

$$\dot{x} = y - x$$
$$\dot{y} = xz + x \tag{3.2}$$
$$\dot{z} = xy - az,$$

where $a = 0.08$.

Note that with six terms, Eq. (3.2) should have two independent parameters through a linear rescaling of the three variables plus time. However, one of the two parameters has a value of ± 1 and can be placed in any of the remaining five terms.

3.3   Equilibria

The system in Eq. (3.2) with $a = 0.08$ has three equilibrium points:

Equilibrium # 1 is an unstable saddle node at $(0, 0, 0)$ with eigenvalues $(-1.618, 0.618, -0.08)$ and a Poincaré index of -1 in the $z = 0$ plane.

Equilibrium # 2 is a stable focus at $(0.2828, 0.2828, 1)$ with eigenvalues $(-1.0702, -0.0049 - 0.3866i, -0.0049 + 0.3866i)$ and a Poincaré index of 0 in the $z = 1$ plane.

Equilibrium # 3 is a stable focus at $(-0.2828, -0.2828, 1)$ with eigenvalues $(-1.0702, -0.0049 - 0.3866i, -0.0049 + 0.3866i)$ and a Poincaré index of 0 in the $z = 1$ plane.

The strange attractor is self-excited, and the system is symmetric under the transformation $x \rightarrow -x$, $y \rightarrow -y$.

3.4   Attractor

Figure 3.1 shows various views of the attractor for Eq. (3.2) with $a = 0.08$ and initial conditions $(-2, 2, 2)$. The rainbow of colors shows the local value of the largest Lyapunov exponent with red indicating the most positive values (regions of worst predictability) and blue indicating the most negative values (regions of best predictability).

3.5   Time Series

Figure 3.2 shows the time series for the three variables along with the local value of the largest Lyapunov exponent (LL) for Eq. (3.2) with $a = 0.08$. Red color in the Lyapunov exponent indicates that the error is growing parallel to the orbit, while blue indicates growth perpendicular to the orbit.

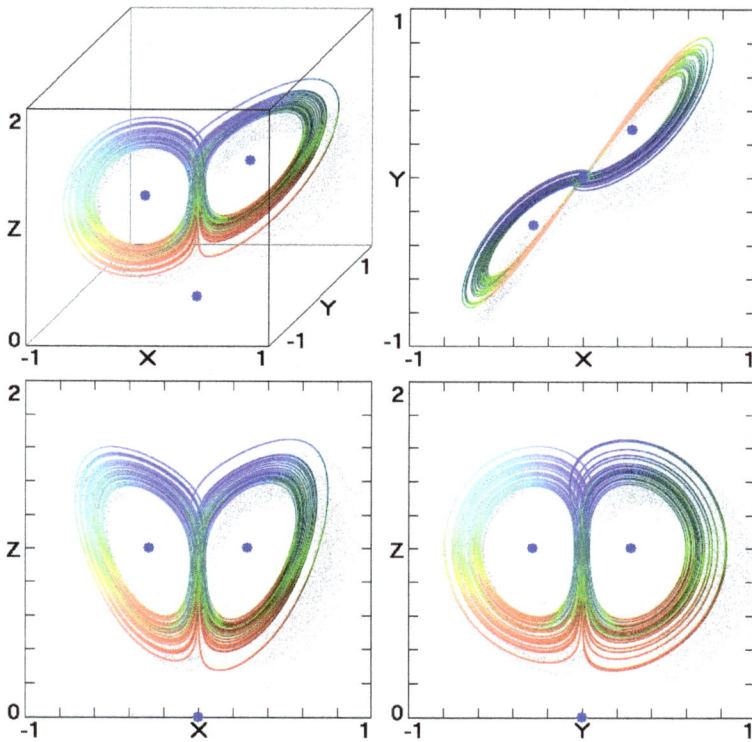

Fig. 3.1   Views of the attractor for Eq. (3.2) with $a = 0.08$ and initial conditions $(-2, 2, 2)$.

Note that the orbit passes through regions where the local Lyapunov exponent is strongly positive and other regions where it is strongly negative as is typical for a chaotic system and is also reflected by the colors in Fig. 3.1.

3.6   Lyapunov Exponents

The global Lyapunov exponents are determined by averaging the local Lyapunov exponents along the orbit. The values typically converge very slowly because of the large variation along the orbit, and an integration time of order 10^8 is required to obtain 4-digit accuracy.

The results of such a calculation for the system in Eq. (3.2) with $a = 0.08$ after a time of 6×10^3 are LE $= (0.0296, 0, -1.1096)$ with a Kaplan–Yorke dimension of 2.0266, where the last digit in the quoted values is only

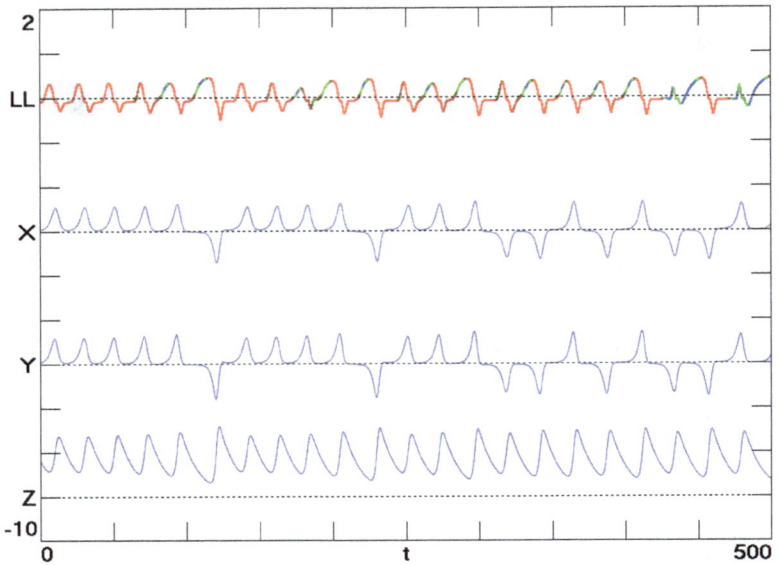

Fig. 3.2 Time series for the variables in Eq. (3.2) with $a = 0.08$ along with the local Lyapunov exponent (LL).

an approximation. The positive value of the largest Lyapunov exponent indicates that the system is chaotic, and the negative sum of the exponents (-1.0800) indicates that the system is dissipative with a strange attractor.

3.7 Basin of Attraction

Figure 3.3 shows (in red) the basin of attraction for Eq. (3.2) with $a = 0.08$ in the $z = 1$ plane. Also shown (in black) is the cross-section of the attractor in the same plane.

3.8 Bifurcations

Figure 3.4 shows the bifurcation diagram for Eq. (3.2) as a function of the parameter a from 0 to 0.16. The initial condition was taken as $(-2, 2, 2)$ for each value of a. Each of the 500 values of a was calculated for a time of about 5×10^4.

The upper plot shows the three Lyapunov exponents. The middle plot shows the Kaplan–Yorke dimension, and the lower plot shows the local

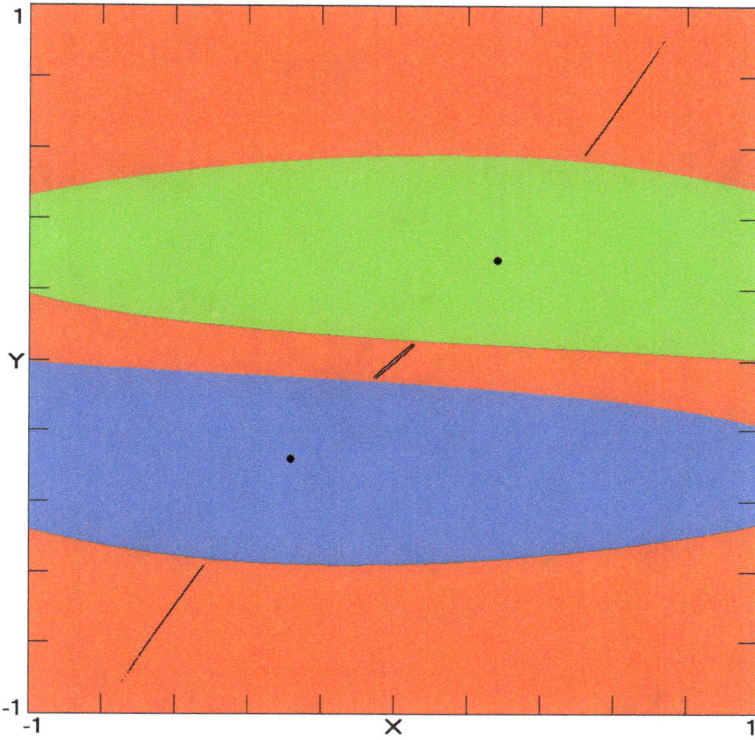

Fig. 3.3 Basin of attraction for Eq. (3.2) with $a = 0.08$ in the $z = 1$ plane.

maxima of x. The chaotic region is in the vicinity of $a = 0.08$, and the route to chaos is clearly shown.

3.9 Robustness

One measure of the robustness of a chaotic system is the amount by which the parameters can be changed from their nominal values before the probability of chaos decreases to 50% [Sprott (2022)]. For the system in Eq. (3.2) with $a = 0.08$ and initial conditions $(-2, 2, 2)$, after 3858 trials, it is estimated that the parameter a can be changed by 48% before the chaos is more likely to be lost than not. Thus the system is somewhat robust. This result is consistent with the data in Fig. 3.4.

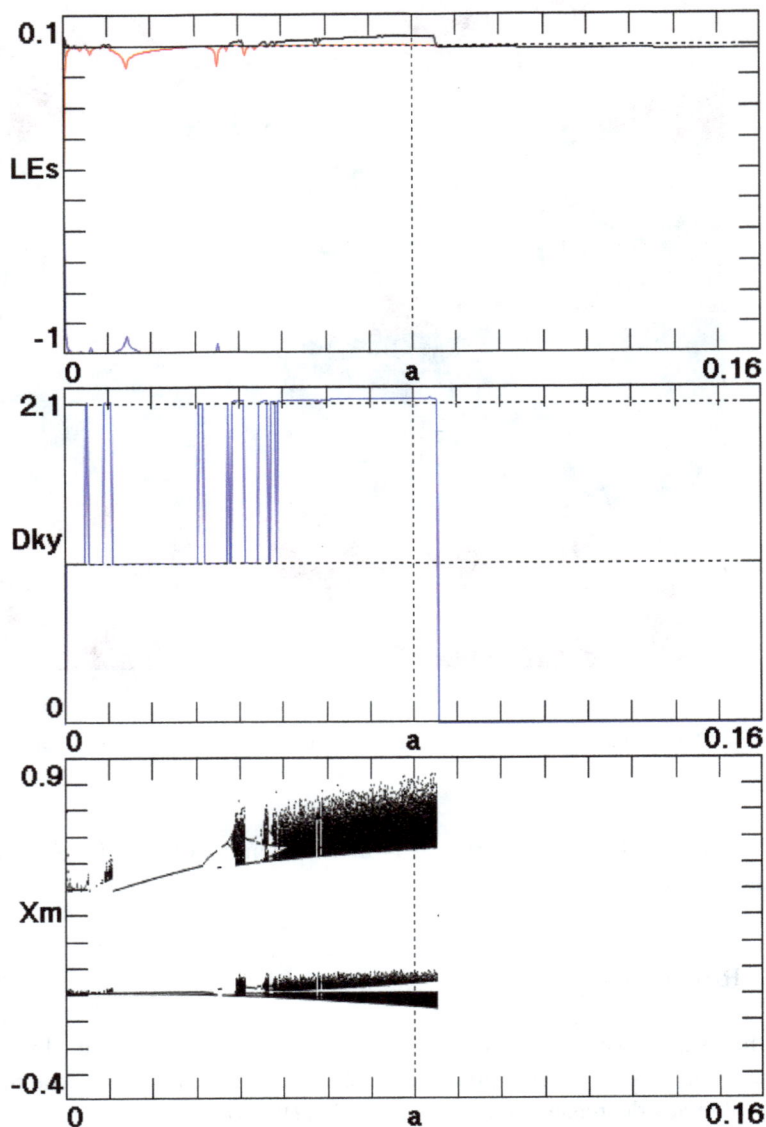

Fig. 3.4   Bifurcation diagram for Eq. (3.2) as a function of the parameter a.

Chapter 4

Rössler System

The Rössler system is nearly as old and well-known as the Lorenz system. It has the same number of terms as the Lorenz system but only a single nonlinearity. Thus it has been widely used as a simple example of chaos.

4.1 Introduction

The Rössler (1976) system written in its most general form with an adjustable coefficient in each of the seven terms is given by

$$\dot{x} = -a_1 y - a_2 z$$
$$\dot{y} = a_3 x + a_4 y \tag{4.1}$$
$$\dot{z} = a_5 + a_6 xz - a_7 z.$$

The usual parameters are $a_1 = a_2 = a_3 = a_6 = 1$, $a_4 = a_5 = 0.2$, and $a_7 = 5.7$.

The following sections were written by the computer program that performed the optimization, carried out the analysis of the resulting system, and produced the corresponding figures, all without human intervention.

4.2 Simplified System

After about 8×10^5 trials, of which 208 were chaotic, simplified parameters for Eq. (4.1) that give chaotic solutions are $a_1 = 1$, $a_2 = 1$, $a_3 = 1$, $a_4 = 0.5$, $a_5 = 1$, $a_6 = 1$, $a_7 = 3$. Thus Eq. (4.1) can be written more compactly as

$$\dot{x} = -y - z$$
$$\dot{y} = x + ay \tag{4.2}$$
$$\dot{z} = 1 + xz - bz,$$

where $a = 0.5$, $b = 3$.

Note that with seven terms, Eq. (4.2) should have three independent parameters through a linear rescaling of the three variables plus time. However, one of the three parameters has a value of ± 1 and can be placed in any of the remaining five terms.

4.3   Equilibria

The system in Eq. (4.2) with $a = 0.5$, $b = 3$ has two equilibrium points:

Equilibrium # 1 is an unstable saddle focus at $(0.1771, -0.3542, 0.3542)$ with eigenvalues $(0.1913 - 0.9702i, 0.1913 + 0.9702i, -2.7055)$ and a Poincaré index of 1 in the $z = 0.3542$ plane.

Equilibrium # 2 is an unstable saddle focus at $(2.8229, -5.6458, 5.6458)$ with eigenvalues $(-0.0393 - 2.5666i, -0.0393 + 2.5666i, 0.4016)$ and a Poincaré index of 1 in the $z = 5.6458$ plane.

The strange attractor is self-excited, and the system has no symmetry.

4.4   Attractor

Figure 4.1 shows various views of the attractor for Eq. (4.2) with $a = 0.5$, $b = 3$ and initial conditions $(3, -2, 1)$. The rainbow of colors shows the local value of the largest Lyapunov exponent with red indicating the most positive values (regions of worst predictability) and blue indicating the most negative values (regions of best predictability).

4.5   Time Series

Figure 4.2 shows the time series for the three variables along with the local value of the largest Lyapunov exponent (LL) for Eq. (4.2) with $a = 0.5$, $b = 3$. Red color in the Lyapunov exponent indicates that the error is growing parallel to the orbit, while blue indicates growth perpendicular to the orbit. Note that the orbit passes through regions where the local Lyapunov exponent is strongly positive and other regions where it is strongly negative as is typical for a chaotic system and is also reflected by the colors in Fig. 4.1.

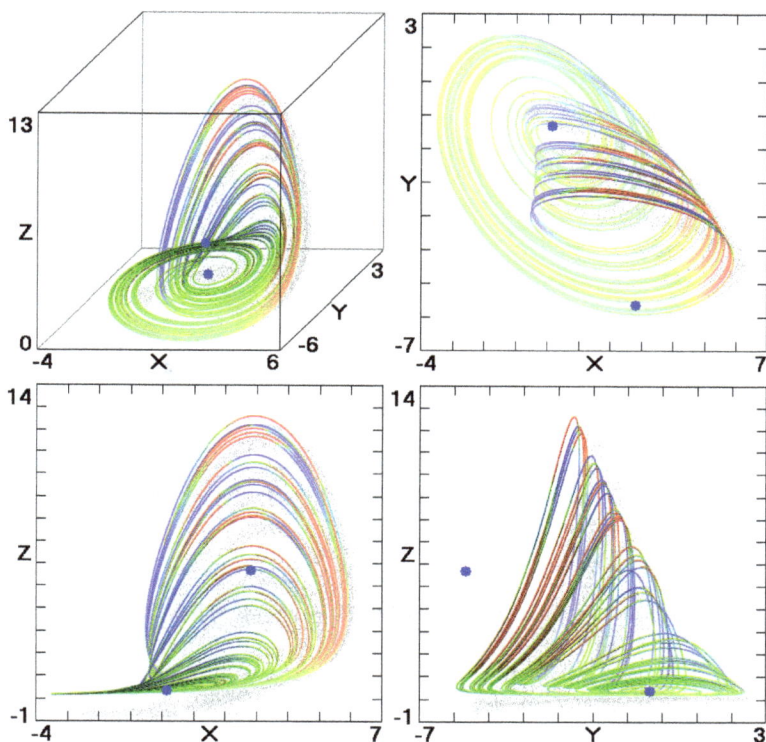

Fig. 4.1 Views of the attractor for Eq. (4.2) with $a = 0.5$, $b = 3$ and initial conditions $(3, -2, 1)$.

4.6 Lyapunov Exponents

The global Lyapunov exponents are determined by averaging the local Lyapunov exponents along the orbit. The values typically converge very slowly because of the large variation along the orbit, and an integration time of order 10^8 is required to obtain 4-digit accuracy.

The results of such a calculation for the system in Eq. (4.2) with $a = 0.5$, $b = 3$ after a time of 2×10^6 are LE = $(0.1174, 0, -1.9381)$ with a Kaplan–Yorke dimension of 2.0605, where the last digit in the quoted values is only an approximation. The positive value of the largest Lyapunov exponent indicates that the system is chaotic, and the negative sum of the exponents (-1.8207) indicates that the system is dissipative with a strange attractor.

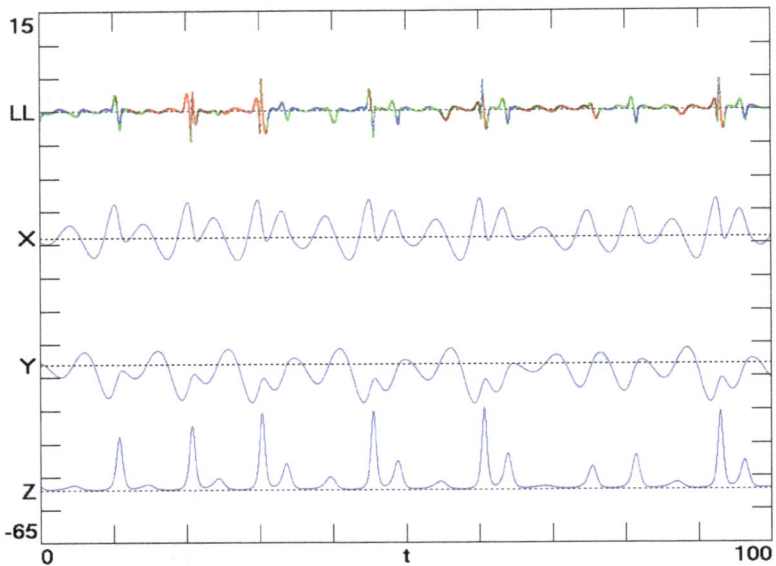

Fig. 4.2 Time series for the variables in Eq. (4.2) with $a = 0.5$, $b = 3$ along with the local Lyapunov exponent (LL).

4.7 Basin of Attraction

Figure 4.3 shows (in red) the basin of attraction for Eq. (4.2) with $a = 0.5$, $b = 3$ in the $z = 5.6458$ plane. Also shown (in black) is the cross-section of the attractor in the same plane.

4.8 Bifurcations

Figure 4.4 shows the bifurcation diagram for Eq. (4.2) as a function of the parameter a from 0 to 1 for $b = 3$. The initial condition was taken as (3, -2, 1) for each value of a. Each of the 500 values of a was calculated for a time of about 2×10^4.

The upper plot shows the three Lyapunov exponents. The middle plot shows the Kaplan–Yorke dimension, and the lower plot shows the local maxima of x. The chaotic region is in the vicinity of $a = 0.5$, $b = 3$, and the route to chaos is clearly shown.

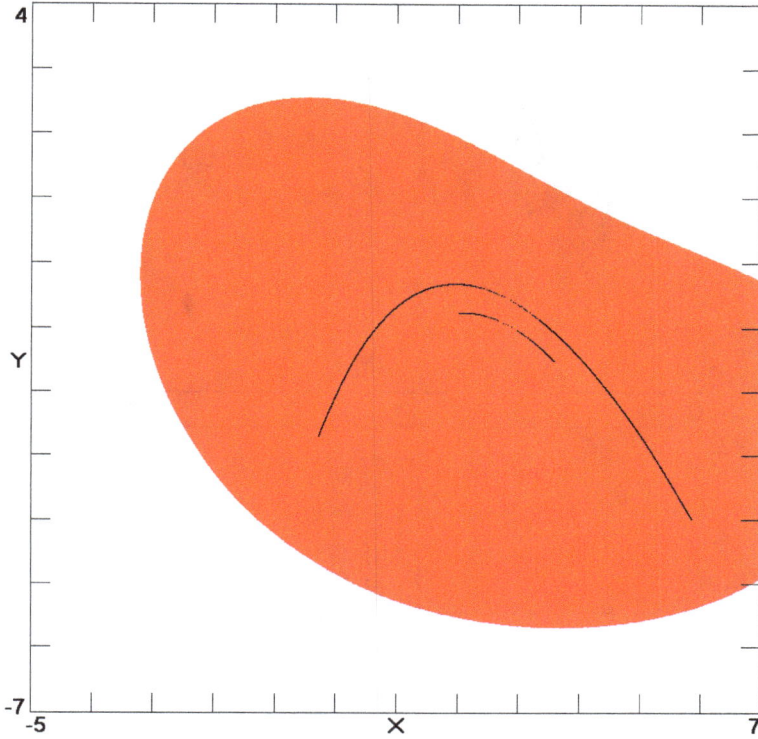

Fig. 4.3   Basin of attraction for Eq. (4.2) with $a = 0.5$, $b = 3$ in the $z = 5.6458$ plane.

4.9   Robustness

One measure of the robustness of a chaotic system is the amount by which the parameters can be changed from their nominal values before the probability of chaos decreases to 50% [Sprott (2022)]. For the system in Eq. (4.2) with $a = 0.5$, $b = 3$ and initial conditions $(3, -2, 1)$, after 11809 trials, it is estimated that the parameters can be changed by 17% before the chaos is more likely to be lost than not. Thus the system is somewhat fragile.

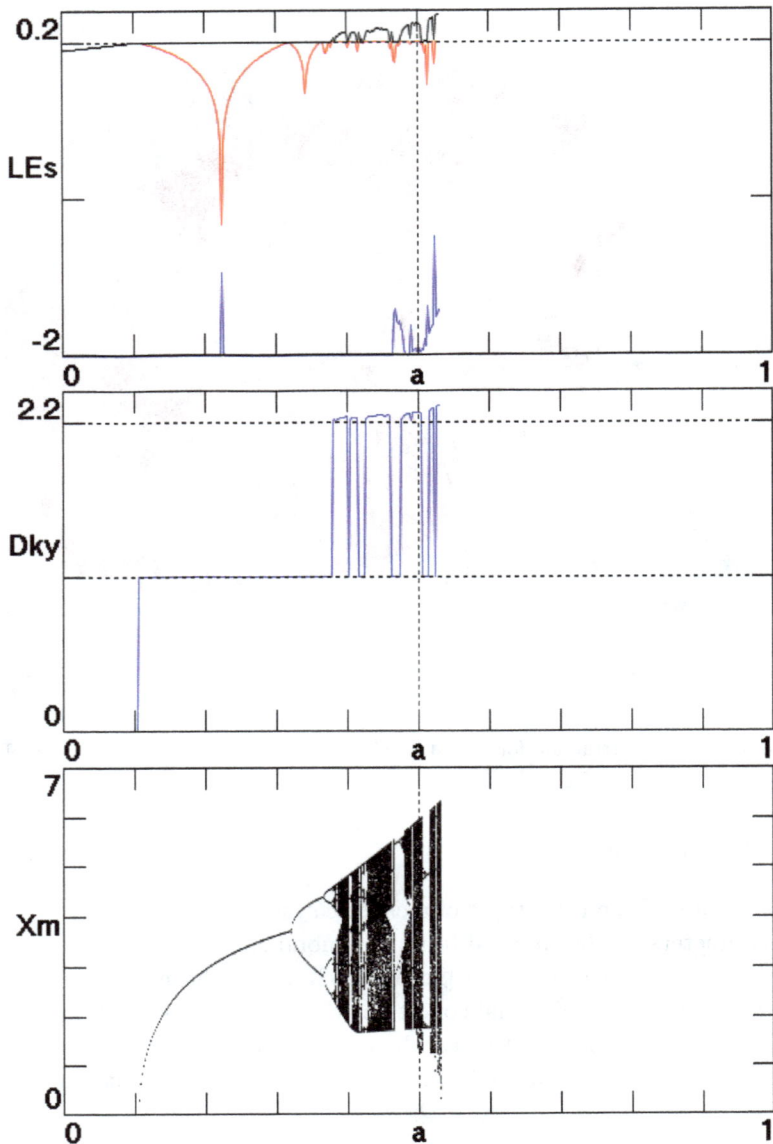

Fig. 4.4   Bifurcation diagram for Eq. (4.2) as a function of the parameter a for $b = 3$.

Chapter 5

Nosé–Hoover System

The Nosé–Hoover system was devised to model a simple harmonic oscillator in thermal equilibrium with a heat bath at constant temperature. Since it is a conservative system, it does not have an attractor, but rather it has a chaotic sea coexisting with regions of quasiperiodicity. With only five terms, it is the simplest example of a thermostatted oscillator, but its dynamics is not ergodic.

5.1 Introduction

The Nosé–Hoover system [Nosé (1984); Hoover (1985)] written in its most general form with an adjustable coefficient in each of the five terms is given by

$$
\begin{aligned}
\dot{x} &= a_1 y \\
\dot{y} &= -a_2 x - a_3 z y \\
\dot{z} &= a_4 y^2 - a_5.
\end{aligned}
\tag{5.1}
$$

The usual parameters are $a_1 = a_2 = a_3 = a_4 = a_5 = 1$, with a_5 corresponding to the 'temperature' of the oscillator. The z variable is called a 'thermostat' since it controls the average energy of the x-y oscillator. This system was independently discovered in a search for the simplest chaotic system with five terms and two quadratic nonlinearities, and thus it is sometimes called the *Sprott A system* [Sprott (1994)].

The following sections were written by the computer program that performed the optimization, carried out the analysis of the resulting system, and produced the corresponding figures, all without human intervention.

5.2   Simplified System

After about 5×10^2 trials, of which 19 were chaotic, simplified parameters for Eq. (5.1) that give chaotic solutions are $a_1 = 1$, $a_2 = 1$, $a_3 = 1$, $a_4 = 1$, $a_5 = 1$. Thus Eq. (5.1) can be written more compactly as

$$\dot{x} = y$$
$$\dot{y} = -x - zy \qquad\qquad (5.2)$$
$$\dot{z} = y^2 - a,$$

where $a = 1$.

Note that with five terms, Eq. (5.2) should have one independent parameter through a linear rescaling of the three variables plus time. However, the parameter a with a value of ± 1 and can be placed in any of the five terms.

5.3   Equilibria

The system in Eq. (5.2) with $a = 1$ has no equilibrium points. The chaotic sea is apparently hidden, and the system is symmetric under the transformation $y \to -y$, $z \to -z$, $t \to -t$.

5.4   Chaotic Sea

Figure 5.1 shows various views of the chaotic sea for Eq. (5.2) with $a = 1$ and initial conditions $(5, 5, 0)$. The rainbow of colors shows the local value of the largest Lyapunov exponent with red indicating the most positive values (regions of worst predictability) and blue indicating the most negative values (regions of best predictability).

5.5   Time Series

Figure 5.2 shows the time series for the three variables along with the local value of the largest Lyapunov exponent (LL) for Eq. (5.2) with $a = 1$. Red color in the Lyapunov exponent indicates that the error is growing parallel to the orbit, while blue indicates growth perpendicular to the orbit. Note that the orbit passes through regions where the local Lyapunov exponent is strongly positive and other regions where it is strongly negative as is typical for a chaotic system and is also reflected by the colors in Fig. 5.1.

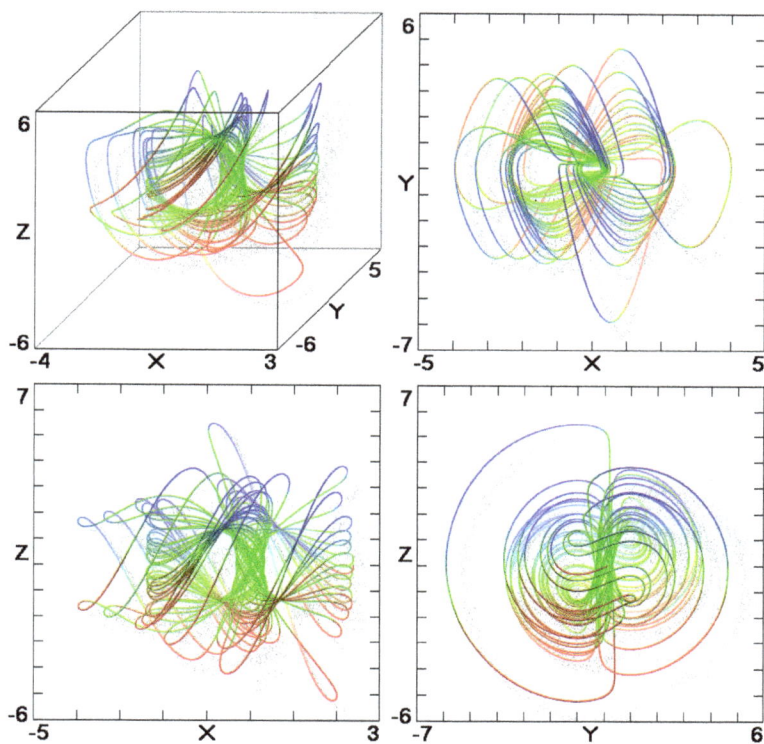

Fig. 5.1   Views of the chaotic sea for Eq. (5.2) with $a = 1$ and initial conditions (5, 5, 0).

5.6   Lyapunov Exponents

The global Lyapunov exponents are determined by averaging the local Lyapunov exponents along the orbit. The values typically converge very slowly because of the large variation along the orbit, and an integration time of order 10^8 is required to obtain 4-digit accuracy.

The results of such a calculation for the system in Eq. (5.2) with $a = 1$ after a time of 3×10^6 are LE $= (0.0140, 0, -0.0140)$ with a Kaplan–Yorke dimension of 3.0, where the last digit in the quoted values is only an approximation. The positive value of the largest Lyapunov exponent indicates that the system is chaotic, and the zero sum of the exponents indicates that the system is conservative with a chaotic sea.

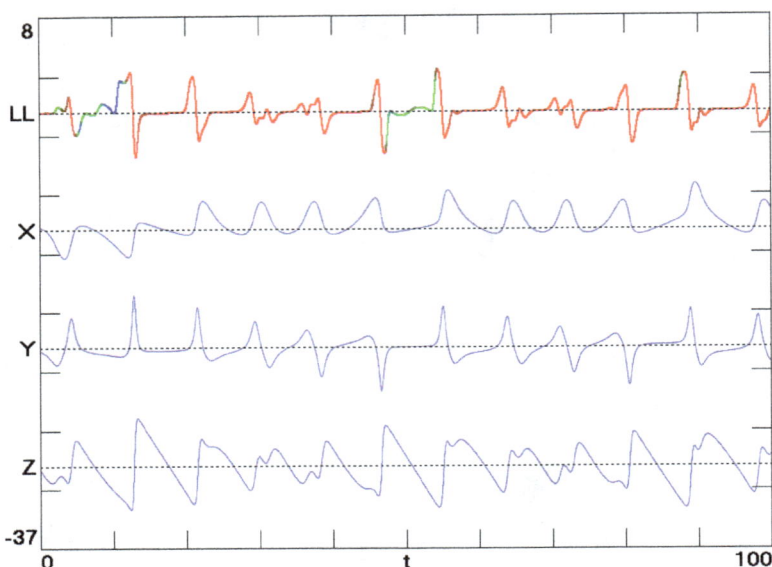

Fig. 5.2   Time series for the variables in Eq. (5.2) with $a = 1$ along with the local Lyapunov exponent (LL).

5.7   Extent of the Chaotic Sea

Figure 5.3 shows (in red) the extent of the chaotic sea for Eq. (5.2) with $a = 1$ in the $z = 0$ plane. Also shown (in black) is the cross-section of the chaotic sea in the same plane.

5.8   Bifurcations

Figure 5.4 shows the bifurcation diagram for Eq. (5.2) as a function of the parameter a from 0 to 2. The initial condition was taken as (5, 5, 0) for each value of a. Each of the 500 values of a was calculated for a time of about 4×10^4.

The upper plot shows the three Lyapunov exponents. The middle plot shows the Kaplan–Yorke dimension, and the lower plot shows the local maxima of x. The chaotic region is in the vicinity of $a = 1$, and the route to chaos is clearly shown.

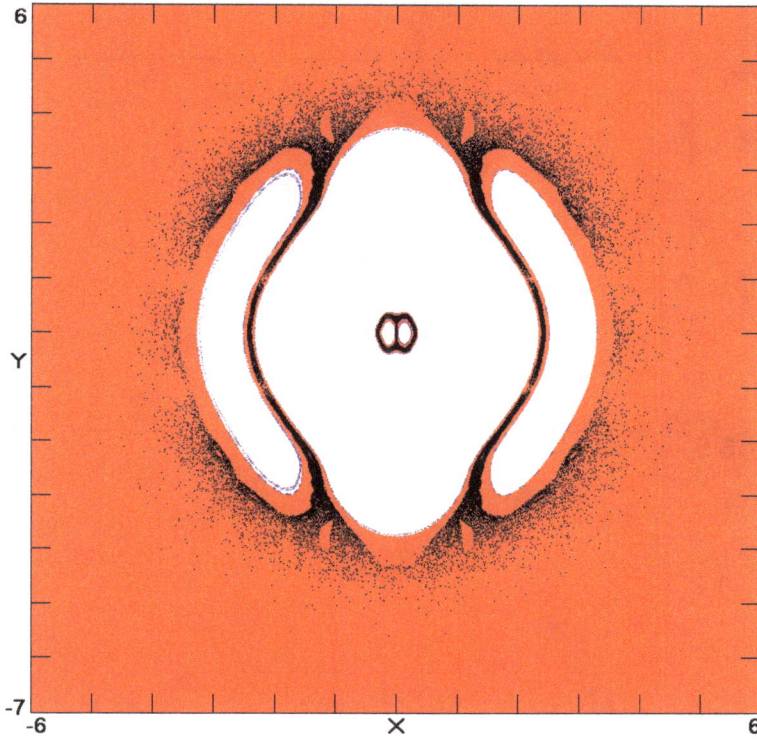

Fig. 5.3   Extent of the chaotic sea for Eq. (5.2) with $a = 1$ in the $z = 0$ plane.

5.9   Robustness

One measure of the robustness of a chaotic system is the amount by which the parameters can be changed from their nominal values before the probability of chaos decreases to 50% [Sprott (2022)]. For the system in Eq. (5.2) with $a = 1$ and initial conditions (5, 5, 0), after 2445 trials, it is estimated that the parameters can be changed by 89% before the chaos is more likely to be lost than not. Thus the system is highly robust.

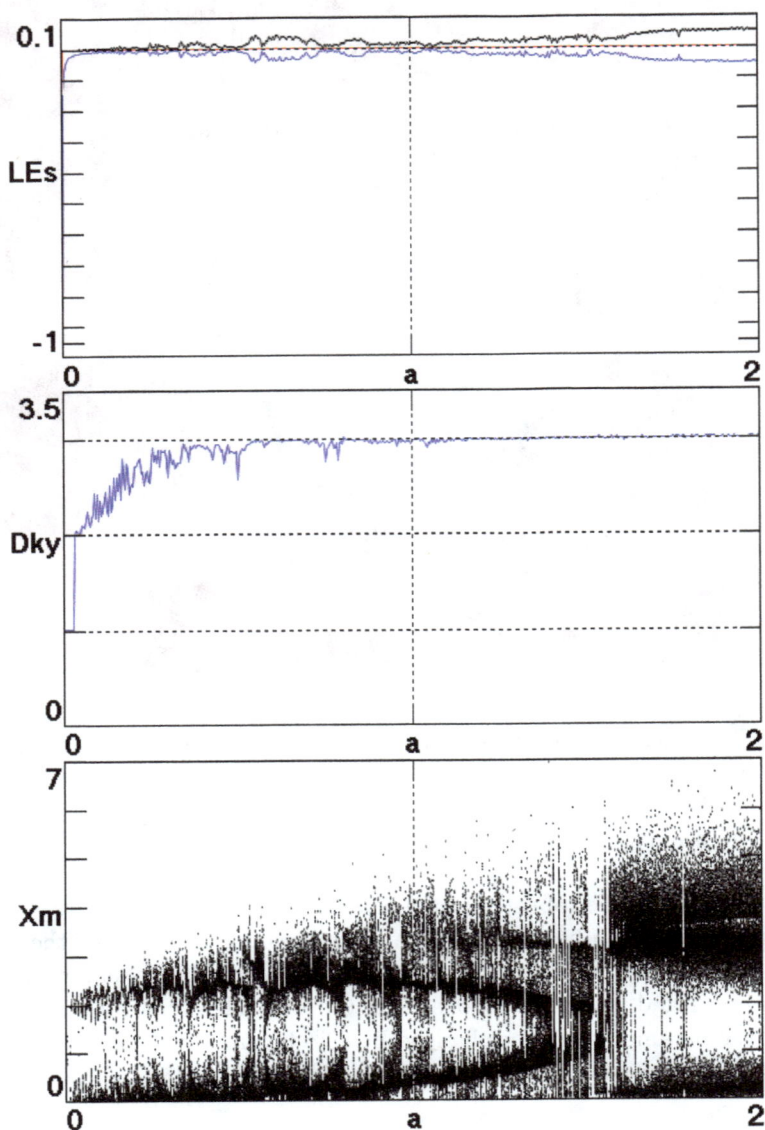

Fig. 5.4   Bifurcation diagram for Eq. (5.2) as a function of the parameter a.

Chapter 6

Diffusionless Lorenz System

The diffusionless Lorenz system is a reduced form of the classical Lorenz system with only five terms and two quadratic nonlinearities. It shares many of the properties of its more familiar relative but with a single parameter.

6.1 Introduction

The diffusionless Lorenz system [van der Schrier and Maas (2000); Munmuangsaen and Srisuchinwong (2009)] written in its most general form with an adjustable coefficient in each of the five terms is given by

$$\dot{x} = a_1 y - a_2 x$$
$$\dot{y} = -a_3 xz \qquad (6.1)$$
$$\dot{z} = a_4 xy - a_5.$$

The usual parameters are $a_1 = a_2 = a_3 = a_4 = a_5 = 1$. This system was earlier discovered in a search for the simplest chaotic system with five terms and two quadratic nonlinearities, and thus it is sometimes called the *Sprott B system* [Sprott (1994)]. A form of this system with an additional nonlinearity was much earlier used by Rikitake (1958) as to model magnetic field reversals in the Earth's dynamo.

The following sections were written by the computer program that performed the optimization, carried out the analysis of the resulting system, and produced the corresponding figures, all without human intervention.

6.2 Simplified System

After about 3×10^1 trials, of which 15 were chaotic, simplified parameters for Eq. (6.1) that give chaotic solutions are $a_1 = 1$, $a_2 = 1$, $a_3 = 1$, $a_4 = 1$,

$a_5 = 1$. Thus Eq. (6.1) can be written more compactly as

$$\dot{x} = y - x$$
$$\dot{y} = -xz \qquad\qquad (6.2)$$
$$\dot{z} = xy - a,$$

where $a = 1$.

Note that with five terms, Eq. (6.2) should have one independent parameter through a linear rescaling of the three variables plus time. However, the parameter a with a value of ± 1 and can be placed in any of the five terms.

6.3   Equilibria

The system in Eq. (6.2) with $a = 1$ has two equilibrium points:

Equilibrium #1 is an unstable saddle focus at $(1, 1, 0)$ with eigenvalues $(-1.3532, 0.1766 - 1.2028i, 0.1766 + 1.2028i)$ and a Poincaré index of 0 in the $z = 0$ plane.

Equilibrium #2 is an unstable saddle focus at $(-1, -1, 0)$ with eigenvalues $(-1.3532, 0.1766 - 1.2028i, 0.1766 + 1.2028i)$ and a Poincaré index of 0 in the $z = 0$ plane.

The strange attractor is self-excited, and the system is symmetric under the transformation $x \rightarrow -x$, $y \rightarrow -y$.

6.4   Attractor

Figure 6.1 shows various views of the attractor for Eq. (6.2) with $a = 1$ and initial conditions $(1, 0, 1)$. The rainbow of colors shows the local value of the largest Lyapunov exponent with red indicating the most positive values (regions of worst predictability) and blue indicating the most negative values (regions of best predictability).

6.5   Time Series

Figure 6.2 shows the time series for the three variables along with the local value of the largest Lyapunov exponent (LL) for Eq. (6.2) with $a = 1$. Red color in the Lyapunov exponent indicates that the error is growing parallel to the orbit, while blue indicates growth perpendicular to the orbit. Note that the orbit passes through regions where the local Lyapunov exponent is strongly positive and other regions where it is strongly negative as is typical for a chaotic system and is also reflected by the colors in Fig. 6.1.

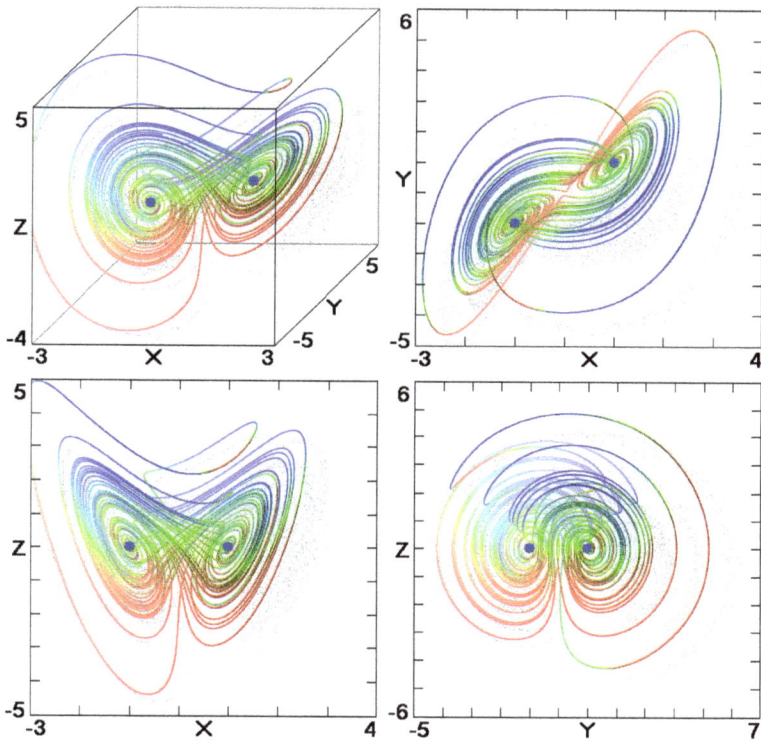

Fig. 6.1 Views of the attractor for Eq. (6.2) with $a = 1$ and initial conditions (1, 0, 1).

6.6 Lyapunov Exponents

The global Lyapunov exponents are determined by averaging the local Lyapunov exponents along the orbit. The values typically converge very slowly because of the large variation along the orbit, and an integration time of order 10^8 is required to obtain 4-digit accuracy.

The results of such a calculation for the system in Eq. (6.2) with $a = 1$ after a time of 2×10^6 are LE = (0.2101, 0, −1.2101) with a Kaplan–Yorke dimension of 2.1736, where the last digit in the quoted values is only an approximation. The positive value of the largest Lyapunov exponent indicates that the system is chaotic, and the negative sum of the exponents (−1) indicates that the system is dissipative with a strange attractor.

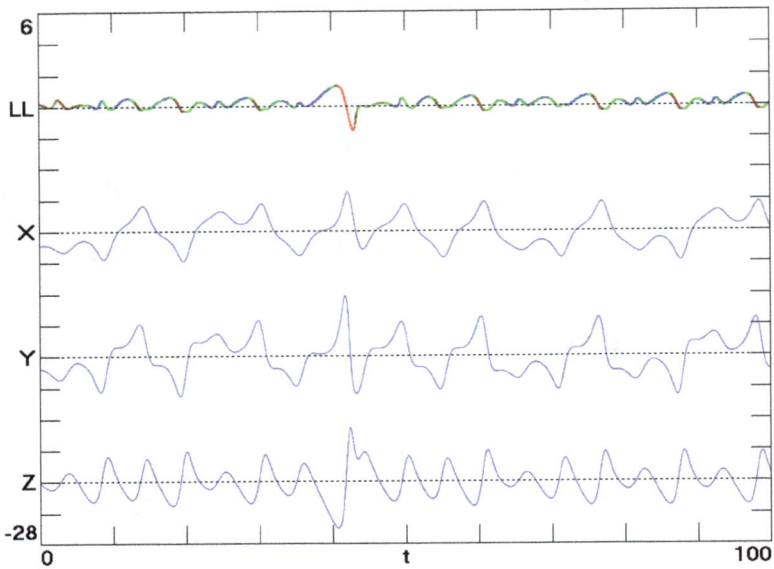

Fig. 6.2   Time series for the variables in Eq. (6.2) with $a = 1$ along with the local Lyapunov exponent (LL).

6.7   Basin of Attraction

Figure 6.3 shows (in red) the basin of attraction for Eq. (6.2) with $a = 1$ in the $z = 0$ plane. Also shown (in black) is the cross-section of the attractor in the same plane.

6.8   Bifurcations

Figure 6.4 shows the bifurcation diagram for Eq. (6.2) as a function of the parameter a from 0 to 2. The initial condition was taken as (1, 0, 1) at $a = 2$ and was not changed as a slowly varied toward $a = 0$. Each of the 500 values of a was calculated for a time of about 1×10^4.

The upper plot shows the three Lyapunov exponents. The middle plot shows the Kaplan–Yorke dimension, and the lower plot shows the local maxima of x. The chaotic region is in the vicinity of $a = 1$, and the route to chaos is clearly shown.

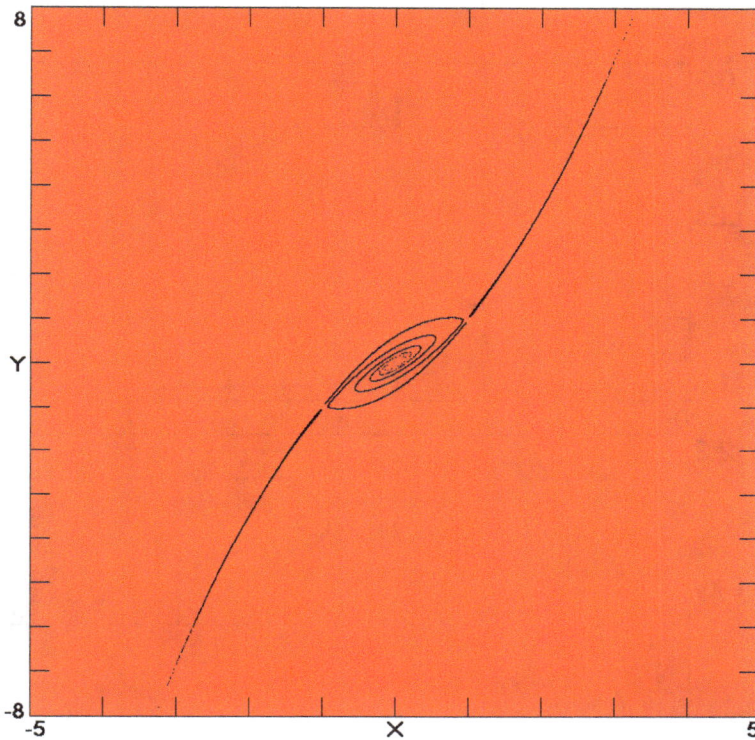

Fig. 6.3   Basin of attraction for Eq. (6.2) with $a = 1$ in the $z = 0$ plane.

6.9   Robustness

One measure of the robustness of a chaotic system is the amount by which the parameters can be changed from their nominal values before the probability of chaos decreases to 50% [Sprott (2022)]. For the system in Eq. (6.2) with $a = 1$ and initial conditions (1, 0, 1), after 1205 trials, it is estimated that the parameters can be changed by 97% before the chaos is more likely to be lost than not. Thus the system is highly robust.

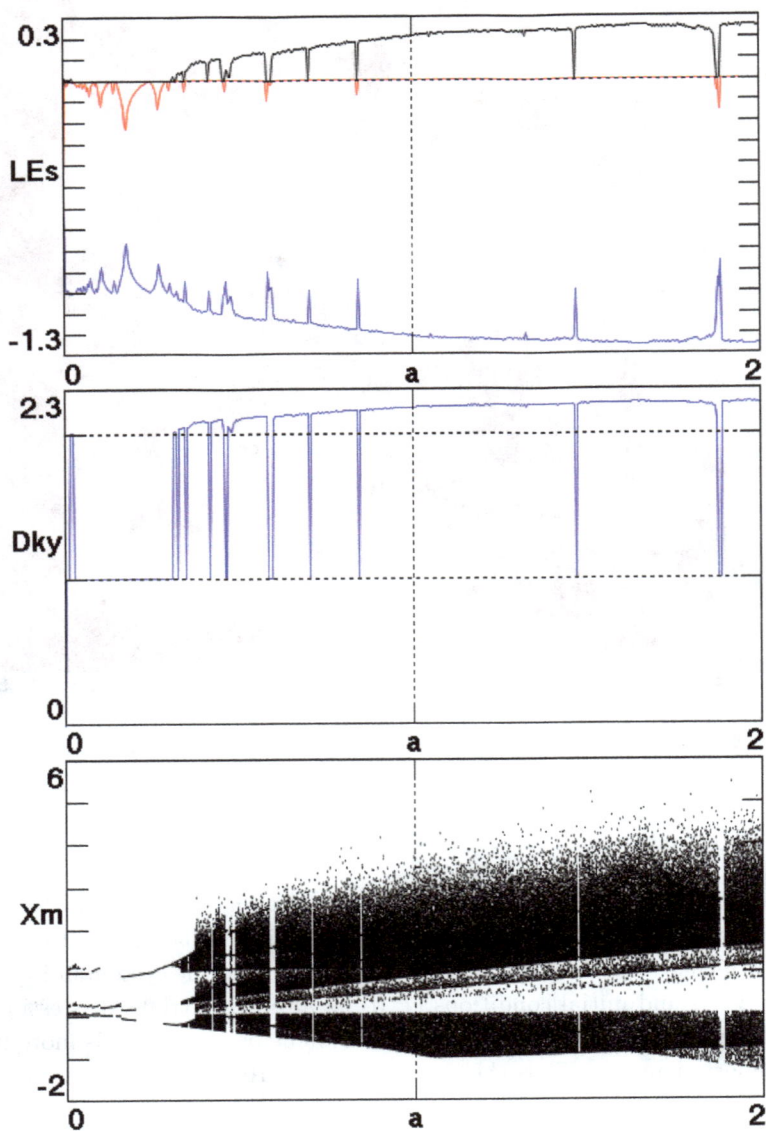

Fig. 6.4   Bifurcation diagram for Eq. (6.2) as a function of the parameter a.

Chapter 7

Sprott C System

The Sprott C system is a slight modification of the diffusionless Lorenz system in which the xy term in the $\dot{z}$ equation is replaced with y^2. Unsurprisingly, the resulting dynamics is very similar.

7.1 Introduction

The Sprott C system [Sprott (1994)] written in its most general form with an adjustable coefficient in each of the five terms is given by

$$\dot{x} = a_1 y - a_2 x$$
$$\dot{y} = -a_3 x z \qquad\qquad (7.1)$$
$$\dot{z} = a_4 y^2 - a_5.$$

The usual parameters are $a_1 = a_2 = a_3 = a_4 = a_5 = 1$. This system was discovered in a search for the simplest chaotic system with five terms and two quadratic nonlinearities. The variables x and y have been interchanged, and the signs have been changed to emphasize the similarity to the Lorenz system.

The following sections were written by the computer program that performed the optimization, carried out the analysis of the resulting system, and produced the corresponding figures, all without human intervention.

7.2 Simplified System

After about 4×10^2 trials, of which 32 were chaotic, simplified parameters for Eq. (7.1) that give chaotic solutions are $a_1 = 1$, $a_2 = 1$, $a_3 = 1$, $a_4 = 1$, $a_5 = 1$. Thus Eq. (7.1) can be written more compactly as

$$\dot{x} = y - x$$
$$\dot{y} = -xz \qquad\qquad (7.2)$$
$$\dot{z} = y^2 - a,$$

where $a = 1$.

Note that with five terms, Eq. (7.2) should have one independent parameter through a linear rescaling of the three variables plus time. However,
the parameter a with a value of ± 1 and can be placed in any of the five
terms.

7.3   Equilibria

The system in Eq. (7.2) with $a = 1$ has two equilibrium points:

Equilibrium # 1 is a neutrally stable non-hyperbolic equilibrium at $(-1,$
$-1, 0)$ with eigenvalues $(-1, 0 - 1.4142i, 0 + 1.4142i)$ and a Poincaré index
of 0 in the $z = 0$ plane.

Equilibrium # 2 is a neutrally stable non-hyperbolic equilibrium at $(1,$
$1, 0)$ with eigenvalues $(-1, 0 - 1.4142i, 0 + 1.4142i)$ and a Poincaré index
of 0 in the $z = 0$ plane.

The strange attractor is self-excited, and the system is symmetric under
the transformation $x \rightarrow -x$, $y \rightarrow -y$.

7.4   Attractor

Figure 7.1 shows various views of the attractor for Eq. (7.2) with $a = 1$ and
initial conditions $(-1, 0, 1)$. The rainbow of colors shows the local value of
the largest Lyapunov exponent with red indicating the most positive values (regions of worst predictability) and blue indicating the most negative
values (regions of best predictability).

7.5   Time Series

Figure 7.2 shows the time series for the three variables along with the local
value of the largest Lyapunov exponent (LL) for Eq. (7.2) with $a = 1$. Red
color in the Lyapunov exponent indicates that the error is growing parallel
to the orbit, while blue indicates growth perpendicular to the orbit. Note
that the orbit passes through regions where the local Lyapunov exponent
is strongly positive and other regions where it is strongly negative as is
typical for a chaotic system and is also reflected by the colors in Fig. 7.1.

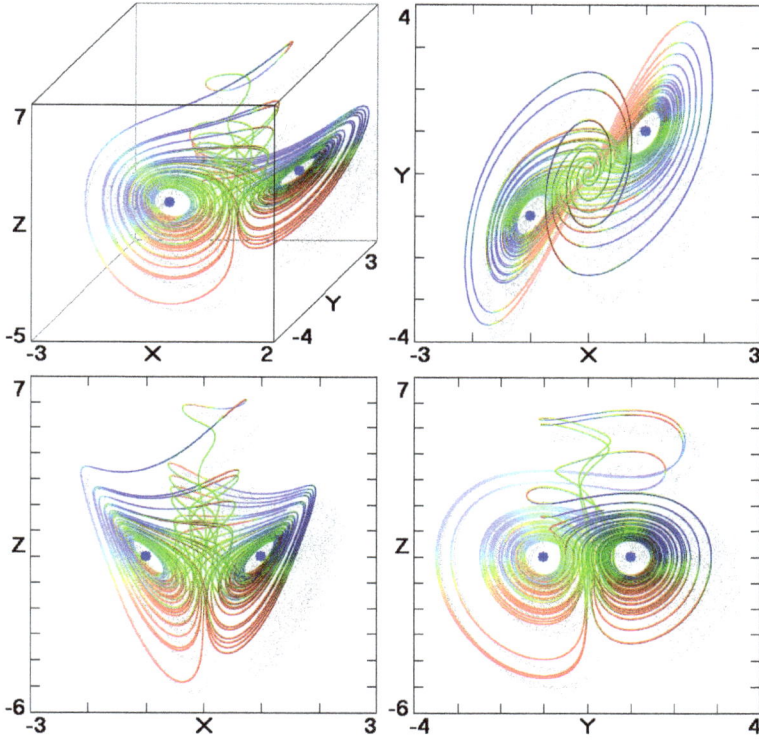

Fig. 7.1   Views of the attractor for Eq. (7.2) with $a = 1$ and initial conditions $(-1, 0, 1)$.

7.6   Lyapunov Exponents

The global Lyapunov exponents are determined by averaging the local Lyapunov exponents along the orbit. The values typically converge very slowly because of the large variation along the orbit, and an integration time of order 10^8 is required to obtain 4-digit accuracy.

The results of such a calculation for the system in Eq. (7.2) with $a = 1$ after a time of 4×10^6 are LE $= (0.1629, 0, -1.1629)$ with a Kaplan–Yorke dimension of 2.1401, where the last digit in the quoted values is only an approximation. The positive value of the largest Lyapunov exponent indicates that the system is chaotic, and the negative sum of the exponents (-1) indicates that the system is dissipative with a strange attractor.

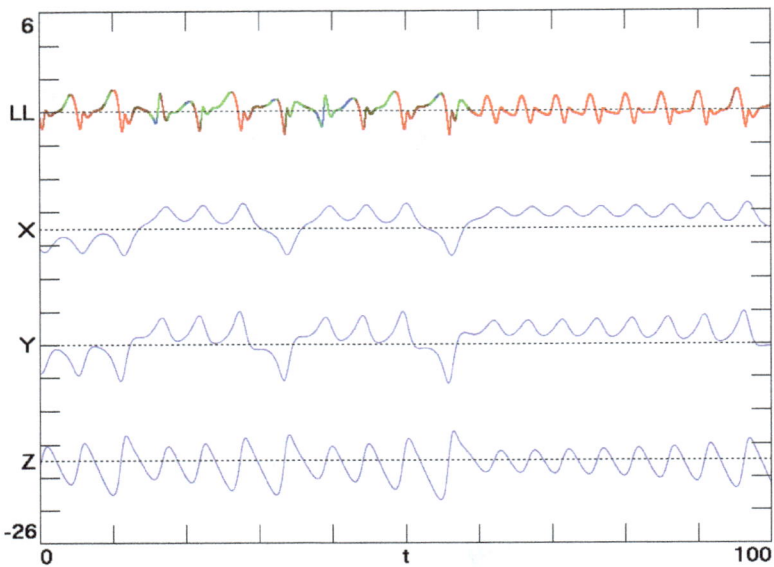

Fig. 7.2   Time series for the variables in Eq. (7.2) with $a = 1$ along with the local Lyapunov exponent (LL).

7.7   Basin of Attraction

Figure 7.3 shows (in red) the basin of attraction for Eq. (7.2) with $a = 1$ in the $z = 0$ plane. Also shown (in black) is the cross-section of the attractor in the same plane.

7.8   Bifurcations

Figure 7.4 shows the bifurcation diagram for Eq. (7.2) as a function of the parameter a from 0 to 2. The initial condition was taken as $(-1, 0, 1)$ at $a = 2$ and was not changed as a slowly varied toward $a = 0$. Each of the 500 values of a was calculated for a time of about 1×10^4.

The upper plot shows the three Lyapunov exponents. The middle plot shows the Kaplan–Yorke dimension, and the lower plot shows the local maxima of x. The chaotic region is in the vicinity of $a = 1$, and the route to chaos is clearly shown.

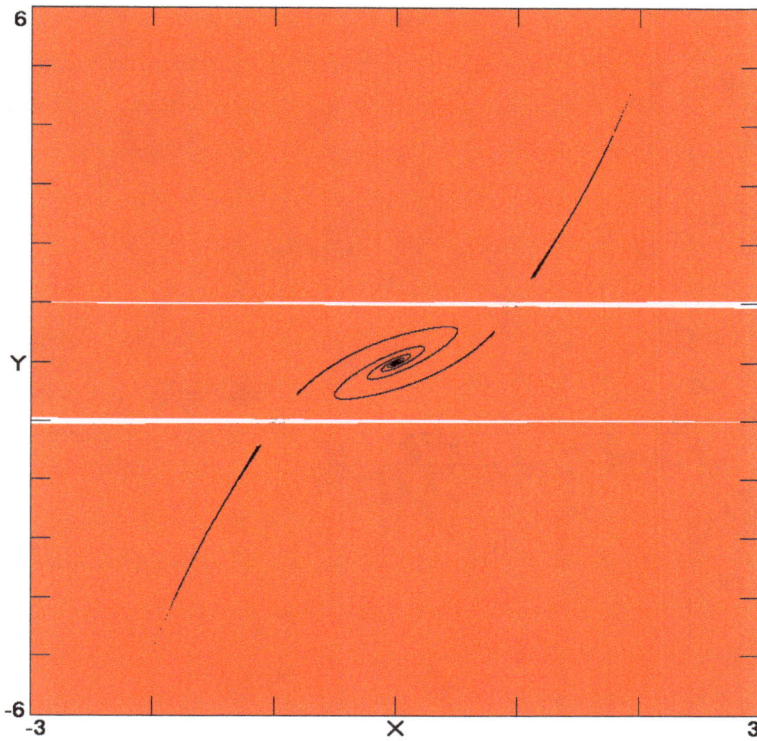

Fig. 7.3　Basin of attraction for Eq. (7.2) with $a = 1$ in the $z = 0$ plane.

7.9　Robustness

One measure of the robustness of a chaotic system is the amount by which the parameters can be changed from their nominal values before the probability of chaos decreases to 50% [Sprott (2022)]. For the system in Eq. (7.2) with $a = 1$ and initial conditions $(-1, 0, 1)$, after 1106 trials, it is estimated that the parameters can be changed by 97% before the chaos is more likely to be lost than not. Thus the system is highly robust.

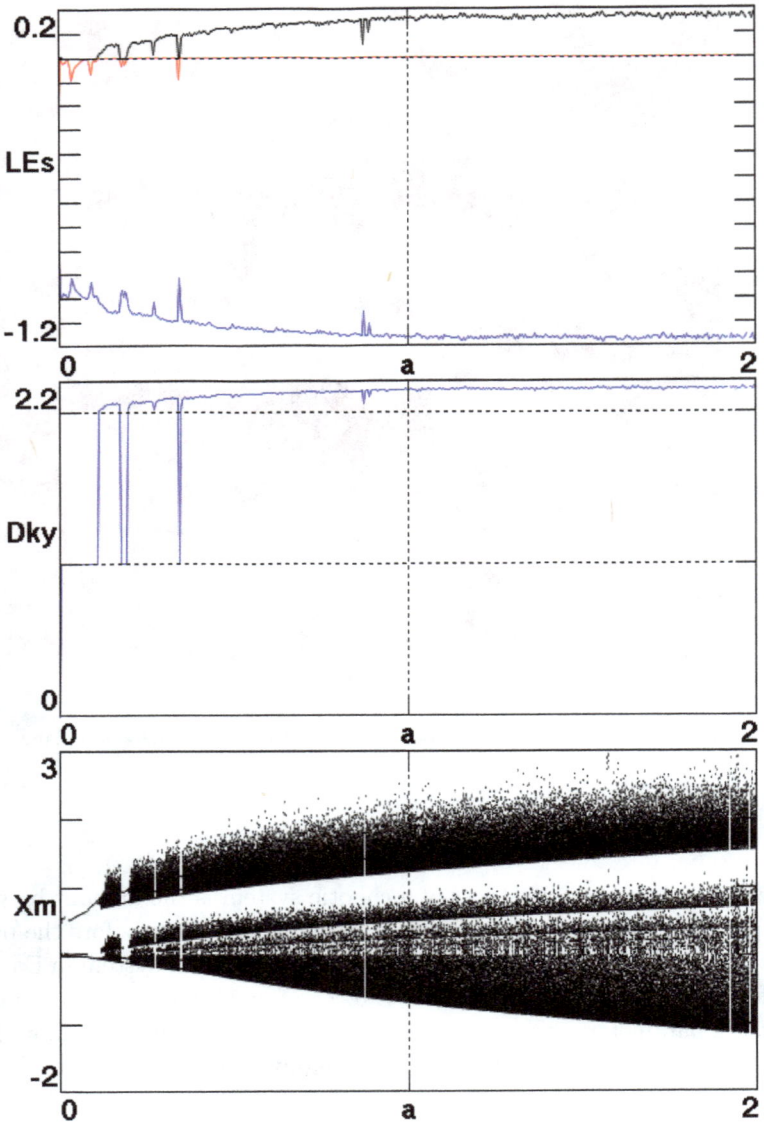

Fig. 7.4   Bifurcation diagram for Eq. (7.2) as a function of the parameter a.

Chapter 8

Sprott D System

The Sprott D system is another system with five terms and two quadratic nonlinearities that was discovered in 1994. It has the unusual feature that it is dissipative but time reversible with a symmetric attractor–repellor pair that exchange roles when time is reversed.

8.1   Introduction

The Sprott D system [Sprott (1994)] written in its most general form with an adjustable coefficient in each of the five terms is given by

$$\begin{aligned}
\dot{x} &= a_1 y \\
\dot{y} &= a_2 x + a_3 z \\
\dot{z} &= a_4 xz + a_5 y^2.
\end{aligned} \tag{8.1}$$

The usual parameters are $a_1 = -1$, $a_2 = a_3 = a_4 = 1$, $a_5 = 3$. This system was discovered in a search for the simplest chaotic system with five terms and two quadratic nonlinearities. It is written here with parameters that can be either positive or negative with a preference for as many positive as possible.

The following sections were written by the computer program that performed the optimization, carried out the analysis of the resulting system, and produced the corresponding figures, all without human intervention.

8.2   Simplified System

After about 3×10^3 trials, of which 29 were chaotic, simplified parameters for Eq. (8.1) that give chaotic solutions are $a_1 = -1$, $a_2 = 1$, $a_3 = 1$, $a_4 = 1$,

49

$a_5 = 3$. Thus Eq. (8.1) can be written more compactly as

$$\dot{x} = -y$$
$$\dot{y} = x + z \tag{8.2}$$
$$\dot{z} = xz + ay^2,$$

where $a = 3$.

Note that with five terms, Eq. (8.2) should have one independent parameter through a linear rescaling of the three variables plus time, and so the dynamics is completely captured by the single parameter a, which could be put in any of the five terms, albeit with a different numerical value.

8.3 Equilibria

The system in Eq. (8.2) with $a = 3$ has a neutrally stable non-hyperbolic equilibrium at $(0, 0, 0)$ with eigenvalues $(0-i, 0+i, 0)$ and a Poincaré index of 1 in the $z = 0$ plane. Despite the neutral linear stability, the equilibrium is nonlinearly unstable.

The strange attractor is self-excited, and the system is symmetric under the transformation $x \to -x$, $z \to -z$, $t \to -t$.

8.4 Attractor

Figure 8.1 shows various views of the attractor for Eq. (8.2) with $a = 3$ and initial conditions $(-1, -1, 2)$. The rainbow of colors shows the local value of the largest Lyapunov exponent with red indicating the most positive values (regions of worst predictability) and blue indicating the most negative values (regions of best predictability).

8.5 Time Series

Figure 8.2 shows the time series for the three variables along with the local value of the largest Lyapunov exponent (LL) for Eq. (8.2) with $a = 3$. Red color in the Lyapunov exponent indicates that the error is growing parallel to the orbit, while blue indicates growth perpendicular to the orbit. Note that the orbit passes through regions where the local Lyapunov exponent is strongly positive and other regions where it is strongly negative as is typical for a chaotic system and is also reflected by the colors in Fig. 8.1.

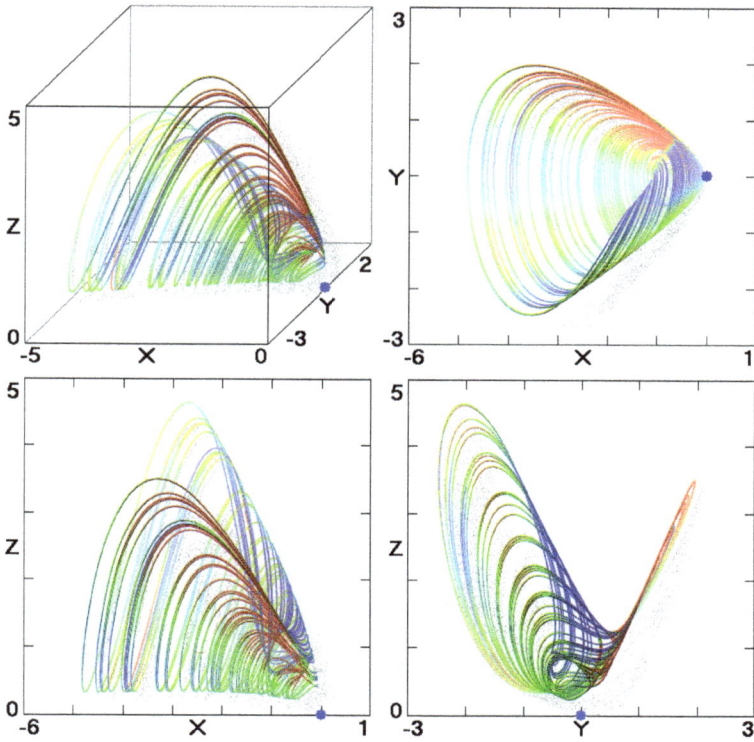

Fig. 8.1   Views of the attractor for Eq. (8.2) with $a = 3$ and initial conditions $(-1, -1, 2)$.

8.6   Lyapunov Exponents

The global Lyapunov exponents are determined by averaging the local Lyapunov exponents along the orbit. The values typically converge very slowly because of the large variation along the orbit, and an integration time of order 10^8 is required to obtain 4-digit accuracy.

The results of such a calculation for the system in Eq. (8.2) with $a = 3$ after a time of 2×10^6 are LE $= (0.1028, 0, -1.3198)$ with a Kaplan–Yorke dimension of 2.0778, where the last digit in the quoted values is only an approximation. The positive value of the largest Lyapunov exponent indicates that the system is chaotic, and the negative sum of the exponents (-1.2170) indicates that the system is dissipative with a strange attractor.

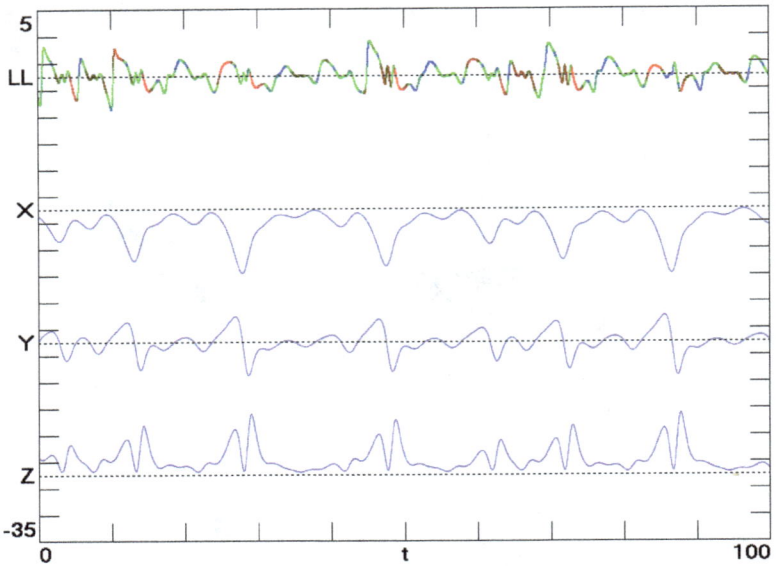

Fig. 8.2   Time series for the variables in Eq. (8.2) with $a = 3$ along with the local Lyapunov exponent (LL).

8.7   Basin of Attraction

Figure 8.3 shows (in red) the basin of attraction for Eq. (8.2) with $a = 3$ in the $z = 2$ plane. Also shown (in black) is the cross-section of the attractor in the same plane.

8.8   Bifurcations

Figure 8.4 shows the bifurcation diagram for Eq. (8.2) as a function of the parameter a from 0 to 6. The initial condition was taken as $(-1, -1, 2)$ for each value of a. Each of the 500 values of a was calculated for a time of about 1×10^4.

The upper plot shows the three Lyapunov exponents. The middle plot shows the Kaplan–Yorke dimension, and the lower plot shows the local maxima of x. The chaotic region is in the vicinity of $a = 3$, and the route to chaos is clearly shown.

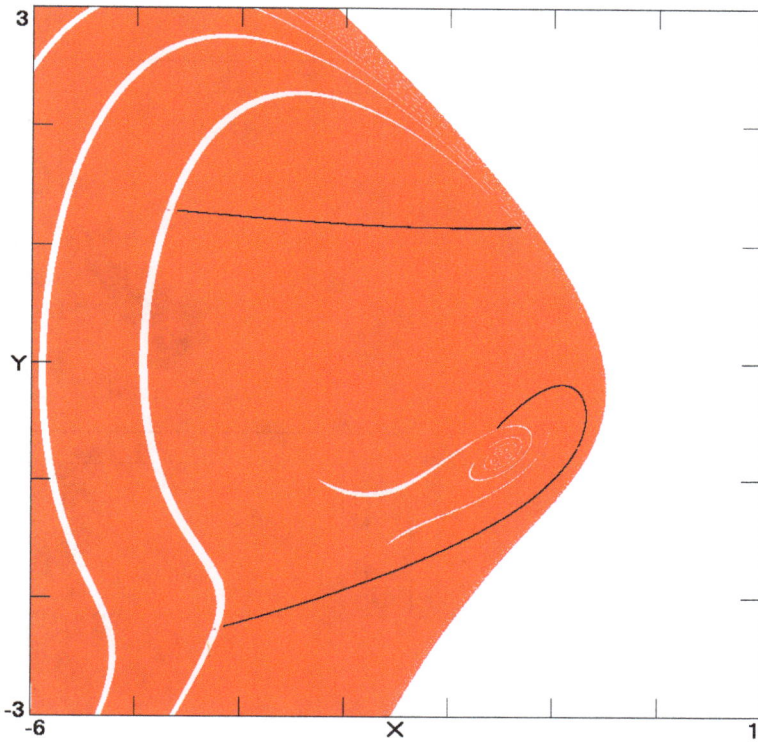

Fig. 8.3   Basin of attraction for Eq. (8.2) with $a = 3$ in the $z = 2$ plane.

8.9   Robustness

One measure of the robustness of a chaotic system is the amount by which the parameters can be changed from their nominal values before the probability of chaos decreases to 50% [Sprott (2022)]. For the system in Eq. (8.2) with $a = 3$ and initial conditions $(-1, -1, 2)$, after 4199 trials, it is estimated that the parameter a can be changed by 48% before the chaos is more likely to be lost than not. Thus the system is somewhat robust. This result is consistent with the data in Fig. 8.4.

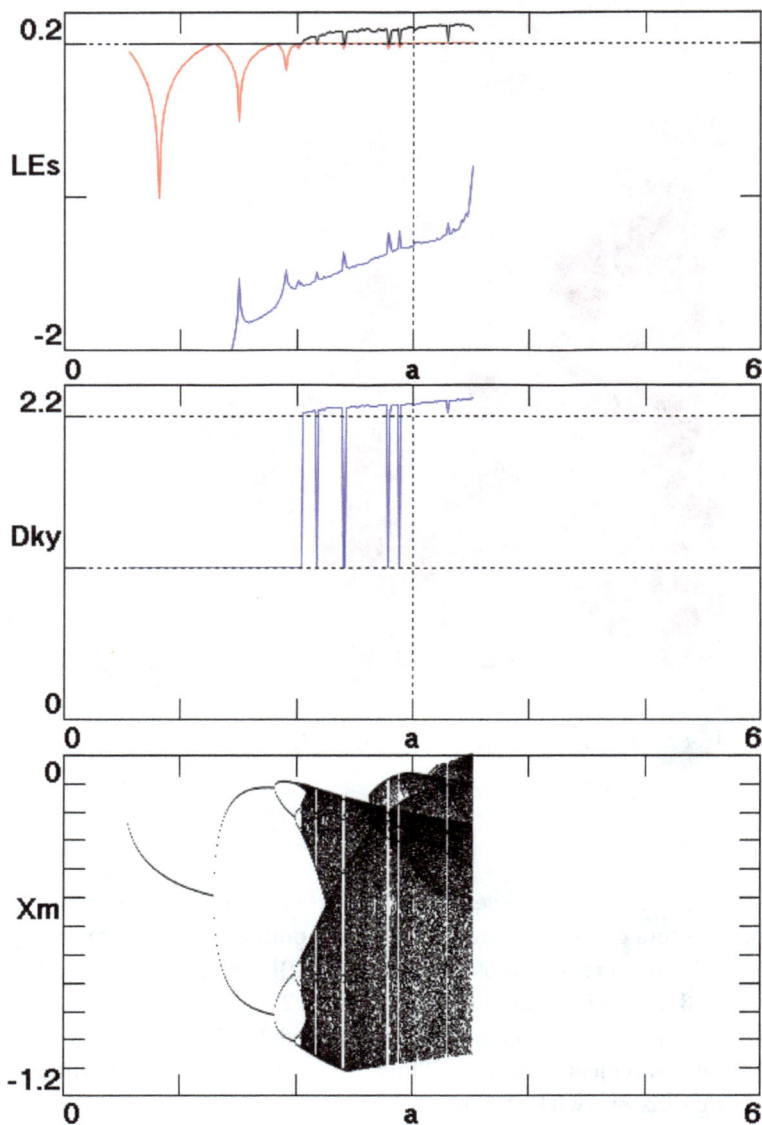

Fig. 8.4　Bifurcation diagram for Eq. (8.2) as a function of the parameter a.

Chapter 9

Sprott E System

The Sprott E system is the final case with five terms and two quadratic nonlinearities that was discovered in 1994. Of the five such cases that were found, it occurred the least frequently. For some choice of the parameters, it produces bursting oscillations.

9.1 Introduction

The Sprott E system [Sprott (1994)] written in its most general form with an adjustable coefficient in each of the five terms is given by

$$\dot{x} = a_1 y z$$
$$\dot{y} = a_2 x^2 + a_3 y \qquad\qquad (9.1)$$
$$\dot{z} = a_4 + a_5 x.$$

The usual parameters are $a_1 = a_2 = a_4 = 1$, $a_3 = -1$, $a_5 = -4$. This system was discovered in a search for the simplest chaotic system with five terms and two quadratic nonlinearities. It is written here with parameters that can be either positive or negative with a preference for as many positive as possible.

The following sections were written by the computer program that performed the optimization, carried out the analysis of the resulting system, and produced the corresponding figures, all without human intervention.

9.2 Simplified System

After about 1×10^4 trials, of which 167 were chaotic, simplified parameters for Eq. (9.1) that give chaotic solutions are $a_1 = 1$, $a_2 = -1$, $a_3 = -1$,

$a_4 = 1$, $a_5 = 6$. Thus Eq. (9.1) can be written more compactly as

$$\dot{x} = yz$$
$$\dot{y} = -x^2 - y \qquad\qquad (9.2)$$
$$\dot{z} = 1 + ax,$$

where $a = 6$.

Note that with five terms, Eq. (9.2) should have one independent parameter through a linear rescaling of the three variables plus time, and so the dynamics is completely captured by the single parameter a, which could be put in any of the five terms, albeit with a different numerical value.

9.3   Equilibria

The system in Eq. (9.2) with $a = 6$ has a neutrally stable non-hyperbolic equilibrium at $(-0.1667, -0.0278, 0)$ with eigenvalues $(0 - 0.4082i, 0 + 0.4082i, -1)$ and a Poincaré index of 0 in the $z = 0$ plane.

The strange attractor is self-excited, and the system has no symmetry.

9.4   Attractor

Figure 9.1 shows various views of the attractor for Eq. (9.2) with $a = 6$ and initial conditions $(1, -2, -2)$. The rainbow of colors shows the local value of the largest Lyapunov exponent with red indicating the most positive values (regions of worst predictability) and blue indicating the most negative values (regions of best predictability).

9.5   Time Series

Figure 9.2 shows the time series for the three variables along with the local value of the largest Lyapunov exponent (LL) for Eq. (9.2) with $a = 6$. Red color in the Lyapunov exponent indicates that the error is growing parallel to the orbit, while blue indicates growth perpendicular to the orbit. Note that the orbit passes through regions where the local Lyapunov exponent is strongly positive and other regions where it is strongly negative as is typical for a chaotic system and is also reflected by the colors in Fig. 9.1.

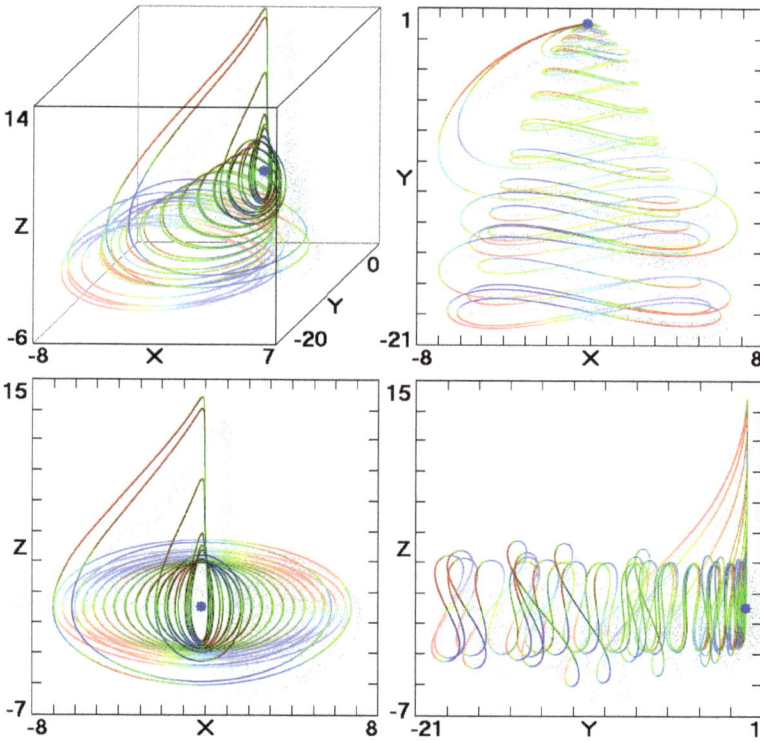

Fig. 9.1   Views of the attractor for Eq. (9.2) with $a = 6$ and initial conditions $(1, -2, -2)$.

9.6   Lyapunov Exponents

The global Lyapunov exponents are determined by averaging the local Lyapunov exponents along the orbit. The values typically converge very slowly because of the large variation along the orbit, and an integration time of order 10^8 is required to obtain 4-digit accuracy.

The results of such a calculation for the system in Eq. (9.2) with $a = 6$ after a time of 3×10^6 are LE $= (0.1129, 0, -1.1129)$ with a Kaplan–Yorke dimension of 2.1014, where the last digit in the quoted values is only an approximation. The positive value of the largest Lyapunov exponent indicates that the system is chaotic, and the negative sum of the exponents (-1) indicates that the system is dissipative with a strange attractor.

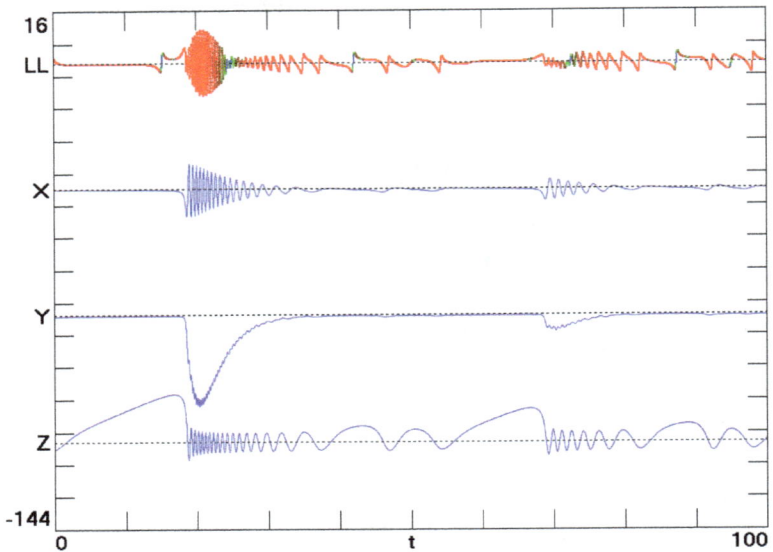

Fig. 9.2 Time series for the variables in Eq. (9.2) with $a = 6$ along with the local Lyapunov exponent (LL).

9.7 Basin of Attraction

Figure 9.3 shows (in red) the basin of attraction for Eq. (9.2) with $a = 6$ in the $z = 0$ plane. Also shown (in black) is the cross-section of the attractor in the same plane.

9.8 Bifurcations

Figure 9.4 shows the bifurcation diagram for Eq. (9.2) as a function of the parameter a from 0 to 12. The initial condition was taken as $(1, -2, -2)$ for each value of a. Each of the 500 values of a was calculated for a time of about 8×10^3.

The upper plot shows the three Lyapunov exponents. The middle plot shows the Kaplan–Yorke dimension, and the lower plot shows the local maxima of x. The chaotic region is in the vicinity of $a = 6$, and the route to chaos is clearly shown.

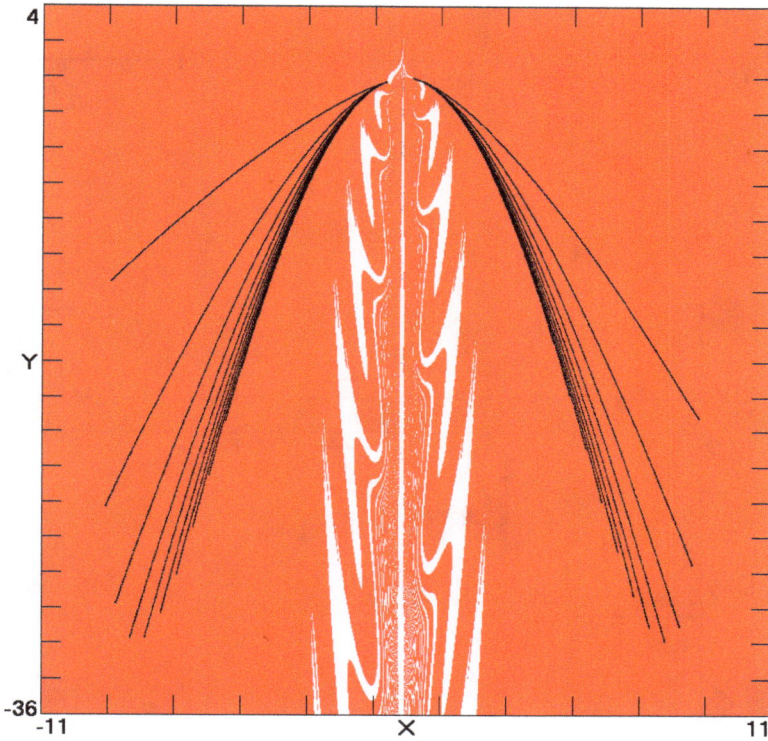

Fig. 9.3   Basin of attraction for Eq. (9.2) with $a = 6$ in the $z = 0$ plane.

9.9   Robustness

One measure of the robustness of a chaotic system is the amount by which the parameters can be changed from their nominal values before the probability of chaos decreases to 50% [Sprott (2022)]. For the system in Eq. (9.2) with $a = 6$ and initial conditions $(1, -2, -2)$, after 3427 trials, it is estimated that the parameter a can be changed by 58% before the chaos is more likely to be lost than not. Thus the system is highly robust. This result is consistent with the data in Fig. 9.4.

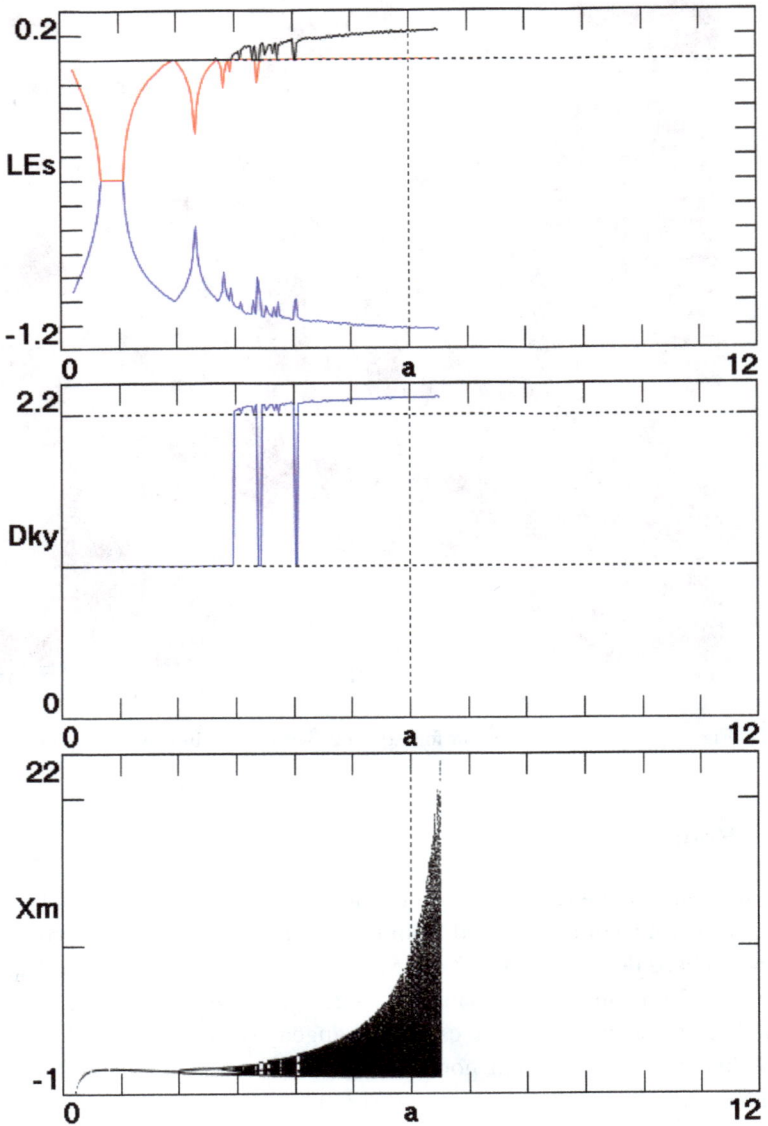

Fig. 9.4   Bifurcation diagram for Eq. (9.2) as a function of the parameter a.

Chapter 10

Sprott F System

The Sprott F system is the first and most frequently found of the fourteen chaotic cases with six terms and one quadratic nonlinearity that were discovered in 1994. The remaining thirteen cases are given in the chapters that follow.

10.1 Introduction

The Sprott F system [Sprott (1994)] written in its most general form with an adjustable coefficient in each of the six terms is given by

$$\dot{x} = a_1 y + a_2 z$$
$$\dot{y} = a_3 x + a_4 y \qquad (10.1)$$
$$\dot{z} = a_5 x^2 + a_6 z.$$

The usual parameters are $a_1 = a_2 = a_5 = 1$, $a_3 = a_6 = -1$, $a_4 = 0.5$. This system was discovered in a search for the simplest chaotic system with six terms and a single quadratic nonlinearity. It is written here with parameters that can be either positive or negative with a preference for as many positive as possible.

The following sections were written by the computer program that performed the optimization, carried out the analysis of the resulting system, and produced the corresponding figures, all without human intervention.

10.2 Simplified System

After about 2×10^4 trials, of which 70 were chaotic, simplified parameters for Eq. (10.1) that give chaotic solutions are $a_1 = -1$, $a_2 = 1$, $a_3 = 1$, $a_4 = 0.5$, $a_5 = 1$, $a_6 = -1$. Thus Eq. (10.1) can be written more compactly

as

$$\dot{x} = -y + z$$
$$\dot{y} = x + ay \tag{10.2}$$
$$\dot{z} = x^2 - z,$$

where $a = 0.5$.

Note that with six terms, Eq. (10.2) should have two independent parameters through a linear rescaling of the three variables plus time. However, one of the two parameters has a value of ± 1 and can be placed in any of the remaining five terms.

10.3   Equilibria

The system in Eq. (10.2) with $a = 0.5$ has two equilibrium points:

Equilibrium # 1 is an unstable saddle focus at $(0, 0, 0)$ with eigenvalues $(0.25 - 0.9682i, 0.25 + 0.9682i, -1)$ and a Poincaré index of 1 in the $z = 0$ plane.

Equilibrium # 2 is an unstable saddle focus at $(-2, 4, 4)$ with eigenvalues $(-0.3574 - 2.1274i, -0.3574 + 2.1274i, 0.2149)$ and a Poincaré index of 1 in the $z = 4$ plane.

The strange attractor is self-excited, and the system has no symmetry.

10.4   Attractor

Figure 10.1 shows various views of the attractor for Eq. (10.2) with $a = 0.5$ and initial conditions $(-1, 2, 3)$. The rainbow of colors shows the local value of the largest Lyapunov exponent with red indicating the most positive values (regions of worst predictability) and blue indicating the most negative values (regions of best predictability).

10.5   Time Series

Figure 10.2 shows the time series for the three variables along with the local value of the largest Lyapunov exponent (LL) for Eq. (10.2) with $a = 0.5$. Red color in the Lyapunov exponent indicates that the error is growing parallel to the orbit, while blue indicates growth perpendicular to the orbit. Note that the orbit passes through regions where the local Lyapunov exponent is strongly positive and other regions where it is strongly negative as is typical for a chaotic system and is also reflected by the colors in Fig. 10.1.

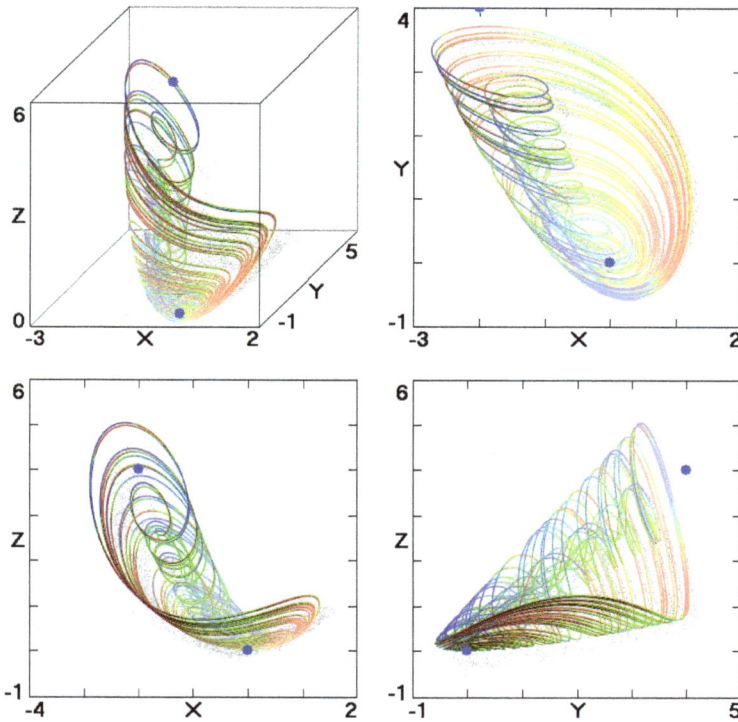

Fig. 10.1   Views of the attractor for Eq. (10.2) with $a = 0.5$ and initial conditions $(-1, 2, 3)$.

10.6   Lyapunov Exponents

The global Lyapunov exponents are determined by averaging the local Lyapunov exponents along the orbit. The values typically converge very slowly because of the large variation along the orbit, and an integration time of order 10^8 is required to obtain 4-digit accuracy.

The results of such a calculation for the system in Eq. (10.2) with $a = 0.5$ after a time of 2×10^6 are LE $= (0.1170, 0, -0.6170)$ with a Kaplan–Yorke dimension of 2.1896, where the last digit in the quoted values is only an approximation. The positive value of the largest Lyapunov exponent indicates that the system is chaotic, and the negative sum of the exponents (-0.5000) indicates that the system is dissipative with a strange attractor.

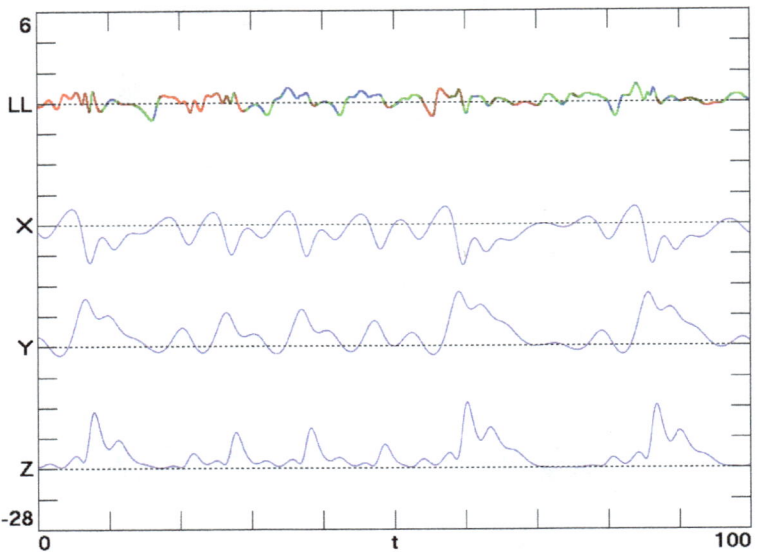

Fig. 10.2   Time series for the variables in Eq. (10.2) with $a = 0.5$ along with the local Lyapunov exponent (LL).

10.7   Basin of Attraction

Figure 10.3 shows (in red) the basin of attraction for Eq. (10.2) with $a = 0.5$ in the $z = 4$ plane. Also shown (in black) is the cross-section of the attractor in the same plane.

10.8   Bifurcations

Figure 10.4 shows the bifurcation diagram for Eq. (10.2) as a function of the parameter a from 0 to 1. The initial condition was taken as $(-1, 2, 3)$ for each value of a. Each of the 500 values of a was calculated for a time of about 8×10^3.

The upper plot shows the three Lyapunov exponents. The middle plot shows the Kaplan–Yorke dimension, and the lower plot shows the local maxima of x. The chaotic region is in the vicinity of $a = 0.5$, and the route to chaos is clearly shown.

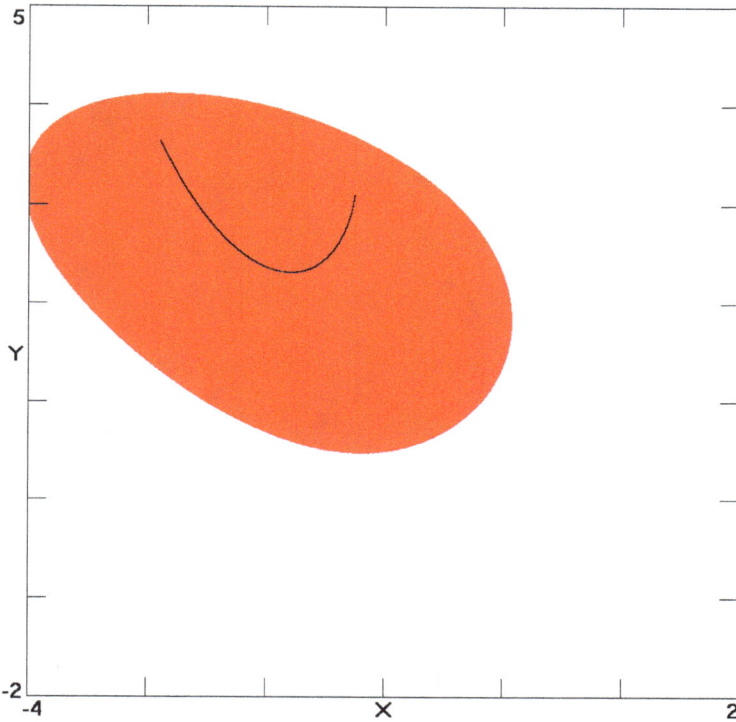

Fig. 10.3   Basin of attraction for Eq. (10.2) with $a = 0.5$ in the $z = 4$ plane.

10.9   Robustness

One measure of the robustness of a chaotic system is the amount by which the parameters can be changed from their nominal values before the probability of chaos decreases to 50% [Sprott (2022)]. For the system in Eq. (10.2) with $a = 0.5$ and initial conditions $(-1, 2, 3)$, after 10681 trials, it is estimated that the parameter a can be changed by 19% before the chaos is more likely to be lost than not. Thus the system is somewhat fragile. This result is consistent with the data in Fig. 10.4.

Elegant Automation

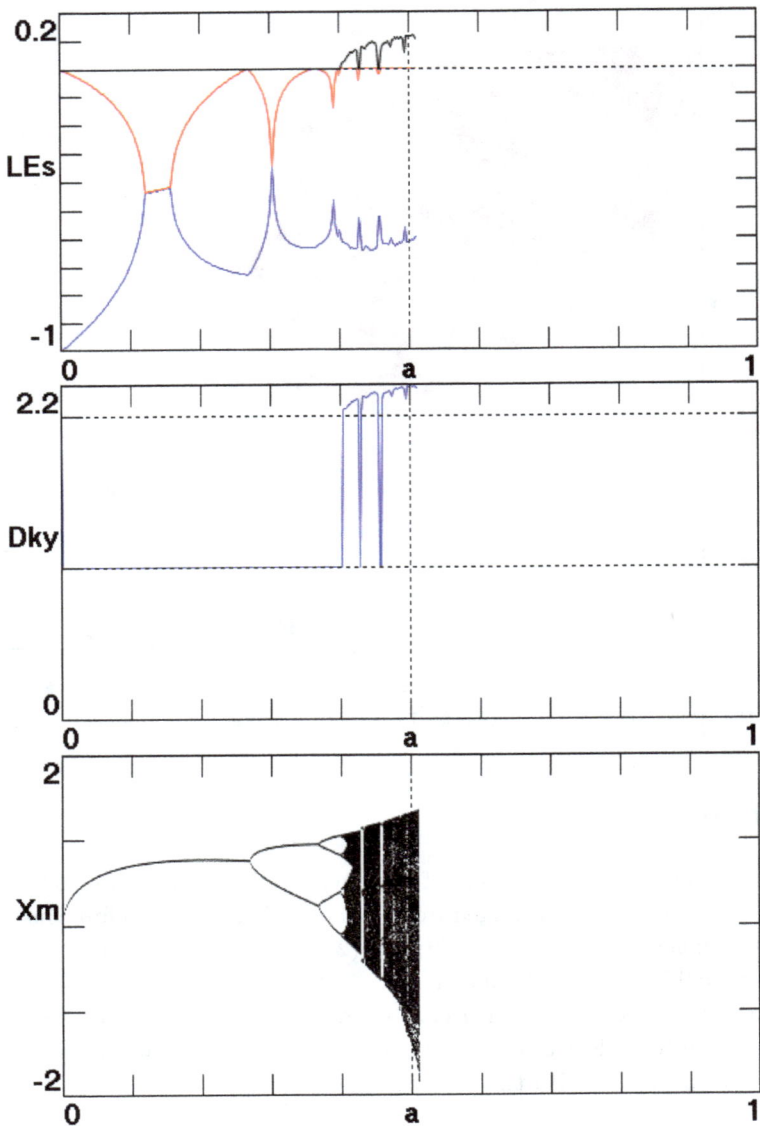

Fig. 10.4   Bifurcation diagram for Eq. (10.2) as a function of the parameter a.

Chapter 11

Sprott G System

The Sprott G system is another case with six terms and one quadratic nonlinearity that was discovered in 1994.

11.1 Introduction

The Sprott G system [Sprott (1994)] written in its most general form with an adjustable coefficient in each of the six terms is given by

$$\dot{x} = a_1 x + a_2 z$$
$$\dot{y} = a_3 xz + a_4 y \qquad (11.1)$$
$$\dot{z} = a_5 x + a_6 y.$$

The usual parameters are $a_1 = 0.4$, $a_2 = a_3 = a_6 = 1$, $a_4 = a_5 = -1$. This system was discovered in a search for the simplest chaotic system with six terms and a single quadratic nonlinearity. It is written here with parameters that can be either positive or negative with a preference for as many positive as possible.

The following sections were written by the computer program that performed the optimization, carried out the analysis of the resulting system, and produced the corresponding figures, all without human intervention.

11.2 Simplified System

After about 6×10^4 trials, of which 345 were chaotic, simplified parameters for Eq. (11.1) that give chaotic solutions are $a_1 = 0.4$, $a_2 = -1$, $a_3 = 1$, $a_4 = -1$, $a_5 = 1$, $a_6 = 1$. Thus Eq. (11.1) can be written more compactly as

$$\dot{x} = ax - z$$
$$\dot{y} = xz - y \qquad (11.2)$$
$$\dot{z} = x + y,$$

where $a = 0.4$.

Note that with six terms, Eq. (11.2) should have two independent parameters through a linear rescaling of the three variables plus time. However, one of the two parameters has a value of ± 1 and can be placed in any of the remaining five terms.

11.3 Equilibria

The system in Eq. (11.2) with $a = 0.4$ has two equilibrium points:

Equilibrium # 1 is an unstable saddle focus at $(0, 0, 0)$ with eigenvalues $(0.2 - 0.9798i, 0.2 + 0.9798i, -1)$ and a Poincaré index of -1 in the $z = 0$ plane.

Equilibrium # 2 is an unstable saddle focus at $(-2.5000, 2.5000, -1)$ with eigenvalues $(0.297, -0.4485 - 1.7791i, -0.4485 + 1.7791i)$ and a Poincaré index of -1 in the $z = -1$ plane.

The strange attractor is self-excited, and the system has no symmetry.

11.4 Attractor

Figure 11.1 shows various views of the attractor for Eq. (11.2) with $a = 0.4$ and initial conditions $(-1, 0, 2)$. The rainbow of colors shows the local value of the largest Lyapunov exponent with red indicating the most positive values (regions of worst predictability) and blue indicating the most negative values (regions of best predictability).

11.5 Time Series

Figure 11.2 shows the time series for the three variables along with the local value of the largest Lyapunov exponent (LL) for Eq. (11.2) with $a = 0.4$. Red color in the Lyapunov exponent indicates that the error is growing parallel to the orbit, while blue indicates growth perpendicular to the orbit. Note that the orbit passes through regions where the local Lyapunov exponent is strongly positive and other regions where it is strongly negative as is typical for a chaotic system and is also reflected by the colors in Fig. 11.1.

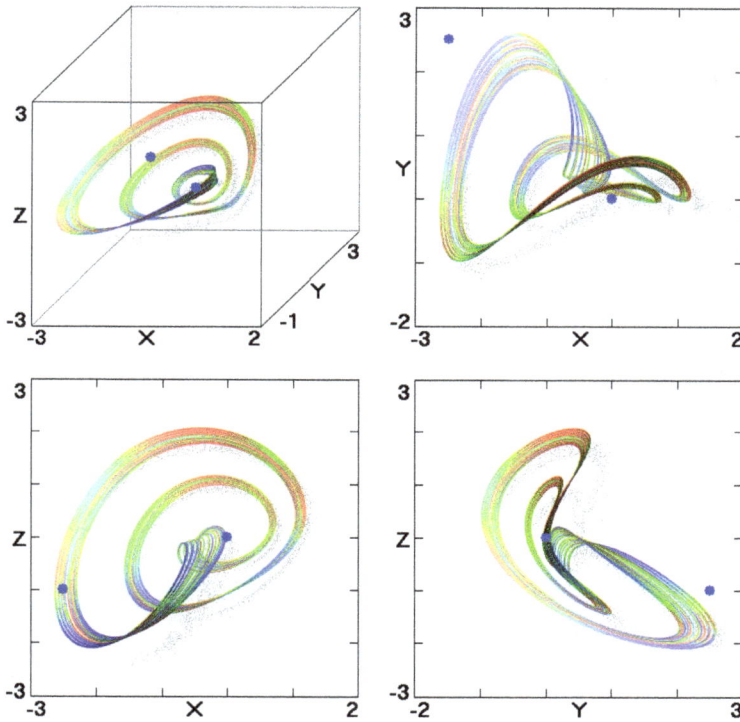

Fig. 11.1 Views of the attractor for Eq. (11.2) with $a = 0.4$ and initial conditions $(-1, 0, 2)$.

11.6 Lyapunov Exponents

The global Lyapunov exponents are determined by averaging the local Lyapunov exponents along the orbit. The values typically converge very slowly because of the large variation along the orbit, and an integration time of order 10^8 is required to obtain 4-digit accuracy.

The results of such a calculation for the system in Eq. (11.2) with $a = 0.4$ after a time of 2×10^6 are LE = $(0.0340, 0, -0.6340)$ with a Kaplan–Yorke dimension of 2.0536, where the last digit in the quoted values is only an approximation. The positive value of the largest Lyapunov exponent indicates that the system is chaotic, and the negative sum of the exponents (-0.6000) indicates that the system is dissipative with a strange attractor.

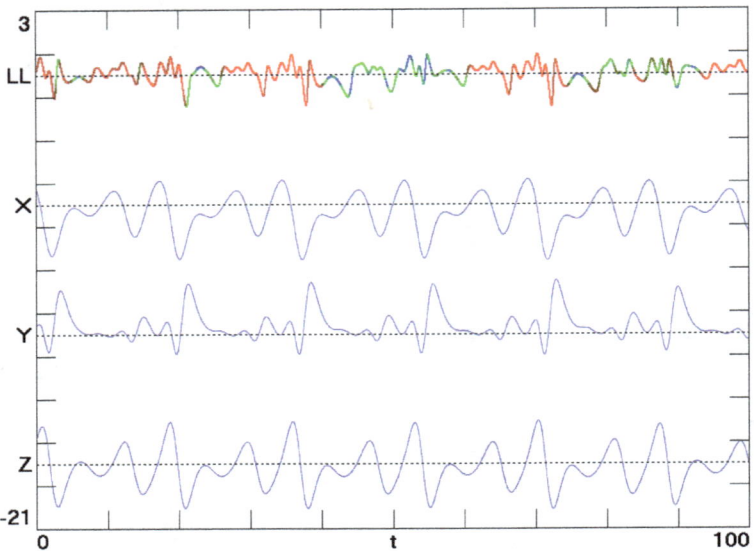

Fig. 11.2   Time series for the variables in Eq. (11.2) with $a = 0.4$ along with the local Lyapunov exponent (LL).

11.7   Basin of Attraction

Figure 11.3 shows (in red) the basin of attraction for Eq. (11.2) with $a = 0.4$ in the $z = 0$ plane. Also shown (in black) is the cross-section of the attractor in the same plane.

11.8   Bifurcations

Figure 11.4 shows the bifurcation diagram for Eq. (11.2) as a function of the parameter a from 0 to 0.8. The initial condition was taken as $(-1, 0, 2)$ for each value of a. Each of the 500 values of a was calculated for a time of about 1×10^4.

The upper plot shows the three Lyapunov exponents. The middle plot shows the Kaplan–Yorke dimension, and the lower plot shows the local maxima of x. The chaotic region is in the vicinity of $a = 0.4$, and the route to chaos is clearly shown.

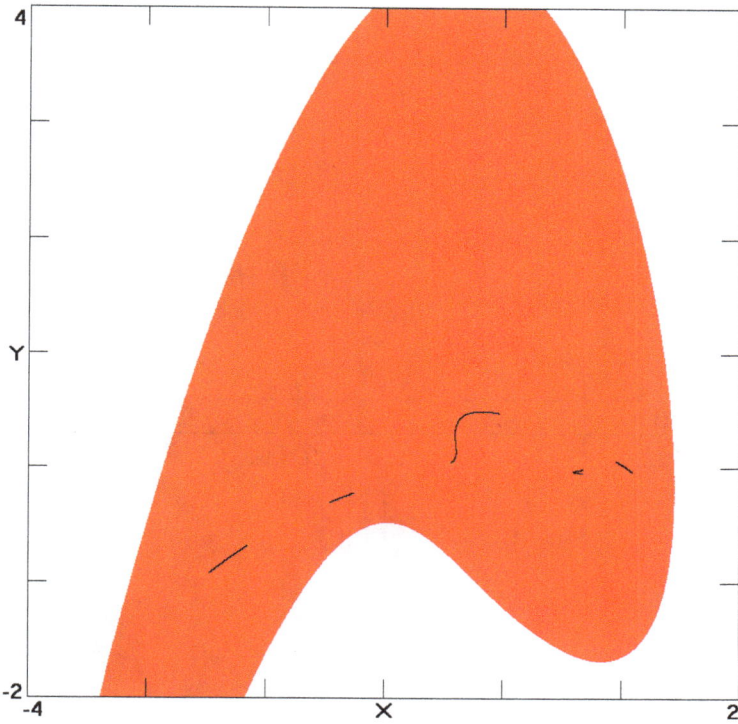

Fig. 11.3   Basin of attraction for Eq. (11.2) with $a = 0.4$ in the $z = 0$ plane.

11.9   Robustness

One measure of the robustness of a chaotic system is the amount by which the parameters can be changed from their nominal values before the probability of chaos decreases to 50% [Sprott (2022)]. For the system in Eq. (11.2) with $a = 0.4$ and initial conditions $(-1, 0, 2)$, after 9713 trials, it is estimated that the parameter a can be changed by 21% before the chaos is more likely to be lost than not. Thus the system is somewhat robust. This result is consistent with the data in Fig. 11.4.

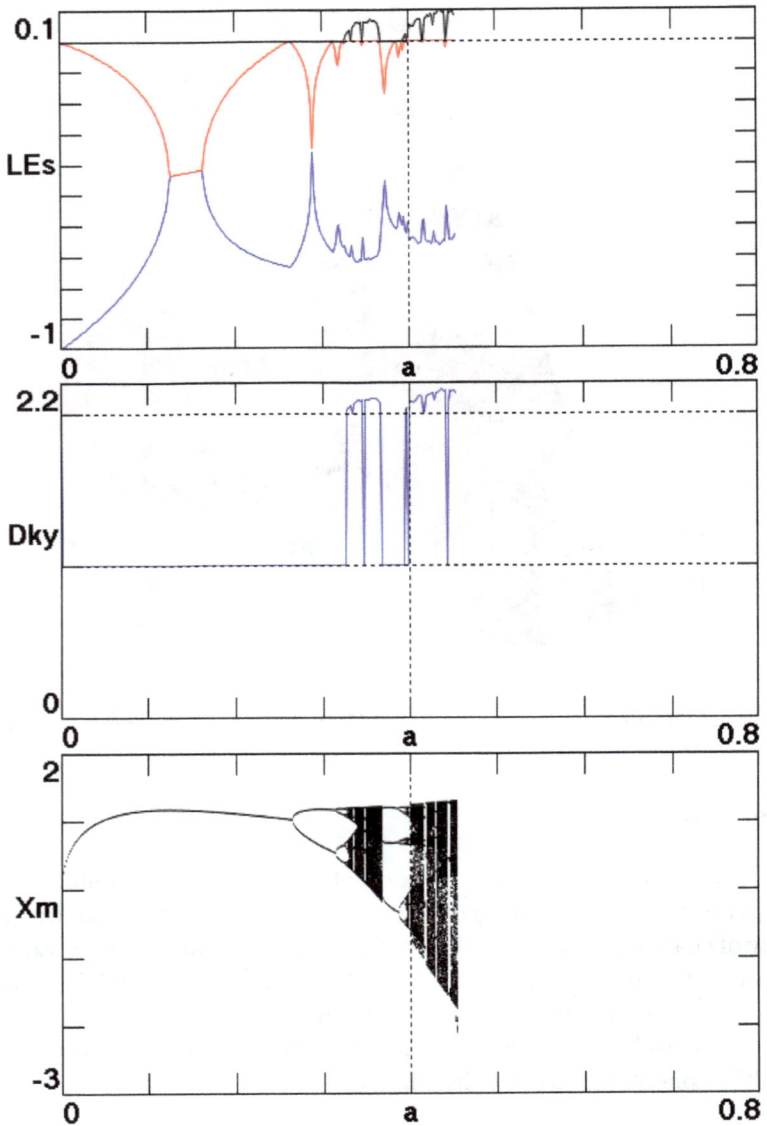

Fig. 11.4   Bifurcation diagram for Eq. (11.2) as a function of the parameter a.

Chapter 12

Sprott H System

The Sprott H system is another case with six terms and one quadratic nonlinearity that was discovered in 1994.

12.1 Introduction

The Sprott H system [Sprott (1994)] written in its most general form with an adjustable coefficient in each of the six terms is given by

$$\dot{x} = a_1 y + a_2 z^2$$
$$\dot{y} = a_3 x + a_4 y \qquad (12.1)$$
$$\dot{z} = a_5 x + a_6 z.$$

The usual parameters are $a_1 = a_6 = -1$, $a_2 = a_3 = a_5 = 1$, $a_4 = 0.5$. This system was discovered in a search for the simplest chaotic system with six terms and a single quadratic nonlinearity. It is written here with parameters that can be either positive or negative with a preference for as many positive as possible.

The following sections were written by the computer program that performed the optimization, carried out the analysis of the resulting system, and produced the corresponding figures, all without human intervention.

12.2 Simplified System

After about 6×10^4 trials, of which 336 were chaotic, simplified parameters for Eq. (12.1) that give chaotic solutions are $a_1 = 1$, $a_2 = 1$, $a_3 = -1$, $a_4 = 0.5$, $a_5 = 1$, $a_6 = -1$. Thus Eq. (12.1) can be written more compactly as

$$\dot{x} = y + z^2$$
$$\dot{y} = -x + ay \qquad (12.2)$$
$$\dot{z} = x - z,$$

where $a = 0.5$.

Note that with six terms, Eq. (12.2) should have two independent parameters through a linear rescaling of the three variables plus time. However, one of the two parameters has a value of ± 1 and can be placed in any of the remaining five terms.

12.3   Equilibria

The system in Eq. (12.2) with $a = 0.5$ has two equilibrium points:

Equilibrium # 1 is an unstable saddle focus at $(-2, -4, -2)$ with eigenvalues $(-0.3574 - 2.1274i, -0.3574 + 2.1274i, 0.2149)$ and a Poincaré index of 1 in the $z = -2$ plane.

Equilibrium # 2 is an unstable saddle focus at $(0, 0, 0)$ with eigenvalues $(0.25 - 0.9682i, 0.25 + 0.9682i, -1)$ and a Poincaré index of 1 in the $z = 0$ plane.

The strange attractor is self-excited, and the system has no symmetry.

12.4   Attractor

Figure 12.1 shows various views of the attractor for Eq. (12.2) with $a = 0.5$ and initial conditions $(1, -2, 1)$. The rainbow of colors shows the local value of the largest Lyapunov exponent with red indicating the most positive values (regions of worst predictability) and blue indicating the most negative values (regions of best predictability).

12.5   Time Series

Figure 12.2 shows the time series for the three variables along with the local value of the largest Lyapunov exponent (LL) for Eq. (12.2) with $a = 0.5$. Red color in the Lyapunov exponent indicates that the error is growing parallel to the orbit, while blue indicates growth perpendicular to the orbit. Note that the orbit passes through regions where the local Lyapunov exponent is strongly positive and other regions where it is strongly negative as is typical for a chaotic system and is also reflected by the colors in Fig. 12.1.

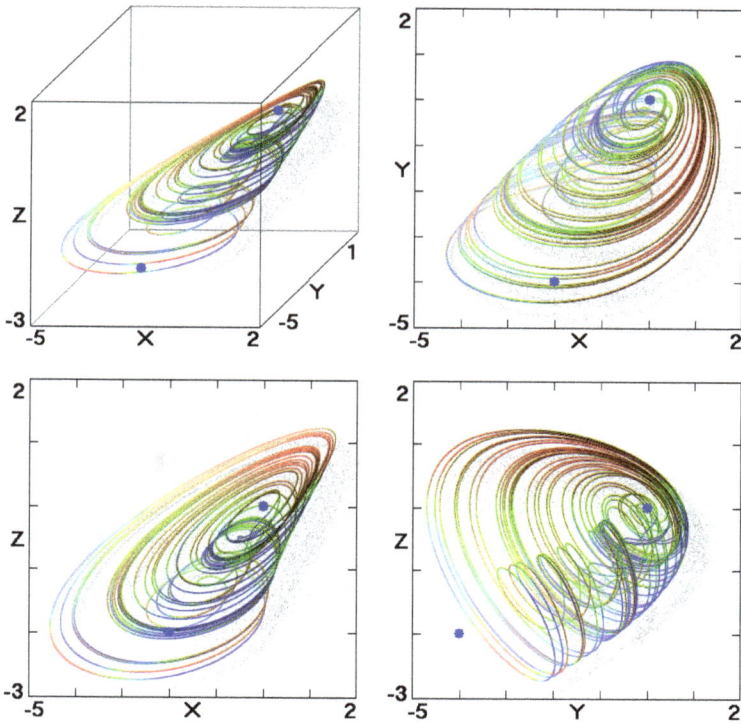

Fig. 12.1   Views of the attractor for Eq. (12.2) with $a = 0.5$ and initial conditions (1, −2, 1).

12.6   Lyapunov Exponents

The global Lyapunov exponents are determined by averaging the local Lyapunov exponents along the orbit. The values typically converge very slowly because of the large variation along the orbit, and an integration time of order 10^8 is required to obtain 4-digit accuracy.

The results of such a calculation for the system in Eq. (12.2) with $a = 0.5$ after a time of 2×10^6 are LE = (0.1170, 0, −0.6170) with a Kaplan–Yorke dimension of 2.1896, where the last digit in the quoted values is only an approximation. The positive value of the largest Lyapunov exponent indicates that the system is chaotic, and the negative sum of the exponents (−0.5000) indicates that the system is dissipative with a strange attractor.

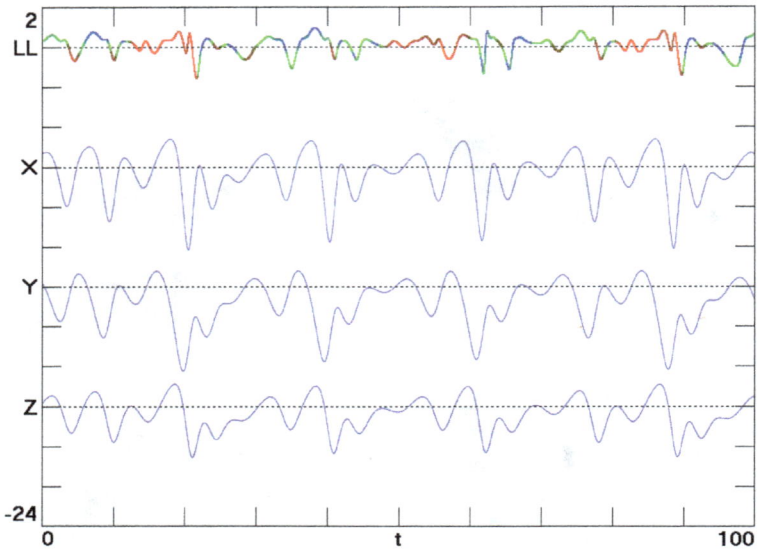

Fig. 12.2   Time series for the variables in Eq. (12.2) with $a = 0.5$ along with the local Lyapunov exponent (LL).

12.7   Basin of Attraction

Figure 12.3 shows (in red) the basin of attraction for Eq. (12.2) with $a = 0.5$ in the $z = 0$ plane. Also shown (in black) is the cross-section of the attractor in the same plane.

12.8   Bifurcations

Figure 12.4 shows the bifurcation diagram for Eq. (12.2) as a function of the parameter a from 0 to 1. The initial condition was taken as $(1, -2, 1)$ for each value of a. Each of the 500 values of a was calculated for a time of about 1×10^4.

The upper plot shows the three Lyapunov exponents. The middle plot shows the Kaplan–Yorke dimension, and the lower plot shows the local maxima of x. The chaotic region is in the vicinity of $a = 0.5$, and the route to chaos is clearly shown.

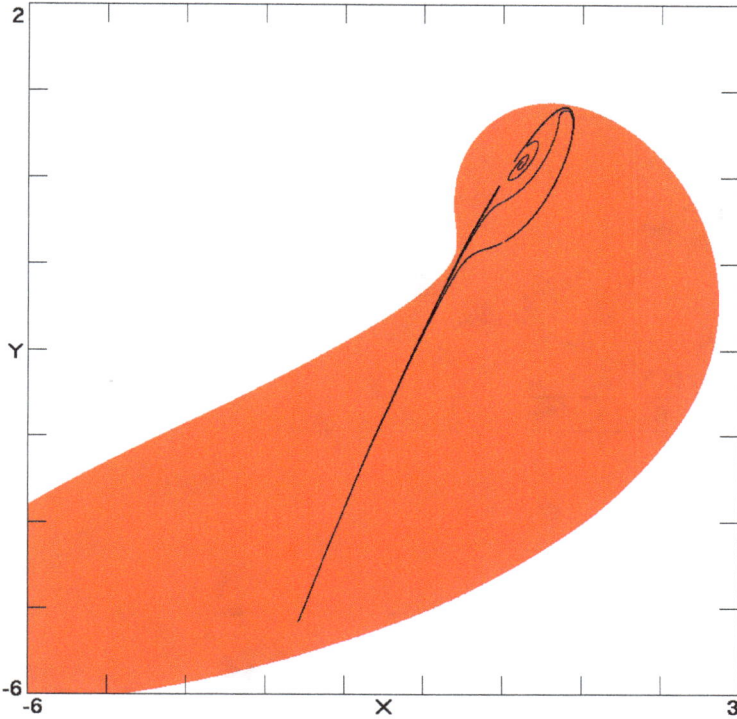

Fig. 12.3 Basin of attraction for Eq. (12.2) with $a = 0.5$ in the $z = 0$ plane.

12.9 Robustness

One measure of the robustness of a chaotic system is the amount by which the parameters can be changed from their nominal values before the probability of chaos decreases to 50% [Sprott (2022)]. For the system in Eq. (12.2) with $a = 0.5$ and initial conditions $(1, -2, 1)$, after 10693 trials, it is estimated that the parameter a can be changed by 19% before the chaos is more likely to be lost than not. Thus the system is somewhat fragile. This result is consistent with the data in Fig. 12.4.

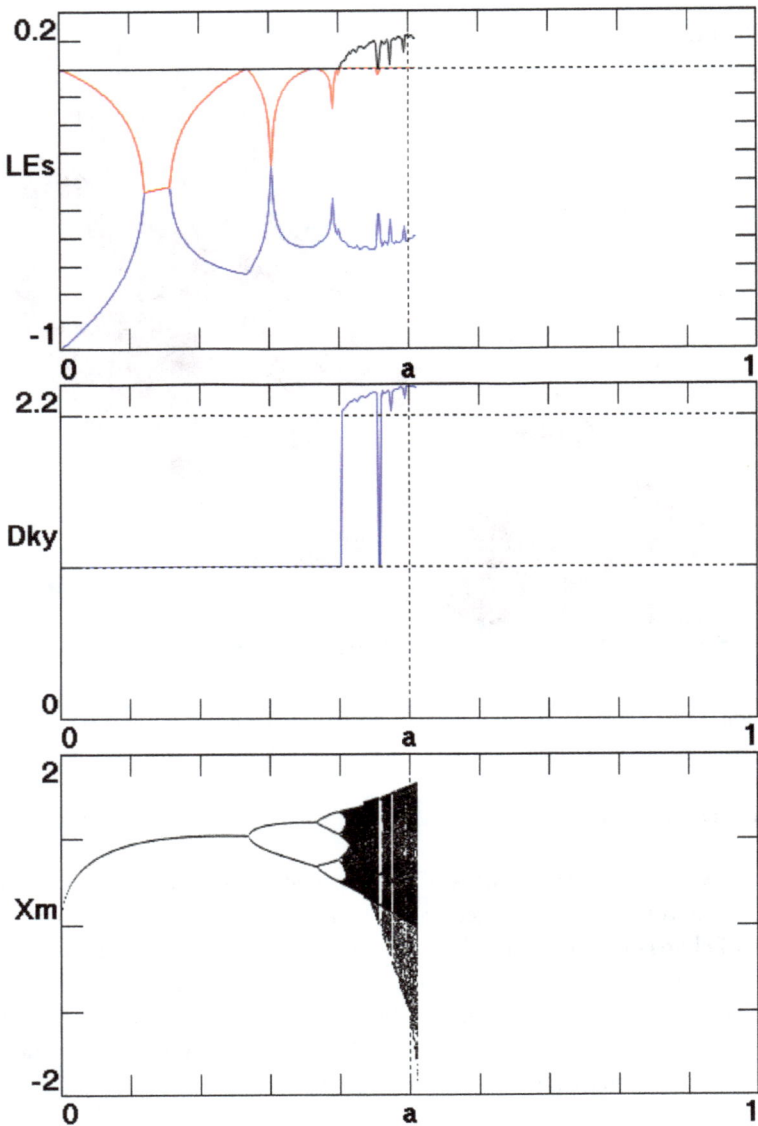

Fig. 12.4 Bifurcation diagram for Eq. (12.2) as a function of the parameter *a*.

Chapter 13

Sprott I System

The Sprott I system is another case with six terms and one quadratic non-linearity that was discovered in 1994.

13.1 Introduction

The Sprott I system [Sprott (1994)] written in its most general form with an adjustable coefficient in each of the six terms is given by

$$\dot{x} = a_1 y$$
$$\dot{y} = a_2 x + a_3 z \qquad (13.1)$$
$$\dot{z} = a_4 x + a_5 y^2 + a_6 z.$$

The usual parameters are $a_1 = -0.2$, $a_2 = a_3 = a_4 = a_5 = 1$, $a_6 = -1$. This system was discovered in a search for the simplest chaotic system with six terms and a single quadratic nonlinearity. It is written here with parameters that can be either positive or negative with a preference for as many positive as possible.

The following sections were written by the computer program that performed the optimization, carried out the analysis of the resulting system, and produced the corresponding figures, all without human intervention.

13.2 Simplified System

After about 3×10^4 trials, of which 227 were chaotic, simplified parameters for Eq. (13.1) that give chaotic solutions are $a_1 = -1$, $a_2 = 1$, $a_3 = 0.84$, $a_4 = 1$, $a_5 = 1$, $a_6 = -1$. Thus Eq. (13.1) can be written more compactly as

$$\dot{x} = -y$$
$$\dot{y} = x + az \qquad (13.2)$$
$$\dot{z} = x + y^2 - z,$$

where $a = 0.84$.

Note that with six terms, Eq. (13.2) should have two independent parameters through a linear rescaling of the three variables plus time. However, one of the two parameters has a value of ± 1 and can be placed in any of the remaining five terms.

13.3 Equilibria

The system in Eq. (13.2) with $a = 0.84$ has an unstable saddle focus at $(0, 0, 0)$ with eigenvalues $(0.1547 - 1.1753i, 0.1547 + 1.1753i, -1.3094)$ and a Poincaré index of 1 in the $z = 0$ plane.

The strange attractor is self-excited, and the system has no symmetry.

13.4 Attractor

Figure 13.1 shows various views of the attractor for Eq. (13.2) with $a = 0.84$ and initial conditions $(-8, 2, 4)$. The rainbow of colors shows the local value of the largest Lyapunov exponent with red indicating the most positive values (regions of worst predictability) and blue indicating the most negative values (regions of best predictability).

13.5 Time Series

Figure 13.2 shows the time series for the three variables along with the local value of the largest Lyapunov exponent (LL) for Eq. (13.2) with $a = 0.84$. Red color in the Lyapunov exponent indicates that the error is growing parallel to the orbit, while blue indicates growth perpendicular to the orbit. Note that the orbit passes through regions where the local Lyapunov exponent is strongly positive and other regions where it is strongly negative as is typical for a chaotic system and is also reflected by the colors in Fig. 13.1.

13.6 Lyapunov Exponents

The global Lyapunov exponents are determined by averaging the local Lyapunov exponents along the orbit. The values typically converge very slowly because of the large variation along the orbit, and an integration time of order 10^8 is required to obtain 4-digit accuracy.

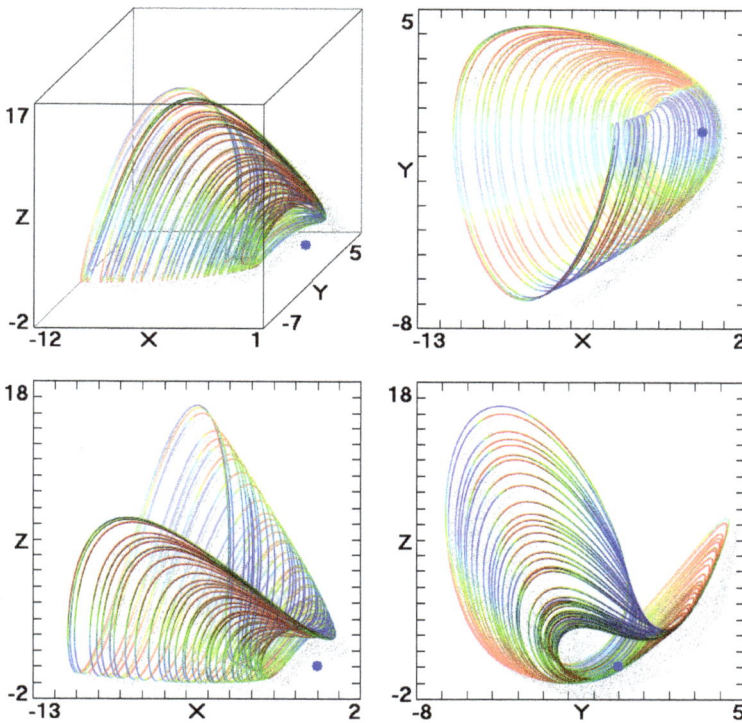

Fig. 13.1 Views of the attractor for Eq. (13.2) with $a = 0.84$ and initial conditions $(-8, 2, 4)$.

The results of such a calculation for the system in Eq. (13.2) with $a = 0.84$ after a time of 2×10^6 are LE $= (0.0860, 0, -1.0860)$ with a Kaplan–Yorke dimension of 2.0791, where the last digit in the quoted values is only an approximation. The positive value of the largest Lyapunov exponent indicates that the system is chaotic, and the negative sum of the exponents (-1) indicates that the system is dissipative with a strange attractor.

13.7 Basin of Attraction

Figure 13.3 shows (in red) the basin of attraction for Eq. (13.2) with $a = 0.84$ in the $z = 0$ plane. Also shown (in black) is the cross-section of the attractor in the same plane.

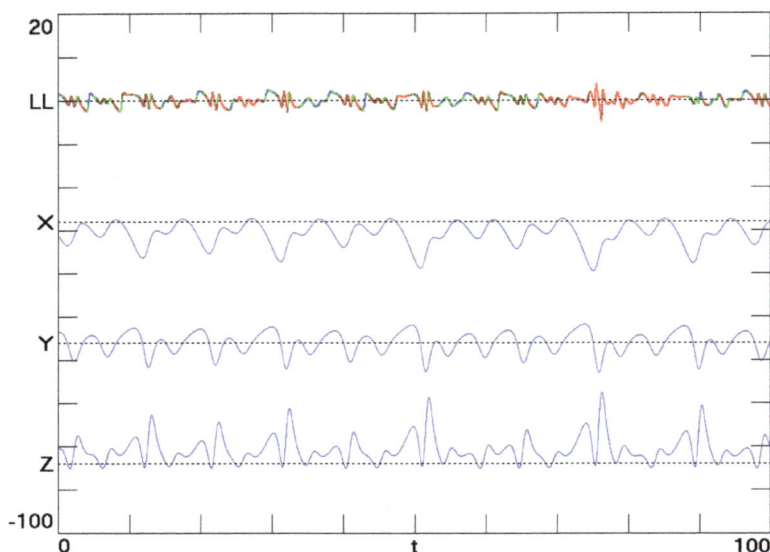

Fig. 13.2   Time series for the variables in Eq. (13.2) with $a = 0.84$ along with the local Lyapunov exponent (LL).

13.8   Bifurcations

Figure 13.4 shows the bifurcation diagram for Eq. (13.2) as a function of the parameter a from 0 to 1.68. The initial condition was taken as $(-8, 2, 4)$ for each value of a. Each of the 500 values of a was calculated for a time of about 1×10^4.

The upper plot shows the three Lyapunov exponents. The middle plot shows the Kaplan–Yorke dimension, and the lower plot shows the local maxima of x. The chaotic region is in the vicinity of $a = 0.84$, and the route to chaos is clearly shown.

13.9   Robustness

One measure of the robustness of a chaotic system is the amount by which the parameters can be changed from their nominal values before the probability of chaos decreases to 50% [Sprott (2022)]. For the system in Eq. (13.2) with $a = 0.84$ and initial conditions $(-8, 2, 4)$, after 50973 trials, it is estimated that the parameter a can be changed by 3.9% before the chaos is

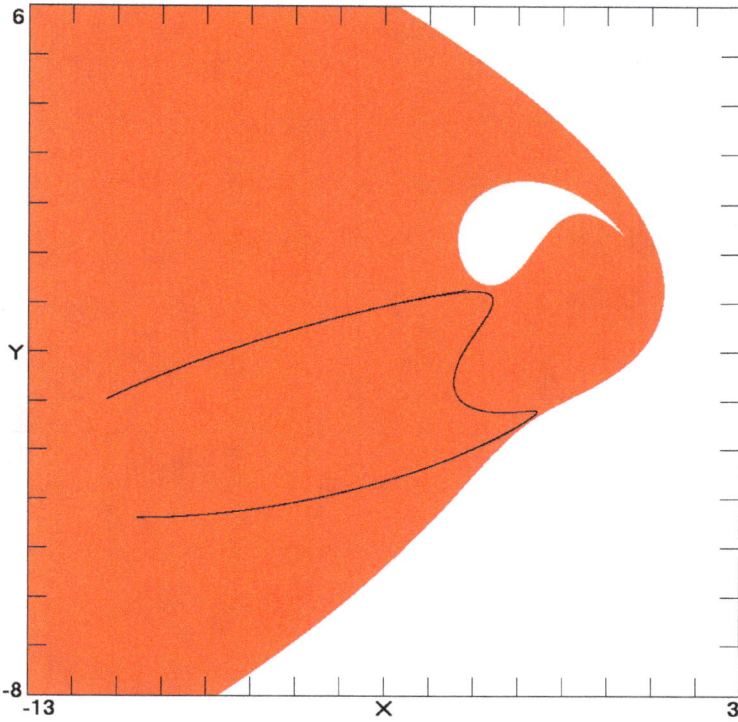

Fig. 13.3   Basin of attraction for Eq. (13.2) with $a = 0.84$ in the $z = 0$ plane.

more likely to be lost than not. Thus the system is somewhat fragile. This result is consistent with the data in Fig. 13.4.

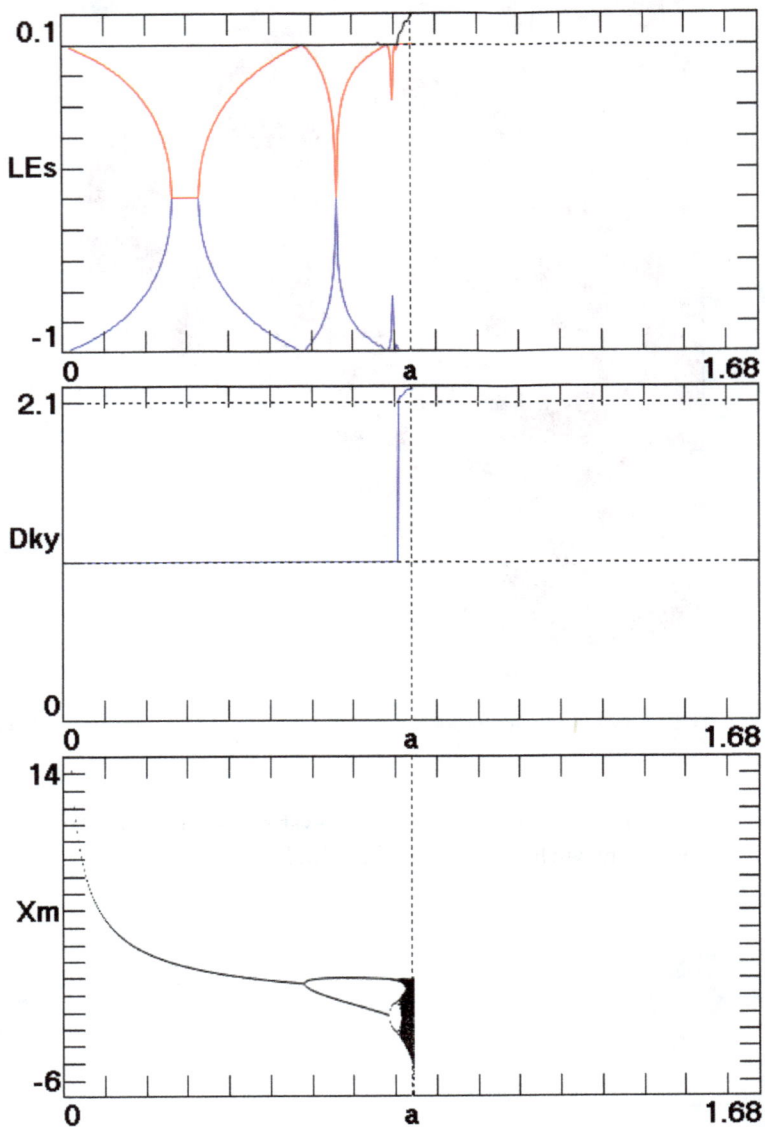

Fig. 13.4   Bifurcation diagram for Eq. (13.2) as a function of the parameter a.

Chapter 14

Sprott J System

The Sprott J system is another case with six terms and one quadratic nonlinearity that was discovered in 1994.

14.1 Introduction

The Sprott J system [Sprott (1994)] written in its most general form with an adjustable coefficient in each of the six terms is given by

$$\dot{x} = a_1 z$$
$$\dot{y} = a_2 y + a_3 z \quad\quad\quad (14.1)$$
$$\dot{z} = a_4 x + a_5 y + a_6 y^2.$$

The usual parameters are $a_1 = 2$, $a_2 = -2$, $a_3 = a_5 = a_6 = 1$, $a_4 = -1$. This system was discovered in a search for the simplest chaotic system with six terms and a single quadratic nonlinearity. It is written here with parameters that can be either positive or negative with a preference for as many positive as possible.

The following sections were written by the computer program that performed the optimization, carried out the analysis of the resulting system, and produced the corresponding figures, all without human intervention.

14.2 Simplified System

After about 3×10^4 trials, of which 195 were chaotic, simplified parameters for Eq. (14.1) that give chaotic solutions are $a_1 = 1$, $a_2 = -1$, $a_3 = 1$, $a_4 = -1$, $a_5 = 0.48$, $a_6 = 1$. Thus Eq. (14.1) can be written more compactly as

$$\dot{x} = z$$
$$\dot{y} = -y + z \quad\quad\quad (14.2)$$
$$\dot{z} = -x + ay + y^2,$$

where $a = 0.48$.

Note that with six terms, Eq. (14.2) should have two independent parameters through a linear rescaling of the three variables plus time. However, one of the two parameters has a value of ± 1 and can be placed in any of the remaining five terms.

14.3   Equilibria

The system in Eq. (14.2) with $a = 0.48$ has an unstable saddle focus at (0, 0, 0) with eigenvalues $(0.1174 - 0.8922i,\ 0.1174 + 0.8922i,\ -1.2348)$ and a Poincaré index of 0 in the $z = 0$ plane.

The strange attractor is self-excited, and the system has no symmetry.

14.4   Attractor

Figure 14.1 shows various views of the attractor for Eq. (14.2) with $a = 0.48$ and initial conditions (4, 1, 0). The rainbow of colors shows the local value of the largest Lyapunov exponent with red indicating the most positive values (regions of worst predictability) and blue indicating the most negative values (regions of best predictability).

14.5   Time Series

Figure 14.2 shows the time series for the three variables along with the local value of the largest Lyapunov exponent (LL) for Eq. (14.2) with $a = 0.48$. Red color in the Lyapunov exponent indicates that the error is growing parallel to the orbit, while blue indicates growth perpendicular to the orbit. Note that the orbit passes through regions where the local Lyapunov exponent is strongly positive and other regions where it is strongly negative as is typical for a chaotic system and is also reflected by the colors in Fig. 14.1.

14.6   Lyapunov Exponents

The global Lyapunov exponents are determined by averaging the local Lyapunov exponents along the orbit. The values typically converge very slowly because of the large variation along the orbit, and an integration time of order 10^8 is required to obtain 4-digit accuracy.

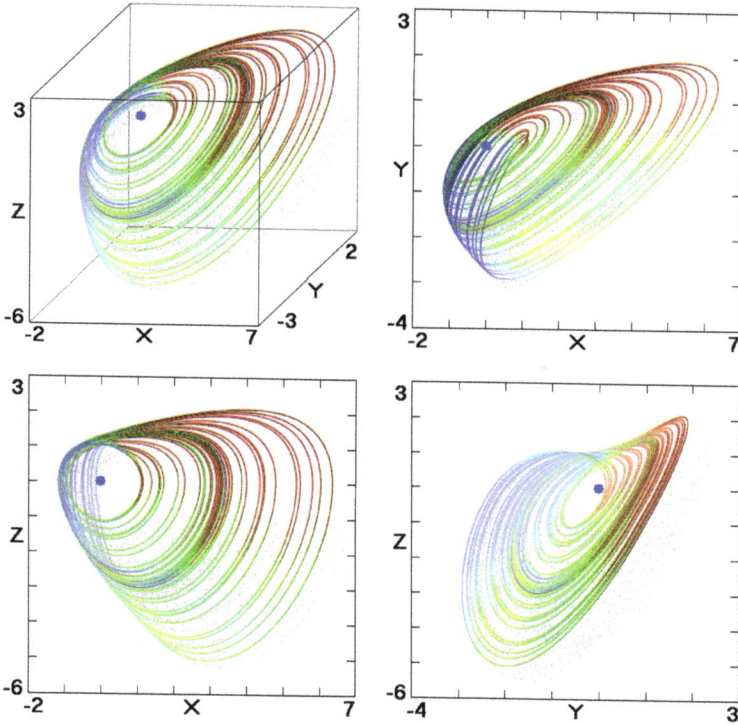

Fig. 14.1   Views of the attractor for Eq. (14.2) with $a = 0.48$ and initial conditions (4, 1, 0).

The results of such a calculation for the system in Eq. (14.2) with $a = 0.48$ after a time of 2×10^6 are LE $= (0.0501, 0, -1.0501)$ with a Kaplan–Yorke dimension of 2.0476, where the last digit in the quoted values is only an approximation. The positive value of the largest Lyapunov exponent indicates that the system is chaotic, and the negative sum of the exponents (-1) indicates that the system is dissipative with a strange attractor.

14.7   Basin of Attraction

Figure 14.3 shows (in red) the basin of attraction for Eq. (14.2) with $a = 0.48$ in the $z = 0$ plane. Also shown (in black) is the cross-section of the attractor in the same plane.

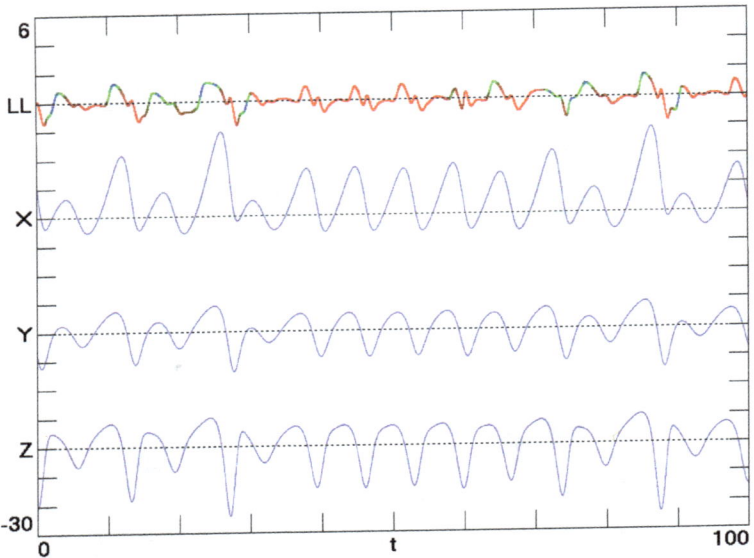

Fig. 14.2   Time series for the variables in Eq. (14.2) with $a = 0.48$ along with the local Lyapunov exponent (LL).

14.8   Bifurcations

Figure 14.4 shows the bifurcation diagram for Eq. (14.2) as a function of the parameter a from 0 to 0.96. The initial condition was taken as (4, 1, 0) for each value of a. Each of the 500 values of a was calculated for a time of about 1×10^4.

The upper plot shows the three Lyapunov exponents. The middle plot shows the Kaplan–Yorke dimension, and the lower plot shows the local maxima of x. The chaotic region is in the vicinity of $a = 0.48$, and the route to chaos is clearly shown.

14.9   Robustness

One measure of the robustness of a chaotic system is the amount by which the parameters can be changed from their nominal values before the probability of chaos decreases to 50% [Sprott (2022)]. For the system in Eq. (14.2) with $a = 0.48$ and initial conditions (4, 1, 0), after 19306 trials, it is estimated that the parameter a can be changed by 10% before the chaos is

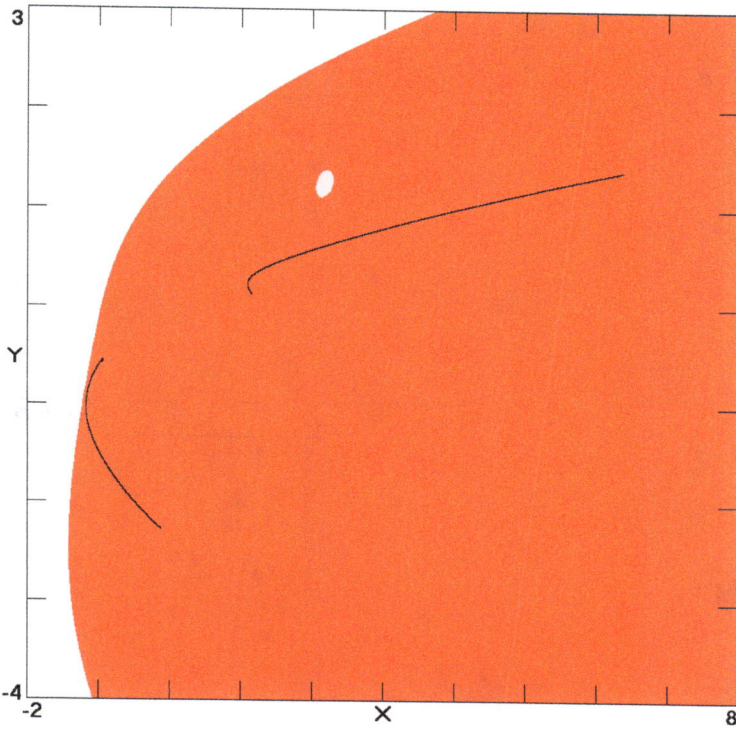

Fig. 14.3   Basin of attraction for Eq. (14.2) with $a = 0.48$ in the $z = 0$ plane.

more likely to be lost than not. Thus the system is somewhat fragile. This result is consistent with the data in Fig. 14.4.

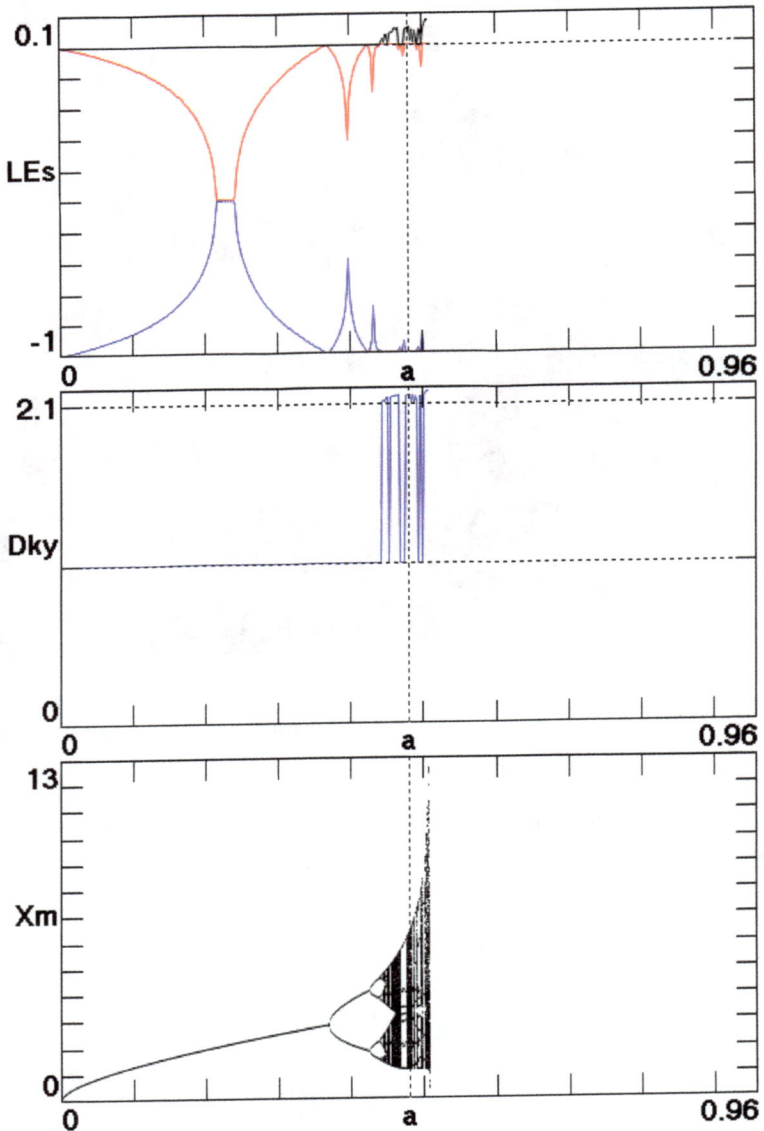

Fig. 14.4   Bifurcation diagram for Eq. (14.2) as a function of the parameter a.

Chapter 15

Sprott K System

The Sprott K system is another case with six terms and one quadratic nonlinearity that was discovered in 1994.

15.1 Introduction

The Sprott K system [Sprott (1994)] written in its most general form with an adjustable coefficient in each of the six terms is given by

$$\dot{x} = a_1 xy + a_2 z$$
$$\dot{y} = a_3 x + a_4 y \qquad (15.1)$$
$$\dot{z} = a_5 x + a_6 z.$$

The usual parameters are $a_1 = a_3 = a_5 = 1$, $a_2 = a_4 = -1$, $a_6 = 0.3$. This system was discovered in a search for the simplest chaotic system with six terms and a single quadratic nonlinearity. It is written here with parameters that can be either positive or negative with a preference for as many positive as possible.

The following sections were written by the computer program that performed the optimization, carried out the analysis of the resulting system, and produced the corresponding figures, all without human intervention.

15.2 Simplified System

After about 2×10^4 trials, of which 228 were chaotic, simplified parameters for Eq. (15.1) that give chaotic solutions are $a_1 = 1$, $a_2 = 1$, $a_3 = 1$, $a_4 = -1$, $a_5 = -1$, $a_6 = 0.3$. Thus Eq. (15.1) can be written more compactly as

$$\dot{x} = xy + z$$
$$\dot{y} = x - y \qquad (15.2)$$
$$\dot{z} = -x + az,$$

where $a = 0.3$.

Note that with six terms, Eq. (15.2) should have two independent parameters through a linear rescaling of the three variables plus time. However, one of the two parameters has a value of ± 1 and can be placed in any of the remaining five terms.

15.3   Equilibria

The system in Eq. (15.2) with $a = 0.3$ has two equilibrium points:

Equilibrium # 1 is an unstable saddle focus at $(0, 0, 0)$ with eigenvalues $(0.15 - 0.9887i, 0.15 + 0.9887i, -1)$ and a Poincaré index of 0 in the $z = 0$ plane.

Equilibrium # 2 is an unstable saddle focus at $(-3.3333, -3.3333, -11.1111)$ with eigenvalues $(-2.0884 - 1.6139i, -2.0884 + 1.6139i, 0.1435)$ and a Poincaré index of 1 in the $z = -11.1111$ plane.

The strange attractor is self-excited, and the system has no symmetry.

15.4   Attractor

Figure 15.1 shows various views of the attractor for Eq. (15.2) with $a = 0.3$ and initial conditions $(-1, -1, 0)$. The rainbow of colors shows the local value of the largest Lyapunov exponent with red indicating the most positive values (regions of worst predictability) and blue indicating the most negative values (regions of best predictability).

15.5   Time Series

Figure 15.2 shows the time series for the three variables along with the local value of the largest Lyapunov exponent (LL) for Eq. (15.2) with $a = 0.3$. Red color in the Lyapunov exponent indicates that the error is growing parallel to the orbit, while blue indicates growth perpendicular to the orbit. Note that the orbit passes through regions where the local Lyapunov exponent is strongly positive and other regions where it is strongly negative as is typical for a chaotic system and is also reflected by the colors in Fig. 15.1.

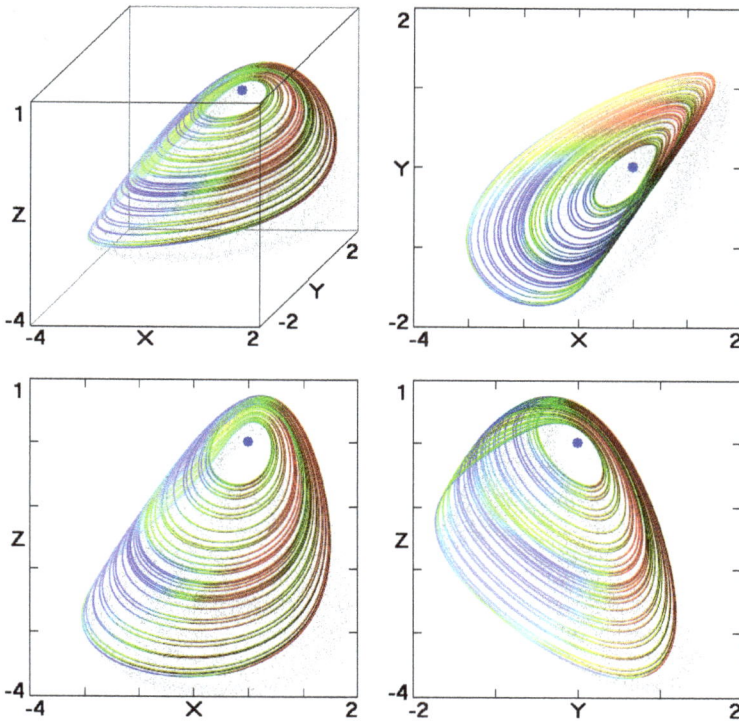

Fig. 15.1 Views of the attractor for Eq. (15.2) with $a = 0.3$ and initial conditions $(-1, -1, 0)$.

15.6   Lyapunov Exponents

The global Lyapunov exponents are determined by averaging the local Lyapunov exponents along the orbit. The values typically converge very slowly because of the large variation along the orbit, and an integration time of order 10^8 is required to obtain 4-digit accuracy.

The results of such a calculation for the system in Eq. (15.2) with $a = 0.3$ after a time of 2×10^6 are LE = $(0.0375, 0, -0.8900)$ with a Kaplan–Yorke dimension of 2.0421, where the last digit in the quoted values is only an approximation. The positive value of the largest Lyapunov exponent indicates that the system is chaotic, and the negative sum of the exponents (-0.8525) indicates that the system is dissipative with a strange attractor.

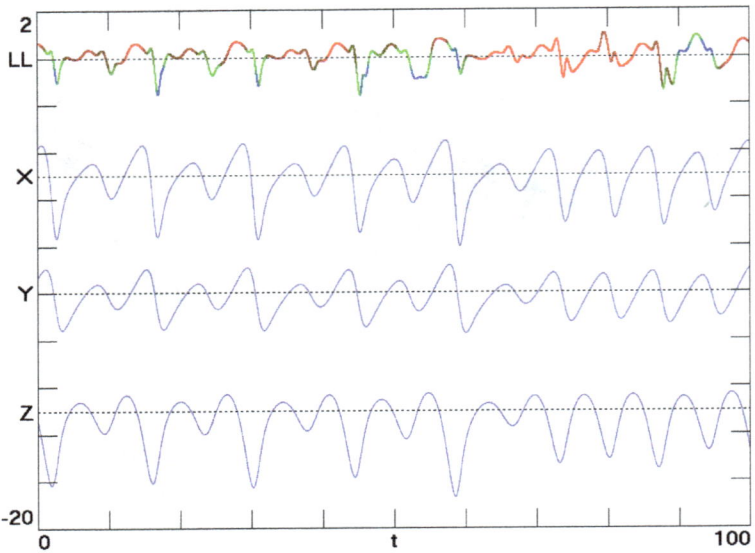

Fig. 15.2 Time series for the variables in Eq. (15.2) with $a = 0.3$ along with the local Lyapunov exponent (LL).

15.7 Basin of Attraction

Figure 15.3 shows (in red) the basin of attraction for Eq. (15.2) with $a = 0.3$ in the $z = 0$ plane. Also shown (in black) is the cross-section of the attractor in the same plane.

15.8 Bifurcations

Figure 15.4 shows the bifurcation diagram for Eq. (15.2) as a function of the parameter a from 0 to 0.6. The initial condition was taken as $(-1, -1, 0)$ for each value of a. Each of the 500 values of a was calculated for a time of about 1×10^4.

The upper plot shows the three Lyapunov exponents. The middle plot shows the Kaplan–Yorke dimension, and the lower plot shows the local maxima of x. The chaotic region is in the vicinity of $a = 0.3$, and the route to chaos is clearly shown.

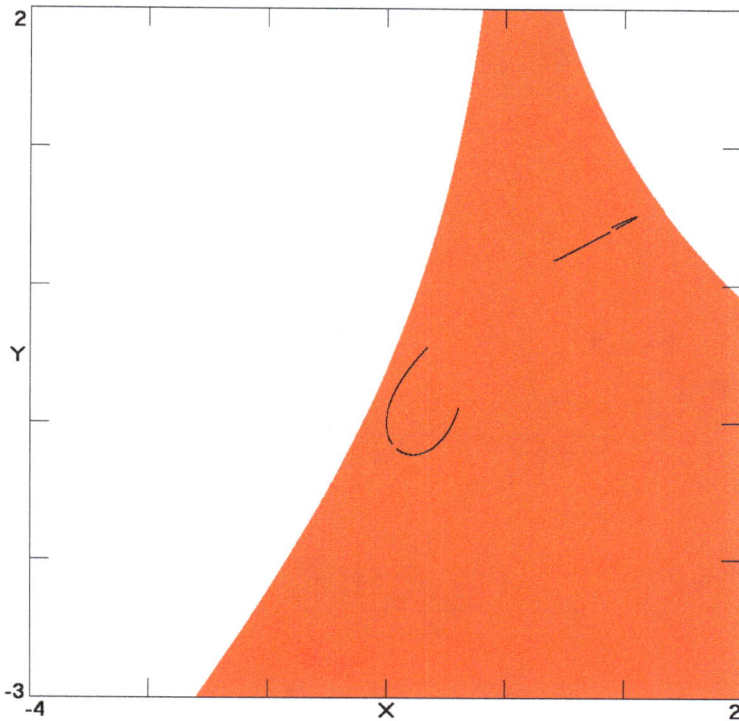

Fig. 15.3   Basin of attraction for Eq. (15.2) with $a = 0.3$ in the $z = 0$ plane.

15.9   Robustness

One measure of the robustness of a chaotic system is the amount by which the parameters can be changed from their nominal values before the probability of chaos decreases to 50% [Sprott (2022)]. For the system in Eq. (15.2) with $a = 0.3$ and initial conditions $(-1, -1, 0)$, after 14722 trials, it is estimated that the parameter a can be changed by 14% before the chaos is more likely to be lost than not. Thus the system is somewhat fragile. This result is consistent with the data in Fig. 15.4.

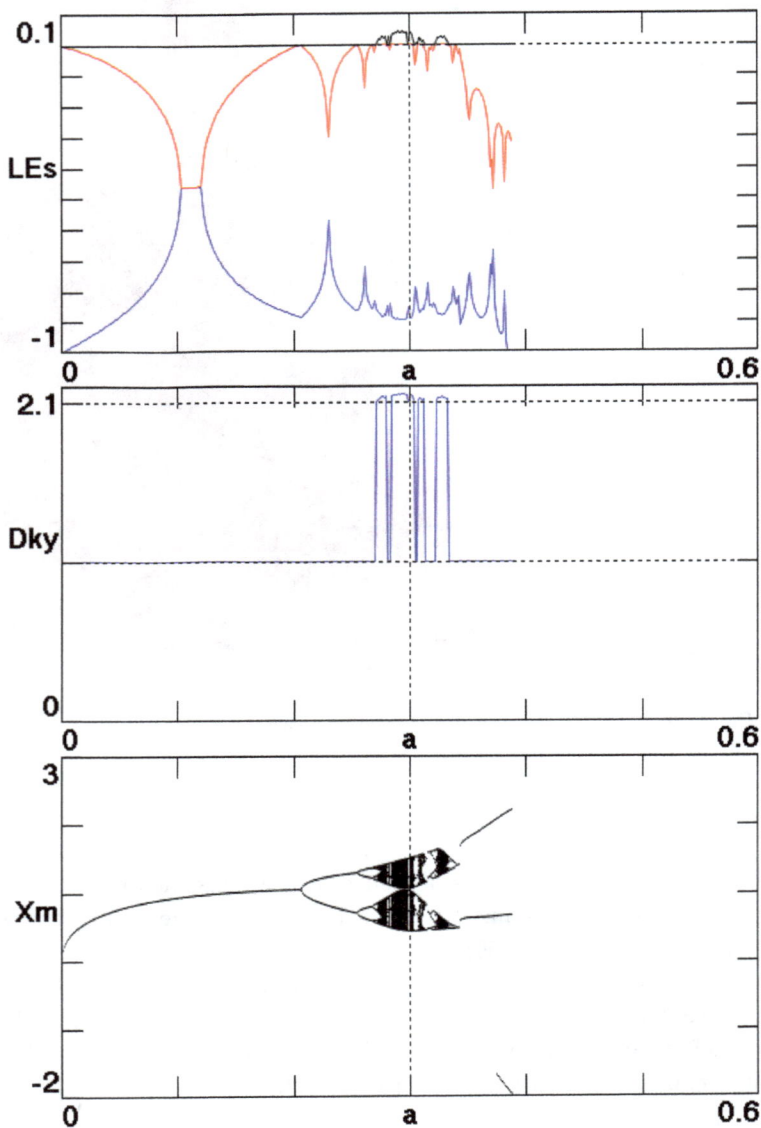

Fig. 15.4   Bifurcation diagram for Eq. (15.2) as a function of the parameter a.

Chapter 16

Sprott L System

The Sprott L system is another case with six terms and one quadratic nonlinearity that was discovered in 1994.

16.1 Introduction

The Sprott L system [Sprott (1994)] written in its most general form with an adjustable coefficient in each of the six terms is given by

$$\dot{x} = a_1 y + a_2 z$$
$$\dot{y} = a_3 x^2 + a_4 y \qquad (16.1)$$
$$\dot{z} = a_5 + a_6 x.$$

The usual parameters are $a_1 = a_5 = 1$, $a_2 = 3.9$, $a_3 = 0.9$, $a_4 = a_6 = -1$. This system was discovered in a search for the simplest chaotic system with six terms and a single quadratic nonlinearity. It is written here with parameters that can be either positive or negative with a preference for as many positive as possible.

The following sections were written by the computer program that performed the optimization, carried out the analysis of the resulting system, and produced the corresponding figures, all without human intervention.

16.2 Simplified System

After about 4×10^4 trials, of which 155 were chaotic, simplified parameters for Eq. (16.1) that give chaotic solutions are $a_1 = 1$, $a_2 = 1$, $a_3 = 0.24$, $a_4 = -1$, $a_5 = 1$, $a_6 = -1$. Thus Eq. (16.1) can be written more compactly as

$$\dot{x} = y + z$$
$$\dot{y} = ax^2 - y \qquad (16.2)$$
$$\dot{z} = 1 - x,$$

where $a = 0.24$.

Note that with six terms, Eq. (16.2) should have two independent parameters through a linear rescaling of the three variables plus time. However, one of the two parameters has a value of ± 1 and can be placed in any of the remaining five terms.

16.3 Equilibria

The system in Eq. (16.2) with $a = 0.24$ has an unstable saddle focus at $(1, 0.2400, -0.2400)$ with eigenvalues $(0.1174 - 0.8922i, 0.1174 + 0.8922i, -1.2348)$ and a Poincaré index of -1 in the $z = -0.24$ plane.

The strange attractor is self-excited, and the system has no symmetry.

16.4 Attractor

Figure 16.1 shows various views of the attractor for Eq. (16.2) with $a = 0.24$ and initial conditions $(-6, 4, -8)$. The rainbow of colors shows the local value of the largest Lyapunov exponent with red indicating the most positive values (regions of worst predictability) and blue indicating the most negative values (regions of best predictability).

16.5 Time Series

Figure 16.2 shows the time series for the three variables along with the local value of the largest Lyapunov exponent (LL) for Eq. (16.2) with $a = 0.24$. Red color in the Lyapunov exponent indicates that the error is growing parallel to the orbit, while blue indicates growth perpendicular to the orbit. Note that the orbit passes through regions where the local Lyapunov exponent is strongly positive and other regions where it is strongly negative as is typical for a chaotic system and is also reflected by the colors in Fig. 16.1.

16.6 Lyapunov Exponents

The global Lyapunov exponents are determined by averaging the local Lyapunov exponents along the orbit. The values typically converge very slowly because of the large variation along the orbit, and an integration time of order 10^8 is required to obtain 4-digit accuracy.

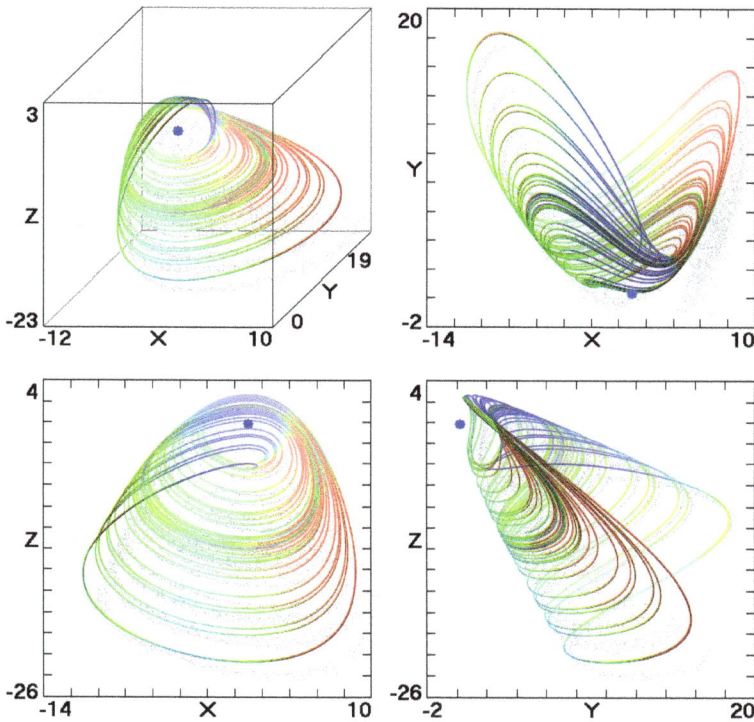

Fig. 16.1   Views of the attractor for Eq. (16.2) with $a = 0.24$ and initial conditions $(-6, 4, -8)$.

The results of such a calculation for the system in Eq. (16.2) with $a = 0.24$ after a time of 2×10^6 are LE $= (0.0502, 0, -1.0502)$ with a Kaplan–Yorke dimension of 2.0477, where the last digit in the quoted values is only an approximation. The positive value of the largest Lyapunov exponent indicates that the system is chaotic, and the negative sum of the exponents (-1) indicates that the system is dissipative with a strange attractor.

16.7   Basin of Attraction

Figure 16.3 shows (in red) the basin of attraction for Eq. (16.2) with $a = 0.24$ in the $z = 0$ plane. Also shown (in black) is the cross-section of the attractor in the same plane.

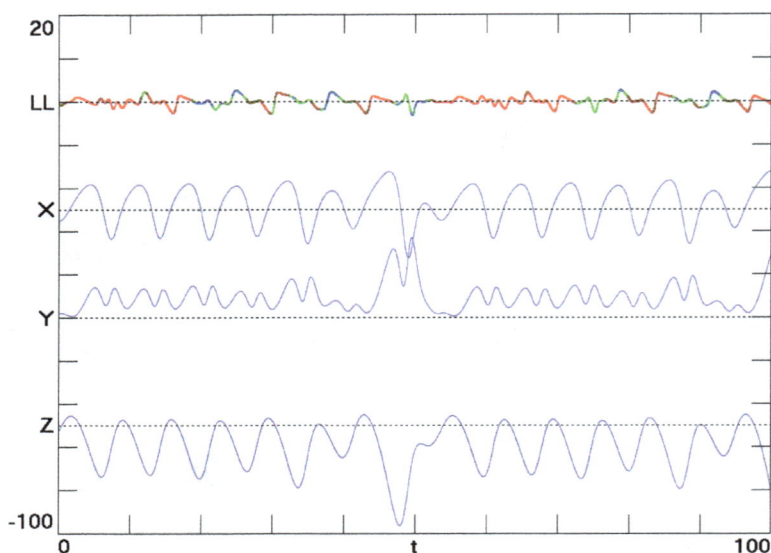

Fig. 16.2   Time series for the variables in Eq. (16.2) with $a = 0.24$ along with the local Lyapunov exponent (LL).

16.8   Bifurcations

Figure 16.4 shows the bifurcation diagram for Eq. (16.2) as a function of the parameter a from 0 to 0.48. The initial condition was taken as $(-6, 4, -8)$ for each value of a. Each of the 500 values of a was calculated for a time of about 2×10^4.

The upper plot shows the three Lyapunov exponents. The middle plot shows the Kaplan–Yorke dimension, and the lower plot shows the local maxima of x. The chaotic region is in the vicinity of $a = 0.24$, and the route to chaos is clearly shown.

16.9   Robustness

One measure of the robustness of a chaotic system is the amount by which the parameters can be changed from their nominal values before the probability of chaos decreases to 50% [Sprott (2022)]. For the system in Eq. (16.2) with $a = 0.24$ and initial conditions $(-6, 4, -8)$, after 19936 trials, it is estimated that the parameter a can be changed by 9.9% before the chaos is

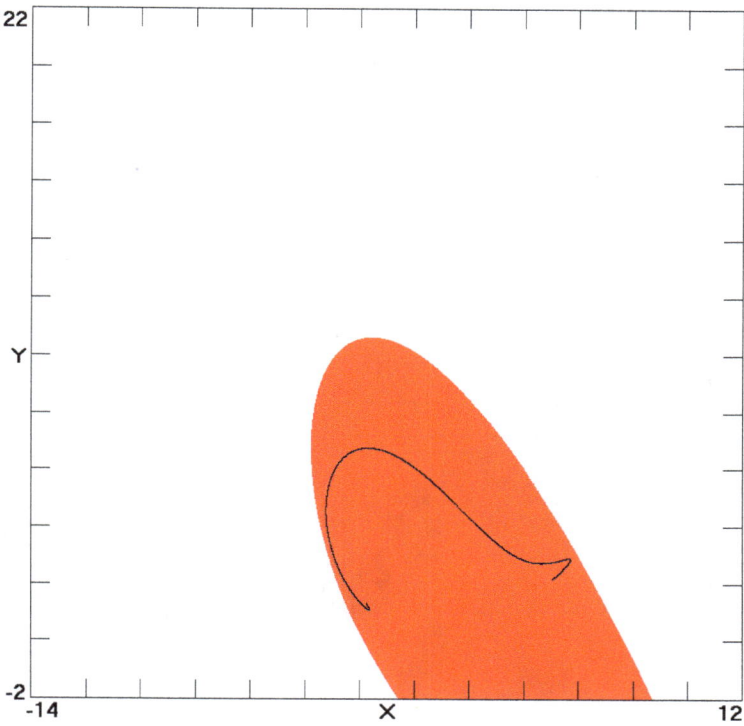

Fig. 16.3   Basin of attraction for Eq. (16.2) with $a = 0.24$ in the $z = 0$ plane.

more likely to be lost than not. Thus the system is somewhat fragile. This result is consistent with the data in Fig. 16.4.

Elegant Automation

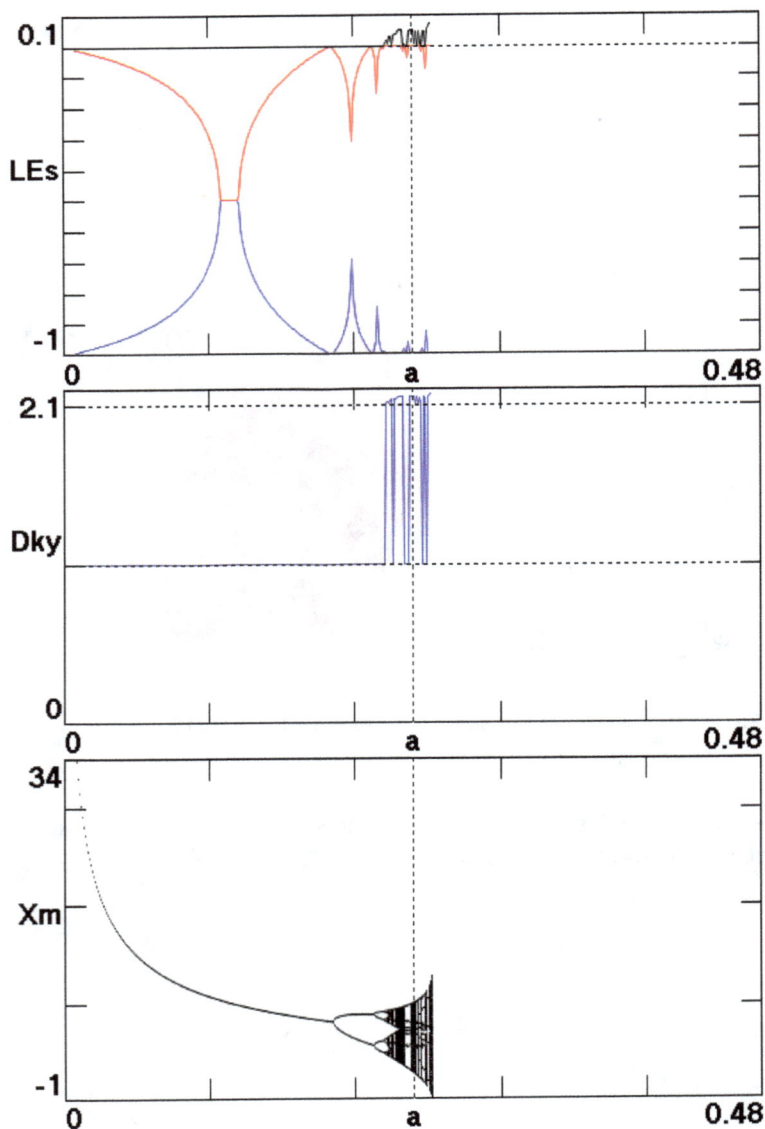

Fig. 16.4 Bifurcation diagram for Eq. (16.2) as a function of the parameter *a*.

Chapter 17

Sprott M System

The Sprott M system is another case with six terms and one quadratic nonlinearity that was discovered in 1994.

17.1 Introduction

The Sprott M system [Sprott (1994)] written in its most general form with an adjustable coefficient in each of the six terms is given by

$$\dot{x} = a_1 z$$
$$\dot{y} = a_2 x^2 + a_3 y \qquad (17.1)$$
$$\dot{z} = a_4 + a_5 x + a_6 y.$$

The usual parameters are $a_1 = a_2 = a_3 = -1$, $a_4 = a_5 = 1.7$, $a_6 = 1$. This system was discovered in a search for the simplest chaotic system with six terms and a single quadratic nonlinearity. It is written here with parameters that can be either positive or negative with a preference for as many positive as possible.

The following sections were written by the computer program that performed the optimization, carried out the analysis of the resulting system, and produced the corresponding figures, all without human intervention.

17.2 Simplified System

After about 8×10^4 trials, of which 23 were chaotic, simplified parameters for Eq. (17.1) that give chaotic solutions are $a_1 = -8$, $a_2 = 1$, $a_3 = -1$, $a_4 = 1$, $a_5 = 1$, $a_6 = -1$. Thus Eq. (17.1) can be written more compactly as

$$\dot{x} = az$$
$$\dot{y} = x^2 - y \qquad (17.2)$$
$$\dot{z} = 1 + x - y,$$

where $a = -8$.

Note that with six terms, Eq. (17.2) should have two independent parameters through a linear rescaling of the three variables plus time. However, one of the two parameters has a value of ± 1 and can be placed in any of the remaining five terms.

17.3   Equilibria

The system in Eq. (17.2) with $a = -8$ has two equilibrium points:

Equilibrium # 1 is an unstable saddle focus at $(1.6180, 2.6180, 0)$ with eigenvalues $(-1.2574 - 3.1982i, -1.2574 + 3.1982i, 1.5148)$ and a Poincaré index of 0 in the $z = 0$ plane.

Equilibrium # 2 is an unstable saddle focus at $(-0.6180, 0.3820, 0)$ with eigenvalues $(0.4311 - 3.0692i, 0.4311 + 3.0692i, -1.8623)$ and a Poincaré index of 0 in the $z = 0$ plane.

The strange attractor is self-excited, and the system has no symmetry.

17.4   Attractor

Figure 17.1 shows various views of the attractor for Eq. (17.2) with $a = -8$ and initial conditions $(-1, 2, 0)$. The rainbow of colors shows the local value of the largest Lyapunov exponent with red indicating the most positive values (regions of worst predictability) and blue indicating the most negative values (regions of best predictability).

17.5   Time Series

Figure 17.2 shows the time series for the three variables along with the local value of the largest Lyapunov exponent (LL) for Eq. (17.2) with $a = -8$. Red color in the Lyapunov exponent indicates that the error is growing parallel to the orbit, while blue indicates growth perpendicular to the orbit. Note that the orbit passes through regions where the local Lyapunov exponent is strongly positive and other regions where it is strongly negative as is typical for a chaotic system and is also reflected by the colors in Fig. 17.1.

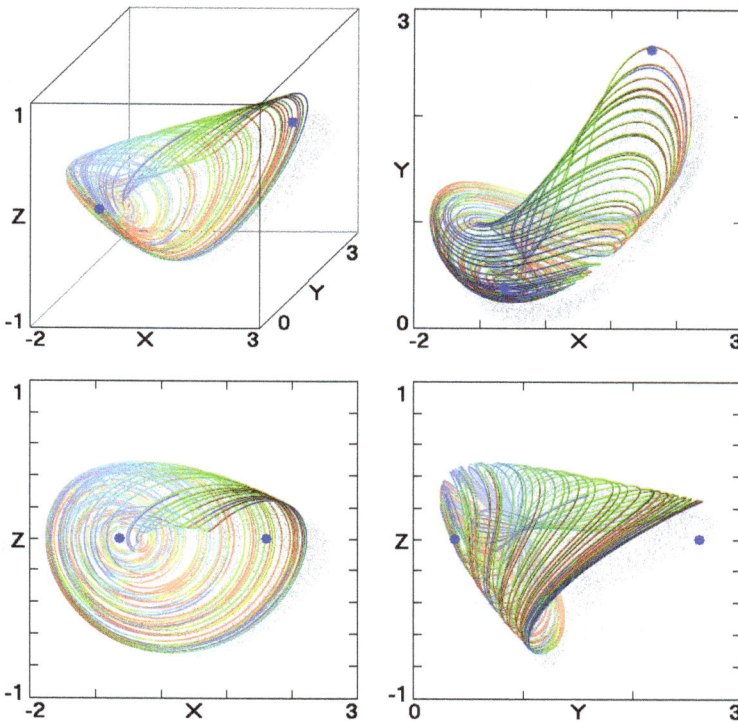

Fig. 17.1   Views of the attractor for Eq. (17.2) with $a = -8$ and initial conditions $(-1, 2, 0)$.

17.6   Lyapunov Exponents

The global Lyapunov exponents are determined by averaging the local Lyapunov exponents along the orbit. The values typically converge very slowly because of the large variation along the orbit, and an integration time of order 10^8 is required to obtain 4-digit accuracy.

The results of such a calculation for the system in Eq. (17.2) with $a = -8$ after a time of 4×10^6 are LE $= (0.2539, 0, -1.2539)$ with a Kaplan–Yorke dimension of 2.2024, where the last digit in the quoted values is only an approximation. The positive value of the largest Lyapunov exponent indicates that the system is chaotic, and the negative sum of the exponents (-1) indicates that the system is dissipative with a strange attractor.

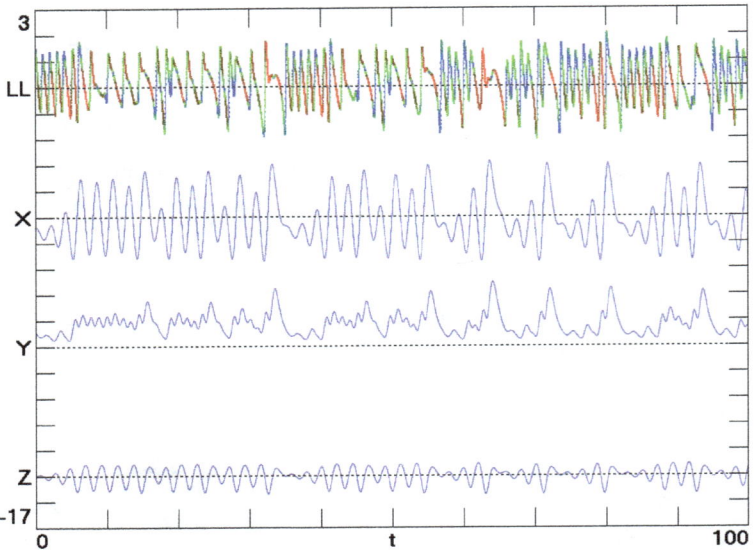

Fig. 17.2   Time series for the variables in Eq. (17.2) with $a = -8$ along with the local Lyapunov exponent (LL).

17.7   Basin of Attraction

Figure 17.3 shows (in red) the basin of attraction for Eq. (17.2) with $a = -8$ in the $z = 0$ plane. Also shown (in black) is the cross-section of the attractor in the same plane.

17.8   Bifurcations

Figure 17.4 shows the bifurcation diagram for Eq. (17.2) as a function of the parameter a from $a = -16$ to $a = 0$. The initial condition was taken as $(-1, 2, 0)$ for each value of a. Each of the 500 values of a was calculated for a time of about 1×10^4.

The upper plot shows the three Lyapunov exponents. The middle plot shows the Kaplan–Yorke dimension, and the lower plot shows the local maxima of x. The chaotic region is in the vicinity of $a = -8$, and the route to chaos is clearly shown.

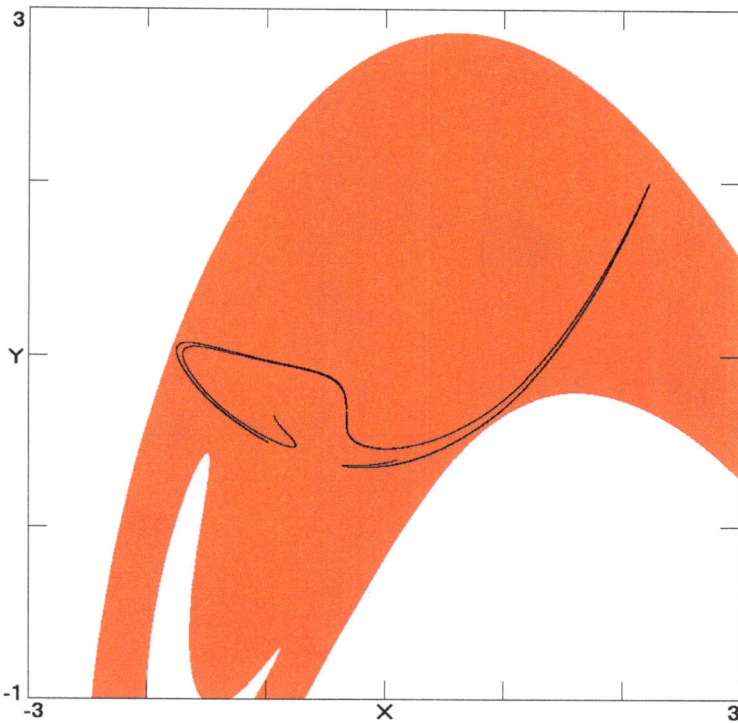

Fig. 17.3   Basin of attraction for Eq. (17.2) with $a = -8$ in the $z = 0$ plane.

17.9   Robustness

One measure of the robustness of a chaotic system is the amount by which the parameters can be changed from their nominal values before the probability of chaos decreases to 50% [Sprott (2022)]. For the system in Eq. (17.2) with $a = -8$ and initial conditions $(-1, 2, 0)$, after 4516 trials, it is estimated that the parameter a can be changed by 46% before the chaos is more likely to be lost than not. Thus the system is somewhat robust. This result is consistent with the data in Fig. 17.4.

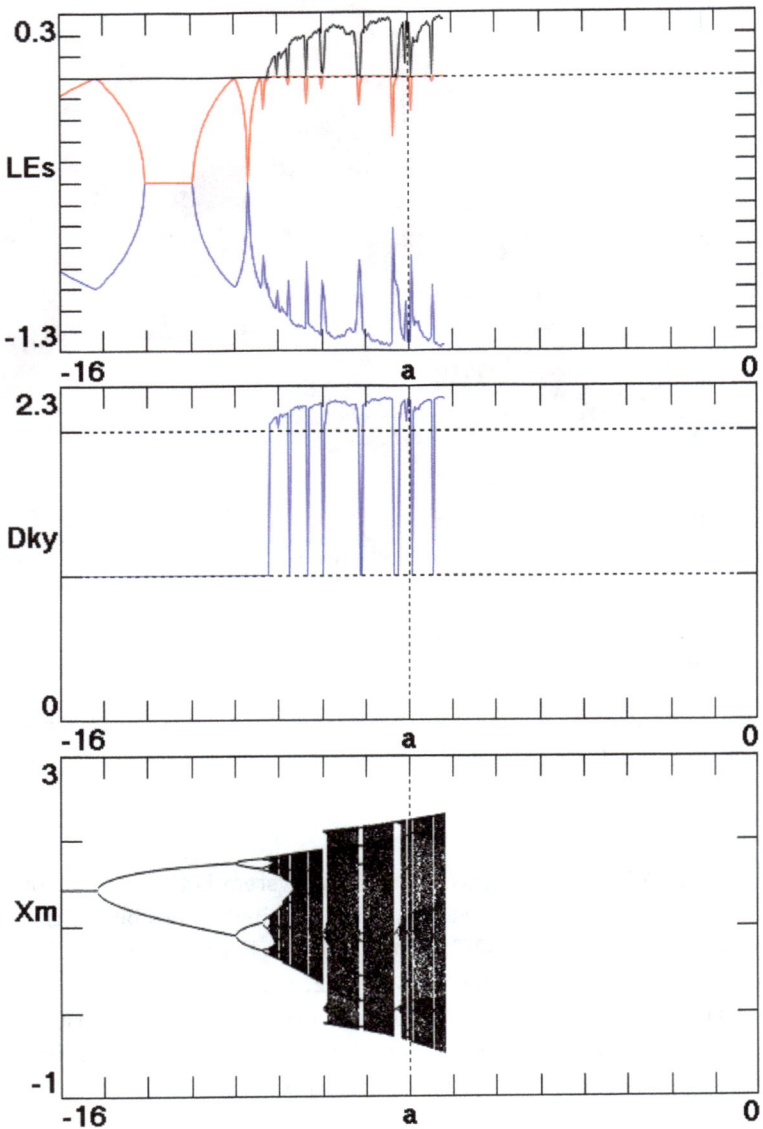

Fig. 17.4 Bifurcation diagram for Eq. (17.2) as a function of the parameter a.

Chapter 18

Sprott N System

The Sprott N system is another case with six terms and one quadratic nonlinearity that was discovered in 1994.

18.1 Introduction

The Sprott N system [Sprott (1994)] written in its most general form with an adjustable coefficient in each of the six terms is given by

$$\dot{x} = a_1 y$$
$$\dot{y} = a_2 x + a_3 z^2 \qquad (18.1)$$
$$\dot{z} = a_4 + a_5 y + a_6 z.$$

The usual parameters are $a_1 = a_6 = -2$, $a_2 = a_3 = a_4 = a_5 = 1$. This system was discovered in a search for the simplest chaotic system with six terms and a single quadratic nonlinearity. It is written here with parameters that can be either positive or negative with a preference for as many positive as possible.

The following sections were written by the computer program that performed the optimization, carried out the analysis of the resulting system, and produced the corresponding figures, all without human intervention.

18.2 Simplified System

After about 3×10^4 trials, of which 80 were chaotic, simplified parameters for Eq. (18.1) that give chaotic solutions are $a_1 = 1$, $a_2 = -1$, $a_3 = 1$, $a_4 = 1$, $a_5 = 0.24$, $a_6 = -1$. Thus Eq. (18.1) can be written more compactly as

$$\dot{x} = y$$
$$\dot{y} = -x + z^2 \qquad (18.2)$$
$$\dot{z} = 1 + ay - z,$$

where $a = 0.24$.

Note that with six terms, Eq. (18.2) should have two independent parameters through a linear rescaling of the three variables plus time. However, one of the two parameters has a value of ± 1 and can be placed in any of the remaining five terms.

18.3 Equilibria

The system in Eq. (18.2) with $a = 0.24$ has an unstable saddle focus at $(1, 0, 1)$ with eigenvalues $(0.1174 - 0.8922i, 0.1174 + 0.8922i, -1.2348)$ and a Poincaré index of 1 in the $z = 1$ plane.

The strange attractor is self-excited, and the system has no symmetry.

18.4 Attractor

Figure 18.1 shows various views of the attractor for Eq. (18.2) with $a = 0.24$ and initial conditions $(28, -42, -2)$. The rainbow of colors shows the local value of the largest Lyapunov exponent with red indicating the most positive values (regions of worst predictability) and blue indicating the most negative values (regions of best predictability).

18.5 Time Series

Figure 18.2 shows the time series for the three variables along with the local value of the largest Lyapunov exponent (LL) for Eq. (18.2) with $a = 0.24$. Red color in the Lyapunov exponent indicates that the error is growing parallel to the orbit, while blue indicates growth perpendicular to the orbit. Note that the orbit passes through regions where the local Lyapunov exponent is strongly positive and other regions where it is strongly negative as is typical for a chaotic system and is also reflected by the colors in Fig. 18.1.

18.6 Lyapunov Exponents

The global Lyapunov exponents are determined by averaging the local Lyapunov exponents along the orbit. The values typically converge very slowly because of the large variation along the orbit, and an integration time of order 10^8 is required to obtain 4-digit accuracy.

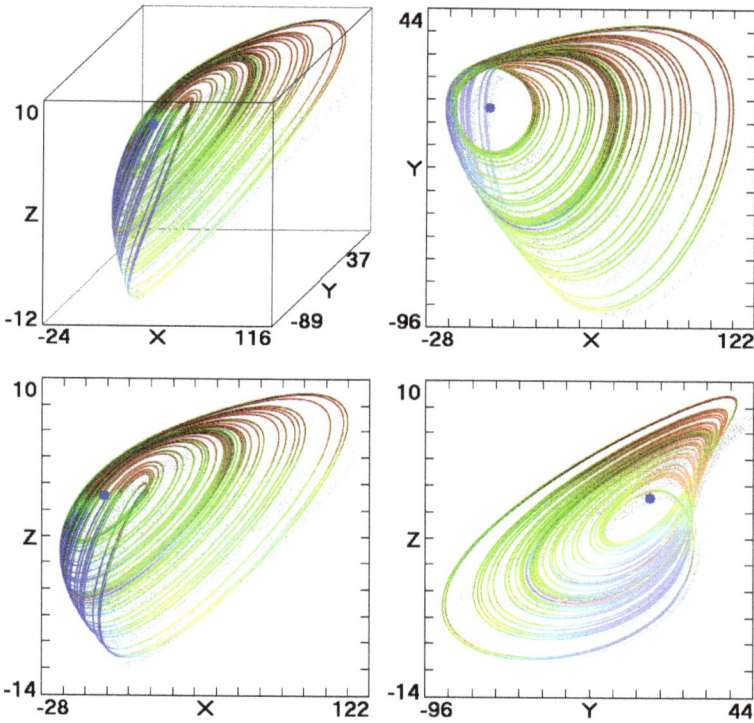

Fig. 18.1   Views of the attractor for Eq. (18.2) with $a = 0.24$ and initial conditions (28, −42, −2).

The results of such a calculation for the system in Eq. (18.2) with $a = 0.24$ after a time of 2×10^6 are LE = (0.0500, 0, −1.0500) with a Kaplan–Yorke dimension of 2.0476, where the last digit in the quoted values is only an approximation. The positive value of the largest Lyapunov exponent indicates that the system is chaotic, and the negative sum of the exponents (−1) indicates that the system is dissipative with a strange attractor.

18.7   Basin of Attraction

Figure 18.3 shows (in red) the basin of attraction for Eq. (18.2) with $a = 0.24$ in the $z = 0$ plane. Also shown (in black) is the cross-section of the attractor in the same plane.

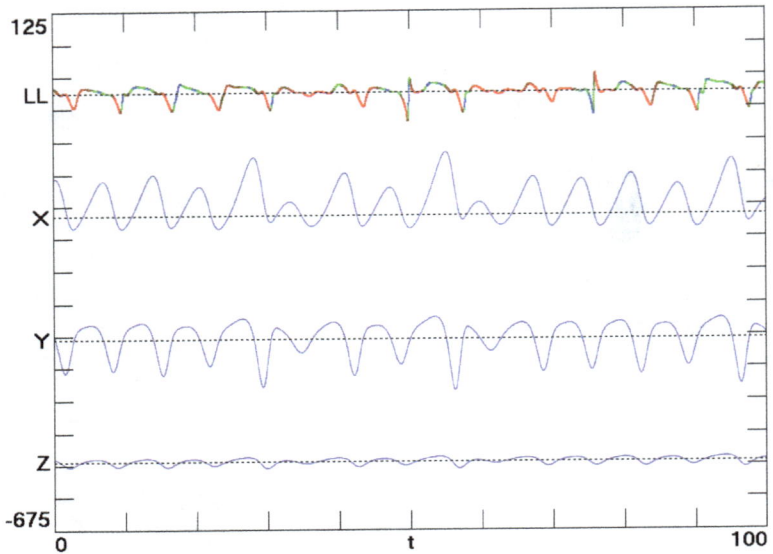

Fig. 18.2   Time series for the variables in Eq. (18.2) with $a = 0.24$ along with the local Lyapunov exponent (LL).

18.8   Bifurcations

Figure 18.4 shows the bifurcation diagram for Eq. (18.2) as a function of the parameter a from 0 to 0.48. The initial condition was taken as (28, −42, −2) for each value of a. Each of the 500 values of a was calculated for a time of about 1×10^4.

The upper plot shows the three Lyapunov exponents. The middle plot shows the Kaplan–Yorke dimension, and the lower plot shows the local maxima of x. The chaotic region is in the vicinity of $a = 0.24$, and the route to chaos is clearly shown.

18.9   Robustness

One measure of the robustness of a chaotic system is the amount by which the parameters can be changed from their nominal values before the probability of chaos decreases to 50% [Sprott (2022)]. For the system in Eq. (18.2) with $a = 0.24$ and initial conditions (28, −42, −2), after 19856 trials, it is estimated that the parameter a can be changed by 10% before the chaos is

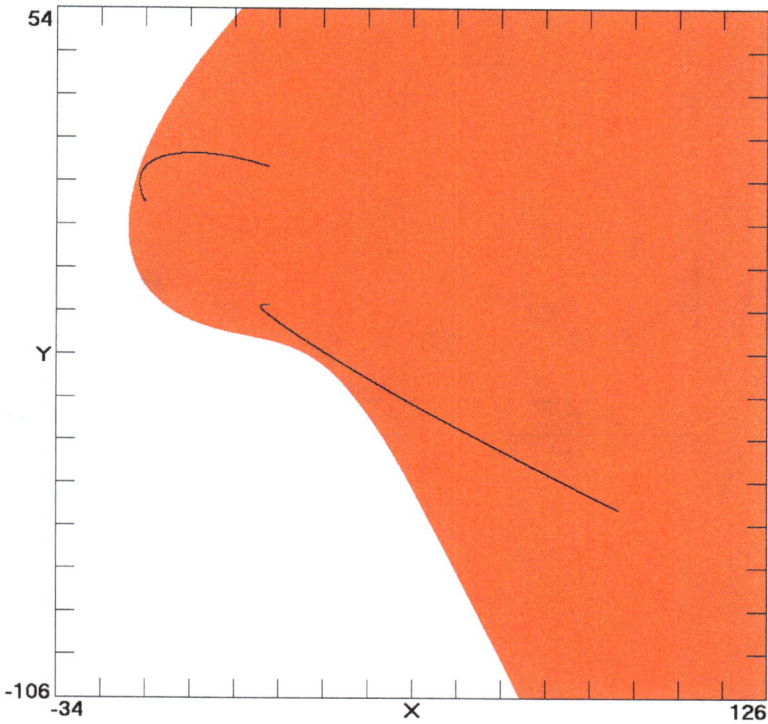

Fig. 18.3   Basin of attraction for Eq. (18.2) with $a = 0.24$ in the $z = 0$ plane.

more likely to be lost than not. Thus the system is somewhat fragile. This result is consistent with the data in Fig. 18.4.

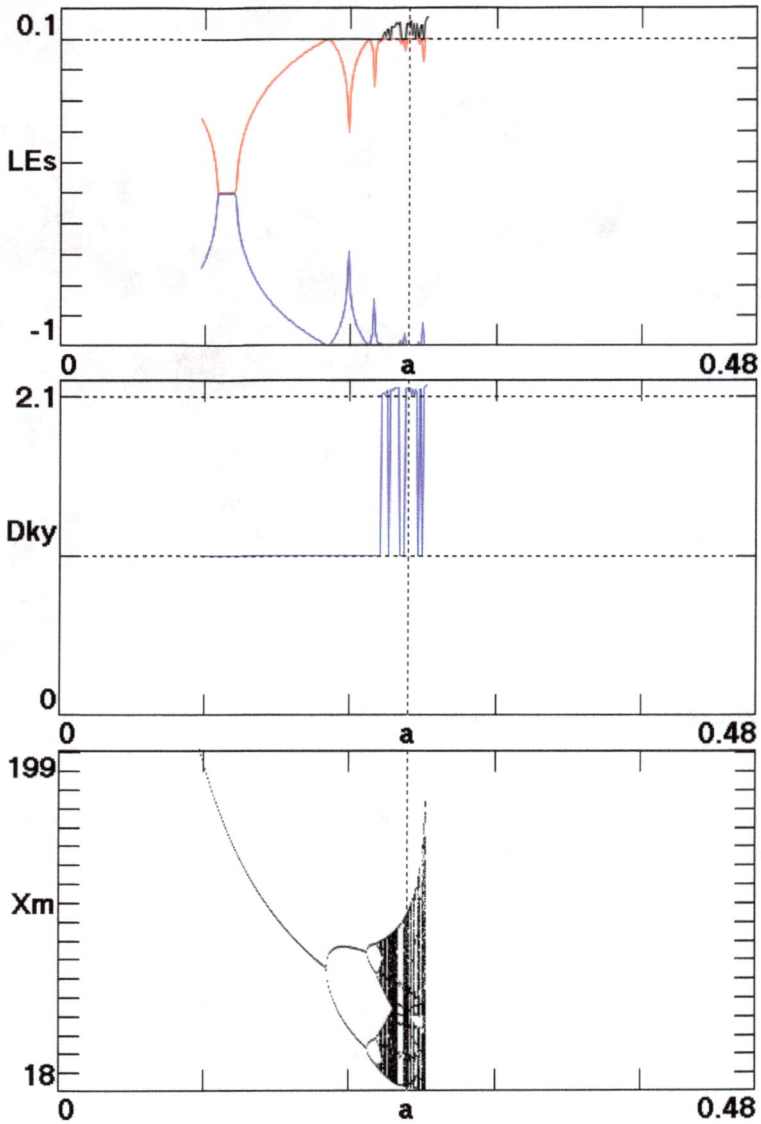

Fig. 18.4   Bifurcation diagram for Eq. (18.2) as a function of the parameter a.

Chapter 19

Sprott O System

The Sprott O system is another case with six terms and one quadratic nonlinearity that was discovered in 1994.

19.1 Introduction

The Sprott O system [Sprott (1994)] written in its most general form with an adjustable coefficient in each of the six terms is given by

$$\dot{x} = a_1 y$$
$$\dot{y} = a_2 x + a_3 z \qquad (19.1)$$
$$\dot{z} = a_4 x + a_5 xz + a_6 y.$$

The usual parameters are $a_1 = a_2 = a_4 = a_5 = 1$, $a_3 = -1$, $a_6 = 2.7$. This system was discovered in a search for the simplest chaotic system with six terms and a single quadratic nonlinearity. It is written here with parameters that can be either positive or negative with a preference for as many positive as possible.

The following sections were written by the computer program that performed the optimization, carried out the analysis of the resulting system, and produced the corresponding figures, all without human intervention.

19.2 Simplified System

After about 6×10^4 trials, of which 68 were chaotic, simplified parameters for Eq. (19.1) that give chaotic solutions are $a_1 = 1$, $a_2 = -1$, $a_3 = 1$, $a_4 = 1$, $a_5 = 1$, $a_6 = 0.3$. Thus Eq. (19.1) can be written more compactly as

$$\dot{x} = y$$
$$\dot{y} = -x + z \qquad (19.2)$$
$$\dot{z} = x + xz + ay,$$

where $a = 0.3$.

Note that with six terms, Eq. (19.2) should have two independent parameters through a linear rescaling of the three variables plus time. However, one of the two parameters has a value of ± 1 and can be placed in any of the remaining five terms.

19.3   Equilibria

The system in Eq. (19.2) with $a = 0.3$ has two equilibrium points:

Equilibrium # 1 is an unstable saddle focus at $(0, 0, 0)$ with eigenvalues $(-0.3859 - 1.0709i, -0.3859 + 1.0709i, 0.7718)$ and a Poincaré index of 1 in the $z = 0$ plane.

Equilibrium # 2 is an unstable saddle focus at $(-1, 0, -1)$ with eigenvalues $(0.0743 - 0.9301i, 0.0743 + 0.9301i, -1.1486)$ and a Poincaré index of 1 in the $z = -1$ plane.

The strange attractor is self-excited, and the system has no symmetry.

19.4   Attractor

Figure 19.1 shows various views of the attractor for Eq. (19.2) with $a = 0.3$ and initial conditions $(-2, 0, -1)$. The rainbow of colors shows the local value of the largest Lyapunov exponent with red indicating the most positive values (regions of worst predictability) and blue indicating the most negative values (regions of best predictability).

19.5   Time Series

Figure 19.2 shows the time series for the three variables along with the local value of the largest Lyapunov exponent (LL) for Eq. (19.2) with $a = 0.3$. Red color in the Lyapunov exponent indicates that the error is growing parallel to the orbit, while blue indicates growth perpendicular to the orbit. Note that the orbit passes through regions where the local Lyapunov exponent is strongly positive and other regions where it is strongly negative as is typical for a chaotic system and is also reflected by the colors in Fig. 19.1.

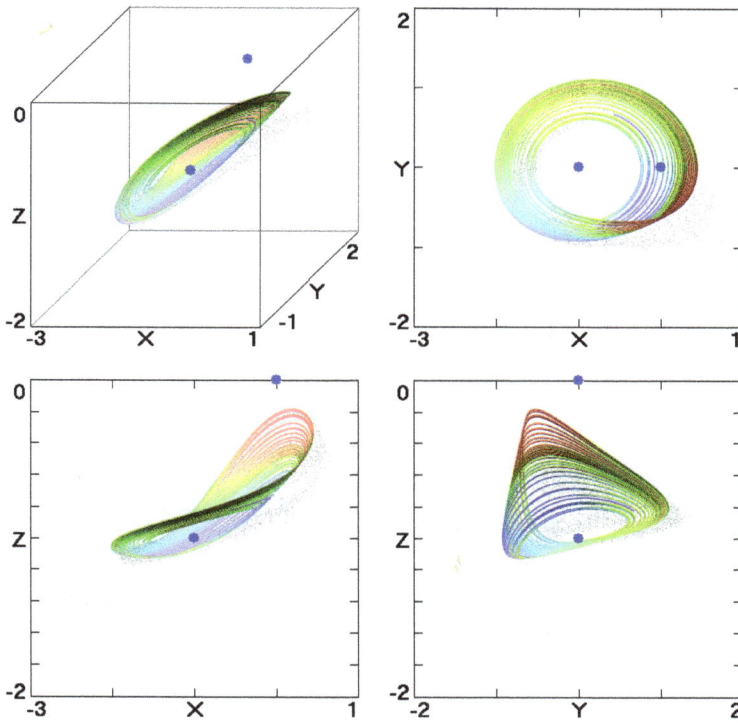

Fig. 19.1   Views of the attractor for Eq. (19.2) with $a = 0.3$ and initial conditions $(-2, 0, -1)$.

19.6   Lyapunov Exponents

The global Lyapunov exponents are determined by averaging the local Lyapunov exponents along the orbit. The values typically converge very slowly because of the large variation along the orbit, and an integration time of order 10^8 is required to obtain 4-digit accuracy.

The results of such a calculation for the system in Eq. (19.2) with $a = 0.3$ after a time of 1×10^6 are LE = $(0.0577, 0, -0.8703)$ with a Kaplan–Yorke dimension of 2.0662, where the last digit in the quoted values is only an approximation. The positive value of the largest Lyapunov exponent indicates that the system is chaotic, and the negative sum of the exponents (-0.8126) indicates that the system is dissipative with a strange attractor.

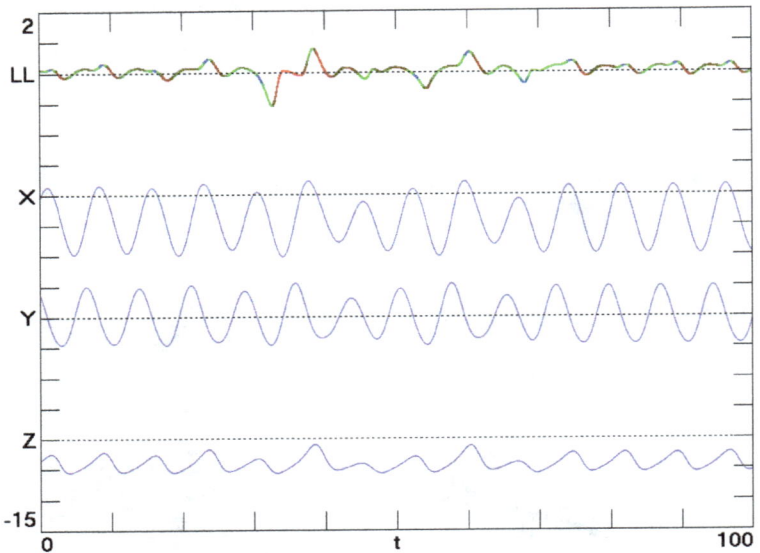

Fig. 19.2 Time series for the variables in Eq. (19.2) with $a = 0.3$ along with the local Lyapunov exponent (LL).

19.7 Basin of Attraction

Figure 19.3 shows (in red) the basin of attraction for Eq. (19.2) with $a = 0.3$ in the $z = -1$ plane. Also shown (in black) is the cross-section of the attractor in the same plane.

19.8 Bifurcations

Figure 19.4 shows the bifurcation diagram for Eq. (19.2) as a function of the parameter a from 0 to 0.6. The initial condition was taken as $(-2, 0, -1)$ for each value of a. Each of the 500 values of a was calculated for a time of about 1×10^4.

The upper plot shows the three Lyapunov exponents. The middle plot shows the Kaplan–Yorke dimension, and the lower plot shows the local maxima of x. The chaotic region is in the vicinity of $a = 0.3$, and the route to chaos is clearly shown.

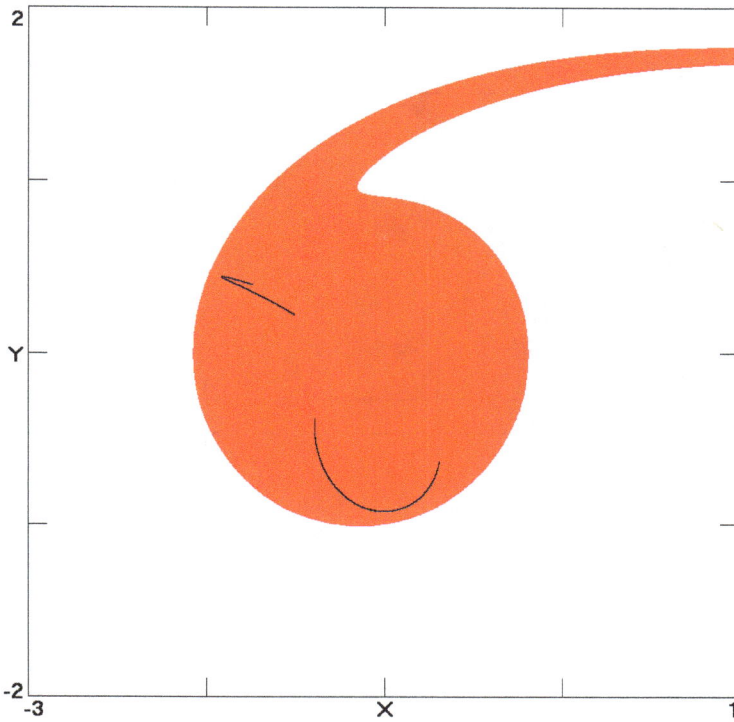

Fig. 19.3   Basin of attraction for Eq. (19.2) with $a = 0.3$ in the $z = -1$ plane.

19.9   Robustness

One measure of the robustness of a chaotic system is the amount by which the parameters can be changed from their nominal values before the probability of chaos decreases to 50% [Sprott (2022)]. For the system in Eq. (19.2) with $a = 0.3$ and initial conditions $(-2, 0, -1)$, after 13986 trials, it is estimated that the parameter a can be changed by 14% before the chaos is more likely to be lost than not. Thus the system is somewhat fragile. This result is consistent with the data in Fig. 19.4.

Elegant Automation

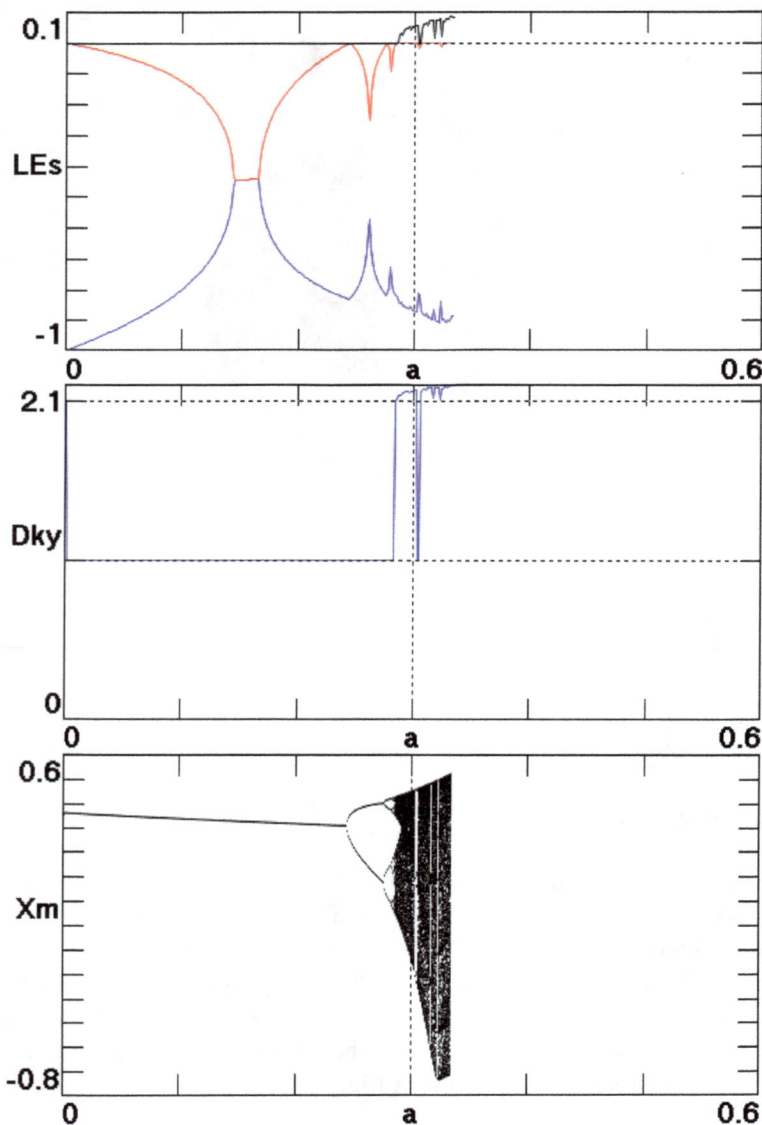

Fig. 19.4   Bifurcation diagram for Eq. (19.2) as a function of the parameter a.

Chapter 20

Sprott P System

The Sprott P system is another case with six terms and one quadratic nonlinearity that was discovered in 1994.

20.1 Introduction

The Sprott P system [Sprott (1994)] written in its most general form with an adjustable coefficient in each of the six terms is given by

$$\dot{x} = a_1 y + a_2 z$$
$$\dot{y} = a_3 x + a_4 y^2 \qquad (20.1)$$
$$\dot{z} = a_5 x + a_6 y.$$

The usual parameters are $a_1 = 2.7$, $a_2 = a_4 = a_5 = a_6 = 1$, $a_3 = -1$. This system was discovered in a search for the simplest chaotic system with six terms and a single quadratic nonlinearity. It is written here with parameters that can be either positive or negative with a preference for as many positive as possible.

The following sections were written by the computer program that performed the optimization, carried out the analysis of the resulting system, and produced the corresponding figures, all without human intervention.

20.2 Simplified System

After about 3×10^5 trials, of which 262 were chaotic, simplified parameters for Eq. (20.1) that give chaotic solutions are $a_1 = 2.7$, $a_2 = 1$, $a_3 = -1$, $a_4 = 1$, $a_5 = 1$, $a_6 = 1$. Thus Eq. (20.1) can be written more compactly as

$$\dot{x} = ay + z$$
$$\dot{y} = -x + y^2 \qquad (20.2)$$
$$\dot{z} = x + y,$$

where $a = 2.7$.

121

Note that with six terms, Eq. (20.2) should have two independent parameters through a linear rescaling of the three variables plus time. However, one of the two parameters has a value of ± 1 and can be placed in any of the remaining five terms.

20.3   Equilibria

The system in Eq. (20.2) with $a = 2.7$ has two equilibrium points:

Equilibrium # 1 is an unstable saddle focus at $(1, -1, 2.7000)$ with eigenvalues $(-1.1914 - 1.0921i, -1.1914 + 1.0921i, 0.3828)$ and a Poincaré index of 1 in the $z = 2.7$ plane.

Equilibrium # 2 is an unstable saddle focus at $(0, 0, 0)$ with eigenvalues $(0.2551 - 1.3767i, 0.2551 + 1.3767i, -0.5101)$ and a Poincaré index of 1 in the $z = 0$ plane.

The strange attractor is self-excited, and the system has no symmetry.

20.4   Attractor

Figure 20.1 shows various views of the attractor for Eq. (20.2) with $a = 2.7$ and initial conditions $(1, 1, -1)$. The rainbow of colors shows the local value of the largest Lyapunov exponent with red indicating the most positive values (regions of worst predictability) and blue indicating the most negative values (regions of best predictability).

20.5   Time Series

Figure 20.2 shows the time series for the three variables along with the local value of the largest Lyapunov exponent (LL) for Eq. (20.2) with $a = 2.7$. Red color in the Lyapunov exponent indicates that the error is growing parallel to the orbit, while blue indicates growth perpendicular to the orbit. Note that the orbit passes through regions where the local Lyapunov exponent is strongly positive and other regions where it is strongly negative as is typical for a chaotic system and is also reflected by the colors in Fig. 20.1.

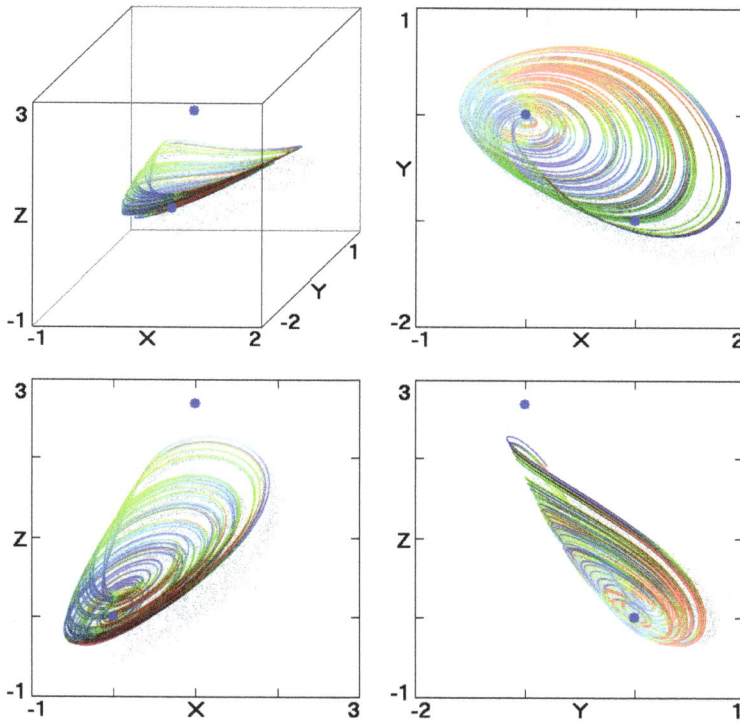

Fig. 20.1   Views of the attractor for Eq. (20.2) with $a = 2.7$ and initial conditions (1, 1, −1).

20.6   Lyapunov Exponents

The global Lyapunov exponents are determined by averaging the local Lyapunov exponents along the orbit. The values typically converge very slowly because of the large variation along the orbit, and an integration time of order 10^8 is required to obtain 4-digit accuracy.

The results of such a calculation for the system in Eq. (20.2) with $a = 2.7$ after a time of 3×10^6 are LE = (0.0871, 0, −0.4812) with a Kaplan–Yorke dimension of 2.1810, where the last digit in the quoted values is only an approximation. The positive value of the largest Lyapunov exponent indicates that the system is chaotic, and the negative sum of the exponents (−0.3940) indicates that the system is dissipative with a strange attractor.

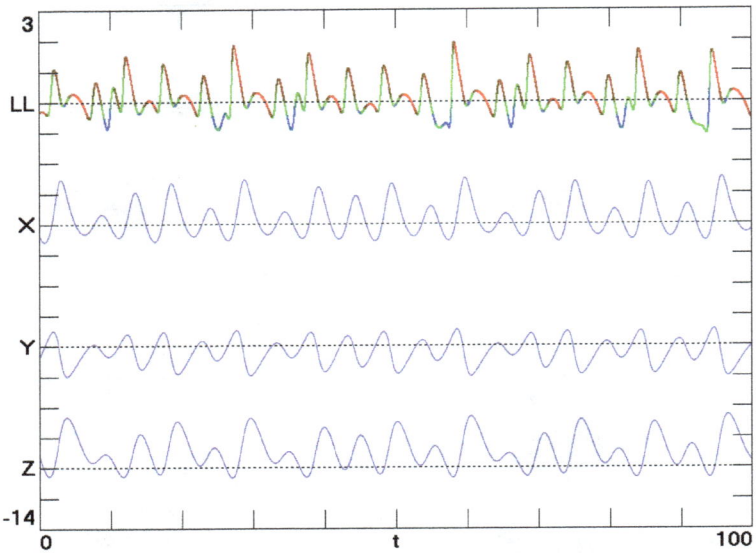

Fig. 20.2   Time series for the variables in Eq. (20.2) with $a = 2.7$ along with the local Lyapunov exponent (LL).

20.7   Basin of Attraction

Figure 20.3 shows (in red) the basin of attraction for Eq. (20.2) with $a = 2.7$ in the $z = 0$ plane. Also shown (in black) is the cross-section of the attractor in the same plane.

20.8   Bifurcations

Figure 20.4 shows the bifurcation diagram for Eq. (20.2) as a function of the parameter a from 0 to 5.4. The initial condition was taken as (1, 1, −1) at $a = 5.4$ and was not changed as a slowly varied toward $a = 0$. Each of the 500 values of a was calculated for a time of about 1×10^4.

The upper plot shows the three Lyapunov exponents. The middle plot shows the Kaplan–Yorke dimension, and the lower plot shows the local maxima of x. The chaotic region is in the vicinity of $a = 2.7$, and the route to chaos is clearly shown.

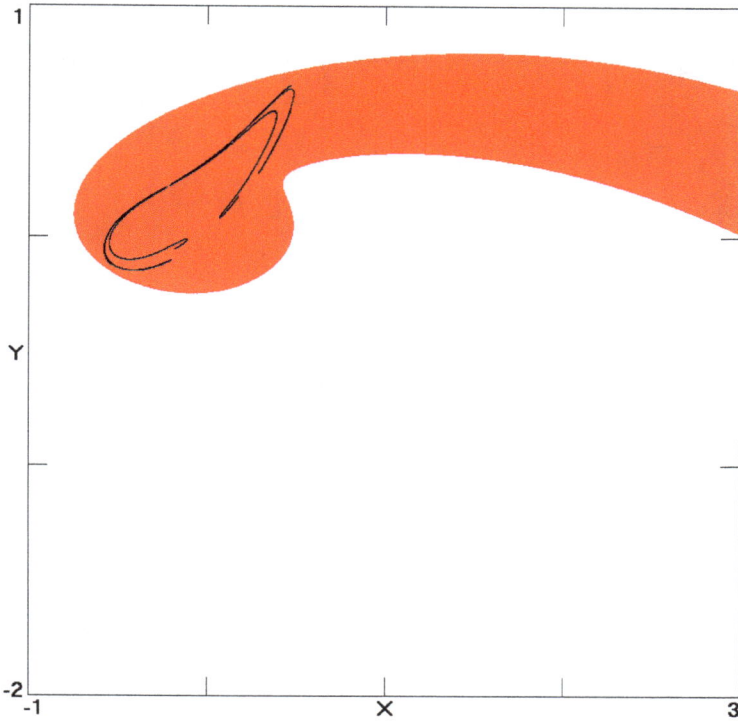

Fig. 20.3   Basin of attraction for Eq. (20.2) with $a = 2.7$ in the $z = 0$ plane.

20.9   Robustness

One measure of the robustness of a chaotic system is the amount by which the parameters can be changed from their nominal values before the probability of chaos decreases to 50% [Sprott (2022)]. For the system in Eq. (20.2) with $a = 2.7$ and initial conditions $(1, 1, -1)$, after 11308 trials, it is estimated that the parameter a can be changed by 4.2% before the chaos is more likely to be lost than not. Thus the system is somewhat fragile. This result is consistent with the data in Fig. 20.4.

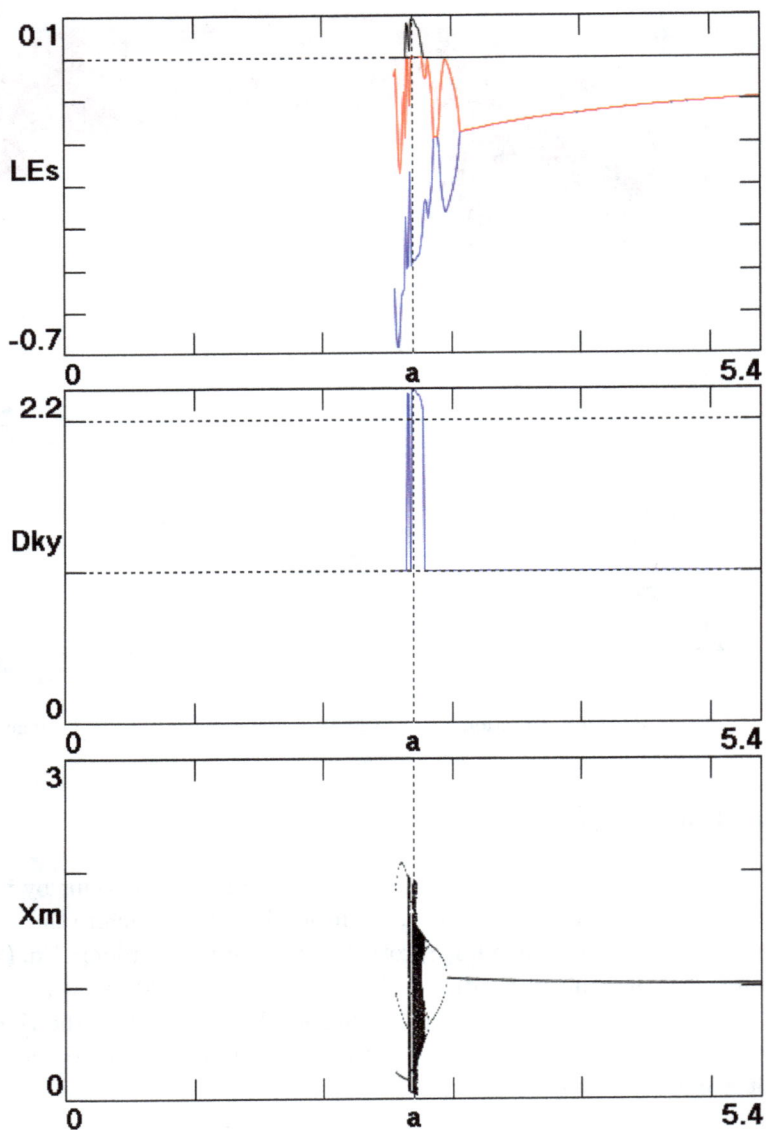

Fig. 20.4 Bifurcation diagram for Eq. (20.2) as a function of the parameter a.

Chapter 21

Sprott Q System

The Sprott Q system is another case with six terms and one quadratic nonlinearity that was discovered in 1994.

21.1 Introduction

The Sprott Q system [Sprott (1994)] written in its most general form with an adjustable coefficient in each of the six terms is given by

$$\dot{x} = a_1 z$$
$$\dot{y} = a_2 x + a_3 y \qquad (21.1)$$
$$\dot{z} = a_4 x + a_5 y^2 + a_6 z.$$

The usual parameters are $a_1 = a_3 = -1$, $a_2 = a_5 = 1$, $a_4 = 3.1$, $a_6 = 0.5$. This system was discovered in a search for the simplest chaotic system with six terms and a single quadratic nonlinearity. It is written here with parameters that can be either positive or negative with a preference for as many positive as possible.

The following sections were written by the computer program that performed the optimization, carried out the analysis of the resulting system, and produced the corresponding figures, all without human intervention.

21.2 Simplified System

After about 7×10^5 trials, of which 86 were chaotic, simplified parameters for Eq. (21.1) that give chaotic solutions are $a_1 = -1$, $a_2 = 1$, $a_3 = -0.5$, $a_4 = 1$, $a_5 = 1$, $a_6 = 0.28$. Thus Eq. (21.1) can be written more compactly as

$$\dot{x} = -z$$
$$\dot{y} = x + ay \qquad (21.2)$$
$$\dot{z} = x + y^2 + bz,$$

where $a = -0.5$, $b = 0.28$.

Note that with six terms, Eq. (21.2) should have two independent parameters through a linear rescaling of the three variables plus time, and so the dynamics is completely captured by the given parameters, which could be put in any of the six terms, albeit with different numerical values.

21.3   Equilibria

The system in Eq. (21.2) with $a = -0.5$, $b = 0.28$ has two equilibrium points:

Equilibrium # 1 is an unstable saddle focus at $(0, 0, 0)$ with eigenvalues $(0.14 - 0.9902i, 0.14 + 0.9902i, -0.5)$ and a Poincaré index of 0 in the $z = 0$ plane.

Equilibrium # 2 is an unstable saddle focus at $(-0.2500, -0.5000, 0)$ with eigenvalues $(-0.3281 - 1.0191i, -0.3281 + 1.0191i, 0.4362)$ and a Poincaré index of 0 in the $z = 0$ plane.

The strange attractor is self-excited, and the system has no symmetry.

21.4   Attractor

Figure 21.1 shows various views of the attractor for Eq. (21.2) with $a = -0.5$, $b = 0.28$ and initial conditions $(0, 1, 0)$. The rainbow of colors shows the local value of the largest Lyapunov exponent with red indicating the most positive values (regions of worst predictability) and blue indicating the most negative values (regions of best predictability).

21.5   Time Series

Figure 21.2 shows the time series for the three variables along with the local value of the largest Lyapunov exponent (LL) for Eq. (21.2) with $a = -0.5$, $b = 0.28$. Red color in the Lyapunov exponent indicates that the error is growing parallel to the orbit, while blue indicates growth perpendicular to the orbit. Note that the orbit passes through regions where the local Lyapunov exponent is strongly positive and other regions where it is strongly negative as is typical for a chaotic system and is also reflected by the colors in Fig. 21.1.

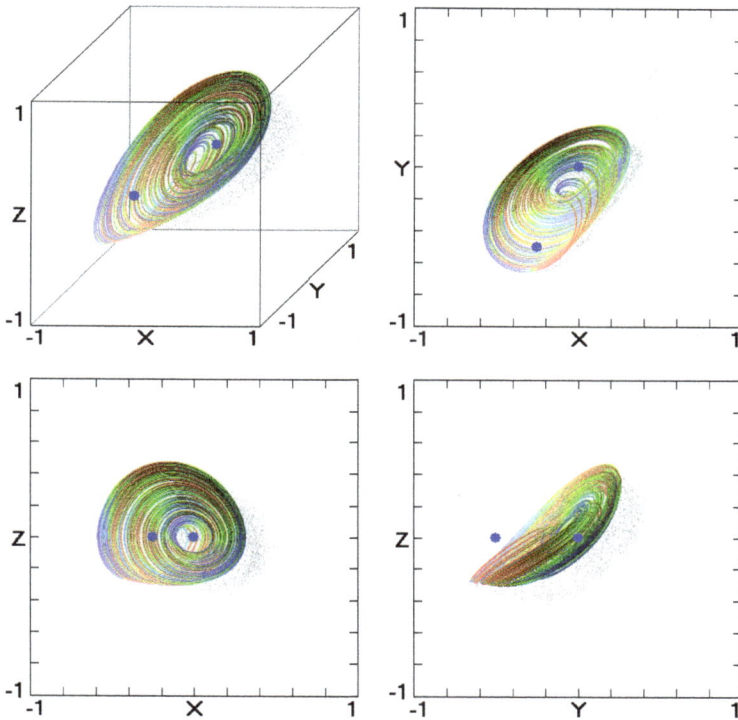

Fig. 21.1 Views of the attractor for Eq. (21.2) with $a = -0.5$, $b = 0.28$ and initial conditions (0, 1, 0).

21.6 Lyapunov Exponents

The global Lyapunov exponents are determined by averaging the local Lyapunov exponents along the orbit. The values typically converge very slowly because of the large variation along the orbit, and an integration time of order 10^8 is required to obtain 4-digit accuracy.

The results of such a calculation for the system in Eq. (21.2) with $a = -0.5$, $b = 0.28$ after a time of 2×10^6 are LE = (0.0670, 0, -0.2870) with a Kaplan–Yorke dimension of 2.2333, where the last digit in the quoted values is only an approximation. The positive value of the largest Lyapunov exponent indicates that the system is chaotic, and the negative sum of the exponents (-0.2200) indicates that the system is dissipative with a strange attractor.

Elegant Automation

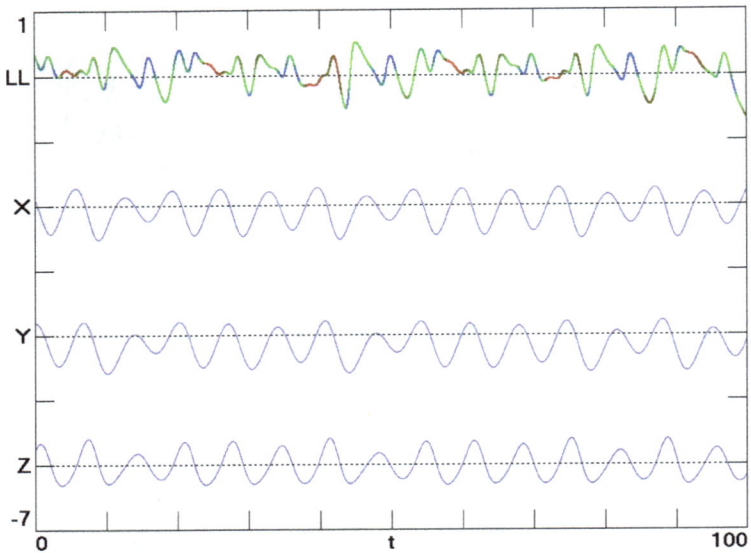

Fig. 21.2   Time series for the variables in Eq. (21.2) with $a = -0.5$, $b = 0.28$ along with the local Lyapunov exponent (LL).

21.7   Basin of Attraction

Figure 21.3 shows (in red) the basin of attraction for Eq. (21.2) with $a = -0.5$, $b = 0.28$ in the $z = 0$ plane. Also shown (in black) is the cross-section of the attractor in the same plane.

21.8   Bifurcations

Figure 21.4 shows the bifurcation diagram for Eq. (21.2) as a function of the parameter a from $a = -1$ to $a = 0$ for $b = 0.28$. The initial condition was taken as $(0, 1, 0)$ at $a = -1$ and was not changed as a slowly varied toward $a = 0$. Each of the 500 values of a was calculated for a time of about 1×10^4.

The upper plot shows the three Lyapunov exponents. The middle plot shows the Kaplan–Yorke dimension, and the lower plot shows the local maxima of x. The chaotic region is in the vicinity of $a = -0.5$, $b = 0.28$, and the route to chaos is clearly shown.

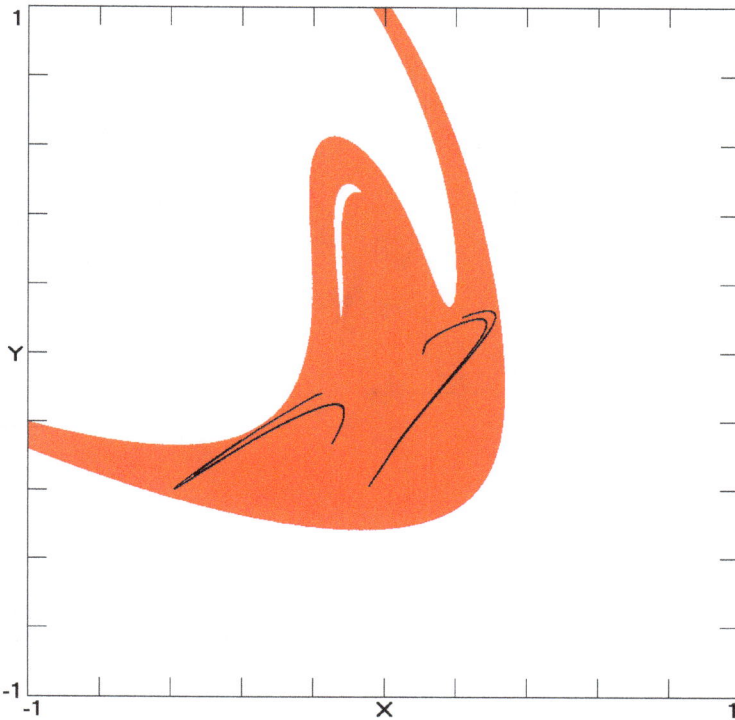

Fig. 21.3   Basin of attraction for Eq. (21.2) with $a = -0.5$, $b = 0.28$ in the $z = 0$ plane.

21.9   Robustness

One measure of the robustness of a chaotic system is the amount by which the parameters can be changed from their nominal values before the probability of chaos decreases to 50% [Sprott (2022)]. For the system in Eq. (21.2) with $a = -0.5$, $b = 0.28$ and initial conditions $(0, 1, 0)$, after 140 trials, it is estimated that the parameters can be changed by 8.3% before the chaos is more likely to be lost than not. Thus the system is somewhat fragile.

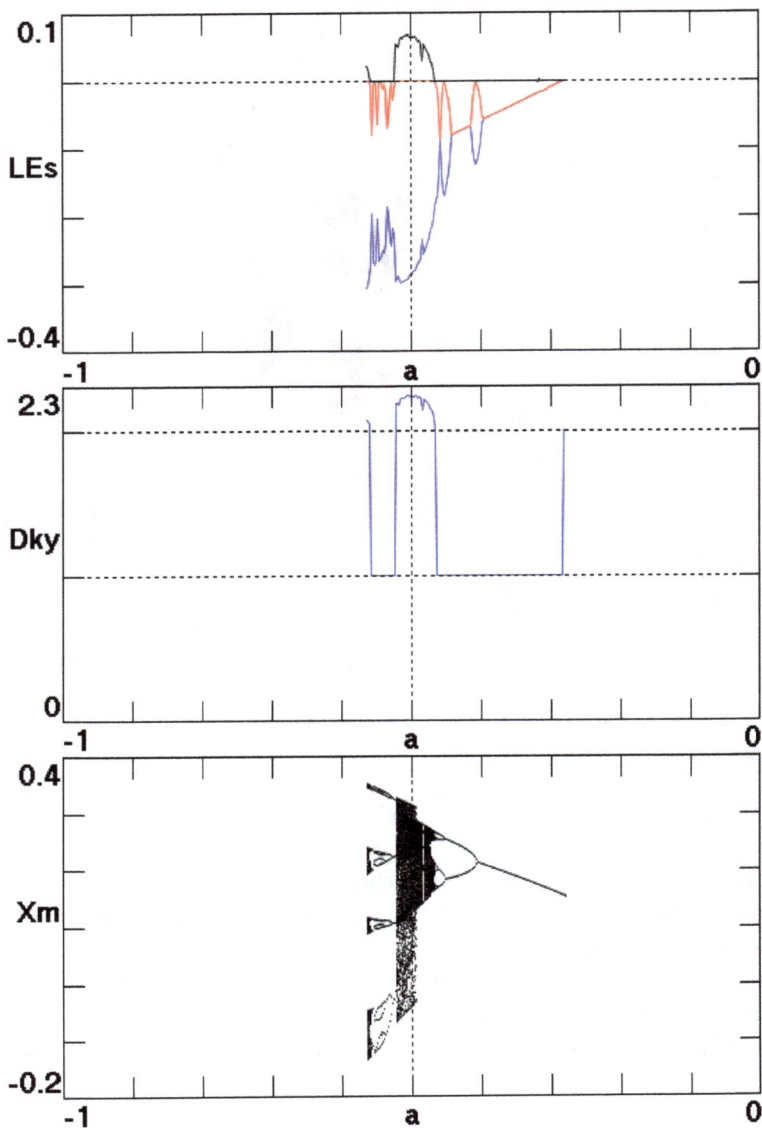

Fig. 21.4   Bifurcation diagram for Eq. (21.2) as a function of the parameter a for $b = 0.28$.

Chapter 22

Sprott R System

The Sprott R system is another case with six terms and one quadratic nonlinearity that was discovered in 1994.

22.1 Introduction

The Sprott R system [Sprott (1994)] written in its most general form with an adjustable coefficient in each of the six terms is given by

$$\dot{x} = a_1 + a_2 y$$
$$\dot{y} = a_3 + a_4 z \qquad (22.1)$$
$$\dot{z} = a_5 xy + a_6 z.$$

The usual parameters are $a_1 = 0.9$, $a_2 = a_6 = -1$, $a_3 = 0.4$, $a_4 = a_5 = 1$. This system was discovered in a search for the simplest chaotic system with six terms and a single quadratic nonlinearity. It is written here with parameters that can be either positive or negative with a preference for as many positive as possible.

The following sections were written by the computer program that performed the optimization, carried out the analysis of the resulting system, and produced the corresponding figures, all without human intervention.

22.2 Simplified System

After about 3×10^4 trials, of which 32 were chaotic, simplified parameters for Eq. (22.1) that give chaotic solutions are $a_1 = 1$, $a_2 = -0.52$, $a_3 = 1$, $a_4 = 1$, $a_5 = 1$, $a_6 = -1$. Thus Eq. (22.1) can be written more compactly as

$$\dot{x} = 1 + ay$$
$$\dot{y} = 1 + z \qquad (22.2)$$
$$\dot{z} = xy - z,$$

where $a = -0.52$.

Elegant Automation

Note that with six terms, Eq. (22.2) should have two independent parameters through a linear rescaling of the three variables plus time. However, one of the two parameters has a value of ± 1 and can be placed in any of the remaining five terms.

22.3   Equilibria

The system in Eq. (22.2) with $a = -0.52$ has an unstable saddle focus at $(-0.5200, 1.9231, -1)$ with eigenvalues $(0.1174 - 0.8922i, 0.1174 + 0.8922i, -1.2348)$ and a Poincaré index of 0 in the $z = -1$ plane.

The strange attractor is self-excited, and the system has no symmetry.

22.4   Attractor

Figure 22.1 shows various views of the attractor for Eq. (22.2) with $a = -0.52$ and initial conditions $(-1, -3, 3)$. The rainbow of colors shows the local value of the largest Lyapunov exponent with red indicating the most positive values (regions of worst predictability) and blue indicating the most negative values (regions of best predictability).

22.5   Time Series

Figure 22.2 shows the time series for the three variables along with the local value of the largest Lyapunov exponent (LL) for Eq. (22.2) with $a = -0.52$. Red color in the Lyapunov exponent indicates that the error is growing parallel to the orbit, while blue indicates growth perpendicular to the orbit. Note that the orbit passes through regions where the local Lyapunov exponent is strongly positive and other regions where it is strongly negative as is typical for a chaotic system and is also reflected by the colors in Fig. 22.1.

22.6   Lyapunov Exponents

The global Lyapunov exponents are determined by averaging the local Lyapunov exponents along the orbit. The values typically converge very slowly because of the large variation along the orbit, and an integration time of order 10^8 is required to obtain 4-digit accuracy.

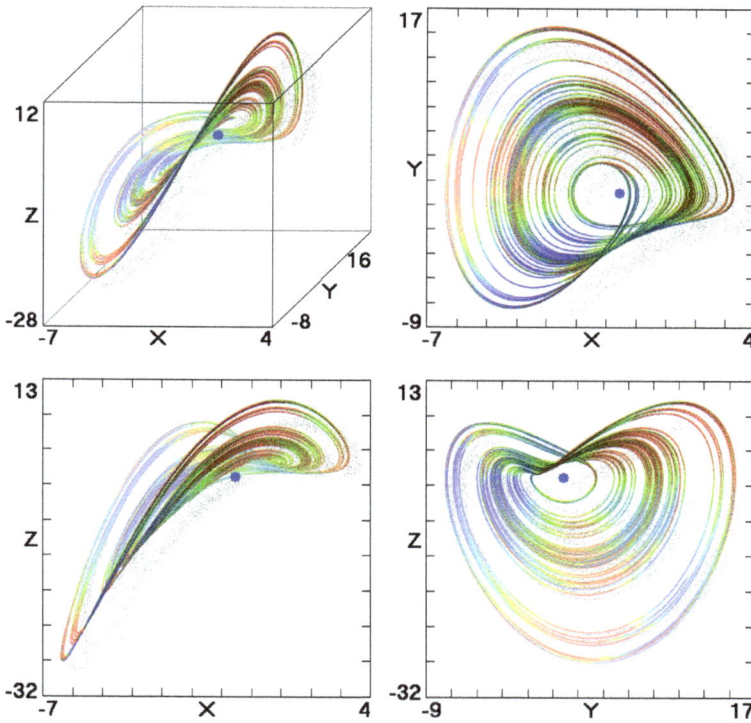

Fig. 22.1   Views of the attractor for Eq. (22.2) with $a = -0.52$ and initial conditions $(-1, -3, 3)$.

The results of such a calculation for the system in Eq. (22.2) with $a = -0.52$ after a time of 3×10^6 are LE $= (0.0501, 0, -1.0501)$ with a Kaplan–Yorke dimension of 2.0476, where the last digit in the quoted values is only an approximation. The positive value of the largest Lyapunov exponent indicates that the system is chaotic, and the negative sum of the exponents (-1) indicates that the system is dissipative with a strange attractor.

22.7   Basin of Attraction

Figure 22.3 shows (in red) the basin of attraction for Eq. (22.2) with $a = -0.52$ in the $z = 0$ plane. Also shown (in black) is the cross-section of the attractor in the same plane.

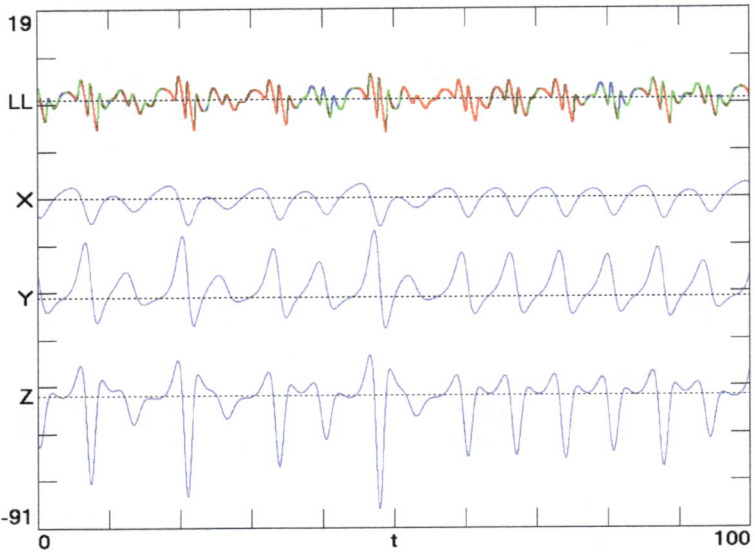

Fig. 22.2   Time series for the variables in Eq. (22.2) with $a = -0.52$ along with the local Lyapunov exponent (LL).

22.8   Bifurcations

Figure 22.4 shows the bifurcation diagram for Eq. (22.2) as a function of the parameter a from $a = -1.04$ to $a = 0$. The initial condition was taken as $(-1, -3, 3)$ at $a = -1.04$ and was not changed as a slowly varied toward $a = 0$. Each of the 500 values of a was calculated for a time of about 1×10^4.

The upper plot shows the three Lyapunov exponents. The middle plot shows the Kaplan–Yorke dimension, and the lower plot shows the local maxima of x. The chaotic region is in the vicinity of $a = -0.52$, and the route to chaos is clearly shown.

22.9   Robustness

One measure of the robustness of a chaotic system is the amount by which the parameters can be changed from their nominal values before the probability of chaos decreases to 50% [Sprott (2022)]. For the system in Eq. (22.2) with $a = -0.52$ and initial conditions $(-1, -3, 3)$, after 3188 trials, it is estimated that the parameter a can be changed by 9.8% before the chaos is

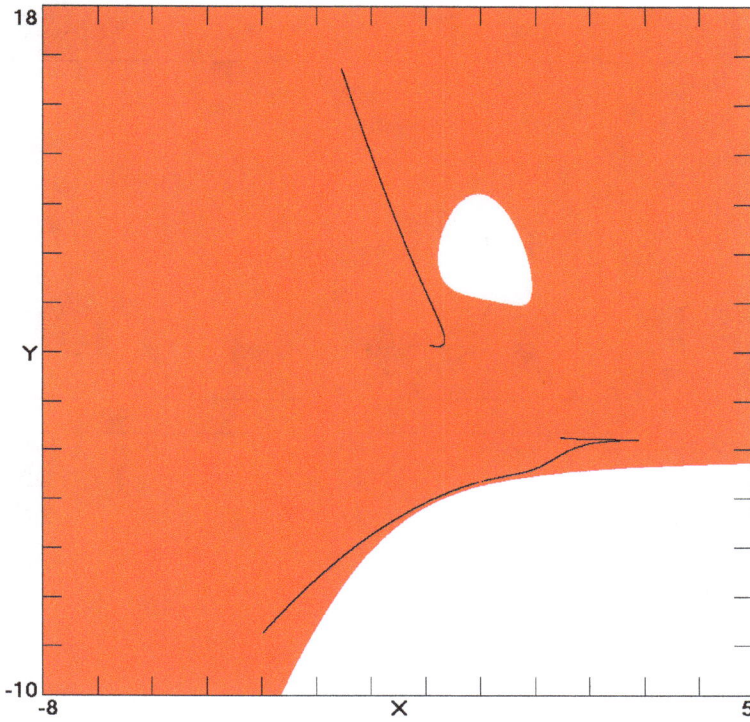

Fig. 22.3   Basin of attraction for Eq. (22.2) with $a = -0.52$ in the $z = 0$ plane.

more likely to be lost than not. Thus the system is somewhat fragile. This result is consistent with the data in Fig. 22.4.

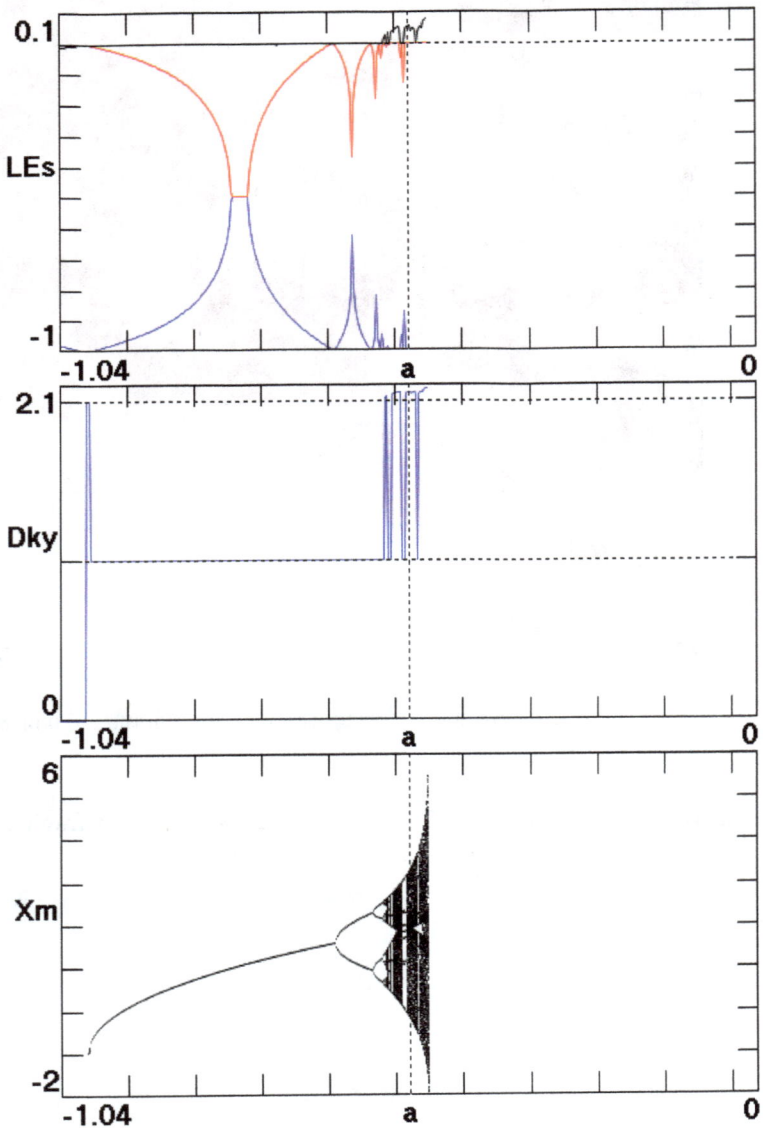

Fig. 22.4   Bifurcation diagram for Eq. (22.2) as a function of the parameter a.

Chapter 23

Sprott S System

The Sprott S system is another case with six terms and one quadratic nonlinearity. Despite being the case from 1994 that occurred the least frequently, this system is highly robust.

23.1 Introduction

The Sprott S system [Sprott (1994)] written in its most general form with an adjustable coefficient in each of the six terms is given by

$$\dot{x} = a_1 x + a_2 y$$
$$\dot{y} = a_3 x + a_4 z^2 \qquad (23.1)$$
$$\dot{z} = a_5 + a_6 x.$$

The usual parameters are $a_1 = -1$, $a_2 = -4$, $a_3 = a_4 = a_5 = a_6 = 1$. This system was discovered in a search for the simplest chaotic system with six terms and a single quadratic nonlinearity. It is written here with parameters that can be either positive or negative with a preference for as many positive as possible.

The following sections were written by the computer program that performed the optimization, carried out the analysis of the resulting system, and produced the corresponding figures, all without human intervention.

23.2 Simplified System

After about 7×10^4 trials, of which 30 were chaotic, simplified parameters for Eq. (23.1) that give chaotic solutions are $a_1 = -1$, $a_2 = -5$, $a_3 = 1$, $a_4 = 1$, $a_5 = 1$, $a_6 = 1$. Thus Eq. (23.1) can be written more compactly as

$$\dot{x} = -x + ay$$
$$\dot{y} = x + z^2 \qquad (23.2)$$
$$\dot{z} = 1 + x,$$

where $a = -5$.

Note that with six terms, Eq. (23.2) should have two independent parameters through a linear rescaling of the three variables plus time. However, one of the two parameters has a value of ± 1 and can be placed in any of the remaining five terms.

23.3   Equilibria

The system in Eq. (23.2) with $a = -5$ has two equilibrium points:

Equilibrium # 1 is an unstable saddle focus at $(-1, 0.2000, 1)$ with eigenvalues $(0.324 - 2.4419i, 0.324 + 2.4419i, -1.648)$ and a Poincaré index of 1 in the $z = 1$ plane.

Equilibrium # 2 is an unstable saddle focus at $(-1, 0.2000, -1)$ with eigenvalues $(-1.1345 - 2.5676i, -1.1345 + 2.5676i, 1.2691)$ and a Poincaré index of 1 in the $z = -1$ plane.

The strange attractor is self-excited, and the system has no symmetry.

23.4   Attractor

Figure 23.1 shows various views of the attractor for Eq. (23.2) with $a = -5$ and initial conditions $(-1, 0, -1)$. The rainbow of colors shows the local value of the largest Lyapunov exponent with red indicating the most positive values (regions of worst predictability) and blue indicating the most negative values (regions of best predictability).

23.5   Time Series

Figure 23.2 shows the time series for the three variables along with the local value of the largest Lyapunov exponent (LL) for Eq. (23.2) with $a = -5$. Red color in the Lyapunov exponent indicates that the error is growing parallel to the orbit, while blue indicates growth perpendicular to the orbit. Note that the orbit passes through regions where the local Lyapunov exponent is strongly positive and other regions where it is strongly negative as is typical for a chaotic system and is also reflected by the colors in Fig. 23.1.

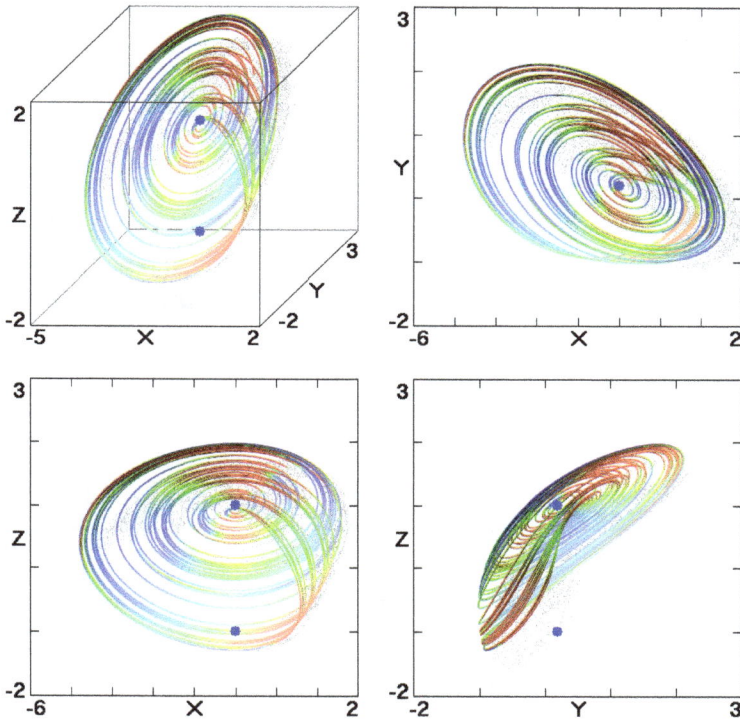

Fig. 23.1   Views of the attractor for Eq. (23.2) with $a = -5$ and initial conditions $(-1, 0, -1)$.

23.6   Lyapunov Exponents

The global Lyapunov exponents are determined by averaging the local Lyapunov exponents along the orbit. The values typically converge very slowly because of the large variation along the orbit, and an integration time of order 10^8 is required to obtain 4-digit accuracy.

The results of such a calculation for the system in Eq. (23.2) with $a = -5$ after a time of 3×10^6 are LE = $(0.1932, 0, -1.1932)$ with a Kaplan–Yorke dimension of 2.1619, where the last digit in the quoted values is only an approximation. The positive value of the largest Lyapunov exponent indicates that the system is chaotic, and the negative sum of the exponents (-1) indicates that the system is dissipative with a strange attractor.

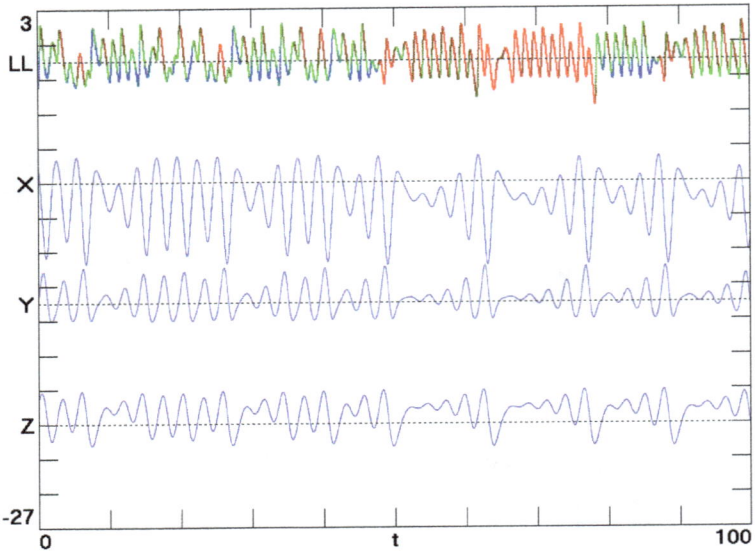

Fig. 23.2   Time series for the variables in Eq. (23.2) with $a = -5$ along with the local Lyapunov exponent (LL).

23.7   Basin of Attraction

Figure 23.3 shows (in red) the basin of attraction for Eq. (23.2) with $a = -5$ in the $z = 0$ plane. Also shown (in black) is the cross-section of the attractor in the same plane.

23.8   Bifurcations

Figure 23.4 shows the bifurcation diagram for Eq. (23.2) as a function of the parameter a from $a = -10$ to $a = 0$. The initial condition was taken as $(-1, 0, -1)$ at $a = -10$ and was not changed as a slowly varied toward $a = 0$. Each of the 500 values of a was calculated for a time of about 1×10^4.

The upper plot shows the three Lyapunov exponents. The middle plot shows the Kaplan–Yorke dimension, and the lower plot shows the local maxima of x. The chaotic region is in the vicinity of $a = -5$, and the route to chaos is clearly shown.

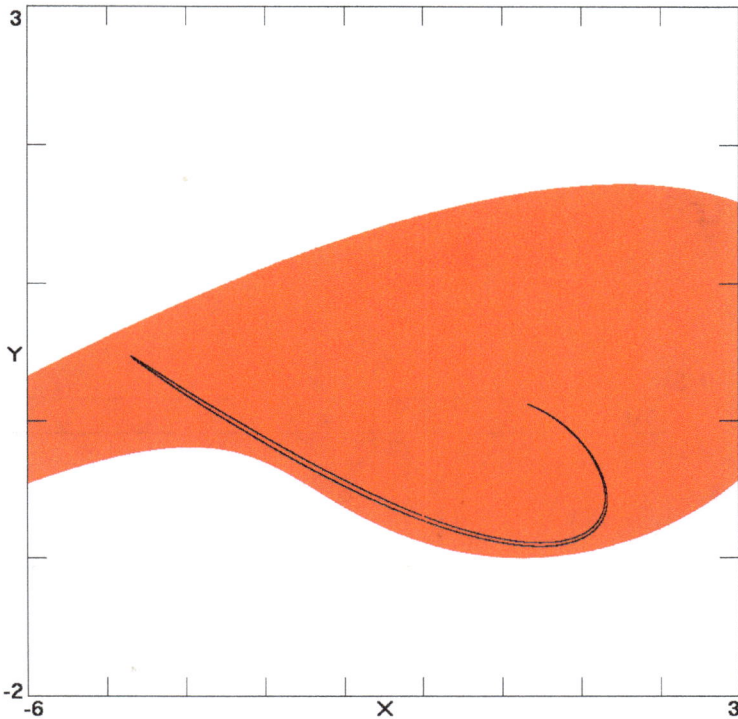

Fig. 23.3   Basin of attraction for Eq. (23.2) with $a = -5$ in the $z = 0$ plane.

23.9   Robustness

One measure of the robustness of a chaotic system is the amount by which the parameters can be changed from their nominal values before the probability of chaos decreases to 50% [Sprott (2022)]. For the system in Eq. (23.2) with $a = -5$ and initial conditions $(-1,\ 0,\ -1)$, after 1248 trials, it is estimated that the parameter a can be changed by 67% before the chaos is more likely to be lost than not. Thus the system is highly robust. This result is consistent with the data in Fig. 23.4.

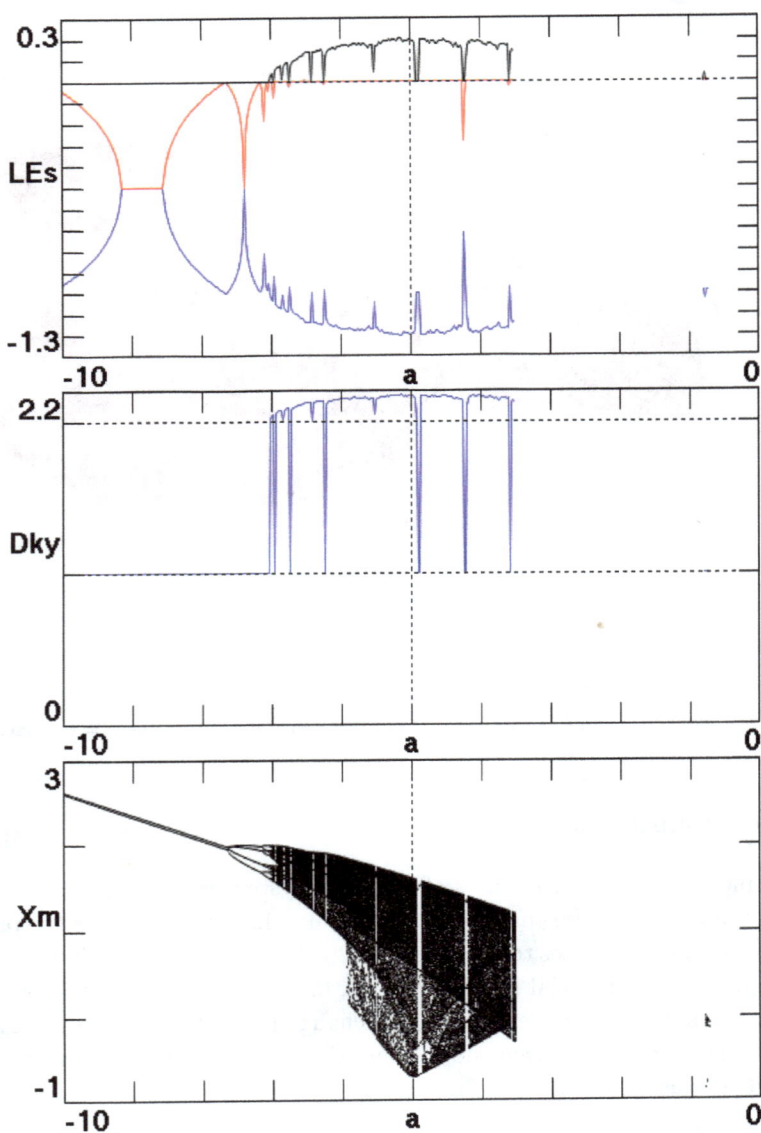

Fig. 23.4   Bifurcation diagram for Eq. (23.2) as a function of the parameter a.

Chapter 24

Rössler Prototype-4 System

This is an old but relatively unknown case with six terms and one quadratic nonlinearity that was somehow missed in the extensive 1994 search for such systems.

24.1 Introduction

The Rössler (1979) prototype-4 system written in its most general form with an adjustable coefficient in each of the six terms is given by

$$\dot{x} = a_1 y + a_2 z$$
$$\dot{y} = a_3 x \qquad\qquad (24.1)$$
$$\dot{z} = a_4 y + a_5 y^2 + a_6 z.$$

The usual parameters are $a_1 = a_2 = -1$, $a_3 = 1$, $a_4 = 0.5$, $a_5 = a_6 = -0.5$. It is written here with parameters that can be either positive or negative with a preference for as many positive as possible.

The following sections were written by the computer program that performed the optimization, carried out the analysis of the resulting system, and produced the corresponding figures, all without human intervention.

24.2 Simplified System

After about 9×10^4 trials, of which 44 were chaotic, simplified parameters for Eq. (24.1) that give chaotic solutions are $a_1 = 1$, $a_2 = 1$, $a_3 = -4$, $a_4 = 1$, $a_5 = 1$, $a_6 = -1$. Thus Eq. (24.1) can be written more compactly as

$$\dot{x} = y + z$$
$$\dot{y} = ax \qquad\qquad (24.2)$$
$$\dot{z} = y + y^2 - z,$$

where $a = -4$.

Note that with six terms, Eq. (24.2) should have two independent parameters through a linear rescaling of the three variables plus time. However, one of the two parameters has a value of ± 1 and can be placed in any of the remaining five terms.

24.3   Equilibria

The system in Eq. (24.2) with $a = -4$ has two equilibrium points:

Equilibrium # 1 is an unstable saddle focus at $(0, -2, 2)$ with eigenvalues $(-1.1015 - 2.3317i, -1.1015 + 2.3317i, 1.203)$ and a Poincaré index of 1 in the $z = 2$ plane.

Equilibrium # 2 is an unstable saddle focus at $(0, 0, 0)$ with eigenvalues $(0.3038 - 2.2101i, 0.3038 + 2.2101i, -1.6075)$ and a Poincaré index of 1 in the $z = 0$ plane.

The strange attractor is self-excited, and the system has no symmetry.

24.4   Attractor

Figure 24.1 shows various views of the attractor for Eq. (24.2) with $a = -4$ and initial conditions $(0, -1, 0)$. The rainbow of colors shows the local value of the largest Lyapunov exponent with red indicating the most positive values (regions of worst predictability) and blue indicating the most negative values (regions of best predictability).

24.5   Time Series

Figure 24.2 shows the time series for the three variables along with the local value of the largest Lyapunov exponent (LL) for Eq. (24.2) with $a = -4$. Red color in the Lyapunov exponent indicates that the error is growing parallel to the orbit, while blue indicates growth perpendicular to the orbit. Note that the orbit passes through regions where the local Lyapunov exponent is strongly positive and other regions where it is strongly negative as is typical for a chaotic system and is also reflected by the colors in Fig. 24.1.

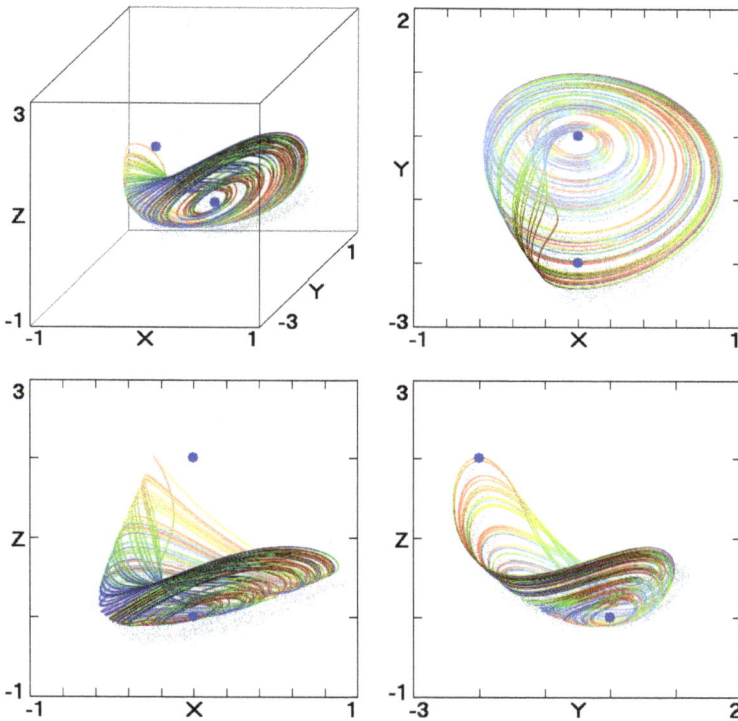

Fig. 24.1   Views of the attractor for Eq. (24.2) with $a = -4$ and initial conditions (0, -1, 0).

24.6   Lyapunov Exponents

The global Lyapunov exponents are determined by averaging the local Lyapunov exponents along the orbit. The values typically converge very slowly because of the large variation along the orbit, and an integration time of order 10^8 is required to obtain 4-digit accuracy.

The results of such a calculation for the system in Eq. (24.2) with $a = -4$ after a time of 3×10^6 are LE = (0.1876, 0, -1.1876) with a Kaplan–Yorke dimension of 2.1579, where the last digit in the quoted values is only an approximation. The positive value of the largest Lyapunov exponent indicates that the system is chaotic, and the negative sum of the exponents (-1) indicates that the system is dissipative with a strange attractor.

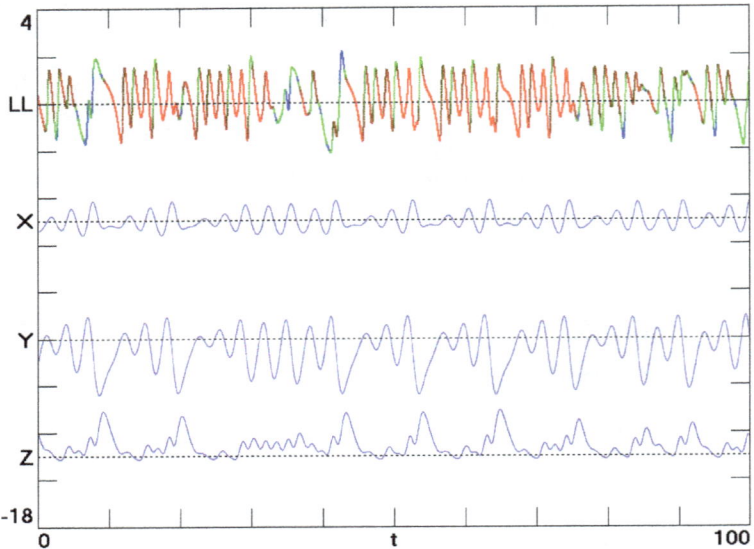

Fig. 24.2   Time series for the variables in Eq. (24.2) with $a = -4$ along with the local Lyapunov exponent (LL).

24.7   Basin of Attraction

Figure 24.3 shows (in red) the basin of attraction for Eq. (24.2) with $a = -4$ in the $z = 0$ plane. Also shown (in black) is the cross-section of the attractor in the same plane.

24.8   Bifurcations

Figure 24.4 shows the bifurcation diagram for Eq. (24.2) as a function of the parameter a from $a = -8$ to $a = 0$. The initial condition was taken as $(0, -1, 0)$ at $a = -8$ and was not changed as a slowly varied toward $a = 0$. Each of the 500 values of a was calculated for a time of about 1×10^4.

The upper plot shows the three Lyapunov exponents. The middle plot shows the Kaplan–Yorke dimension, and the lower plot shows the local maxima of x. The chaotic region is in the vicinity of $a = -4$, and the route to chaos is clearly shown.

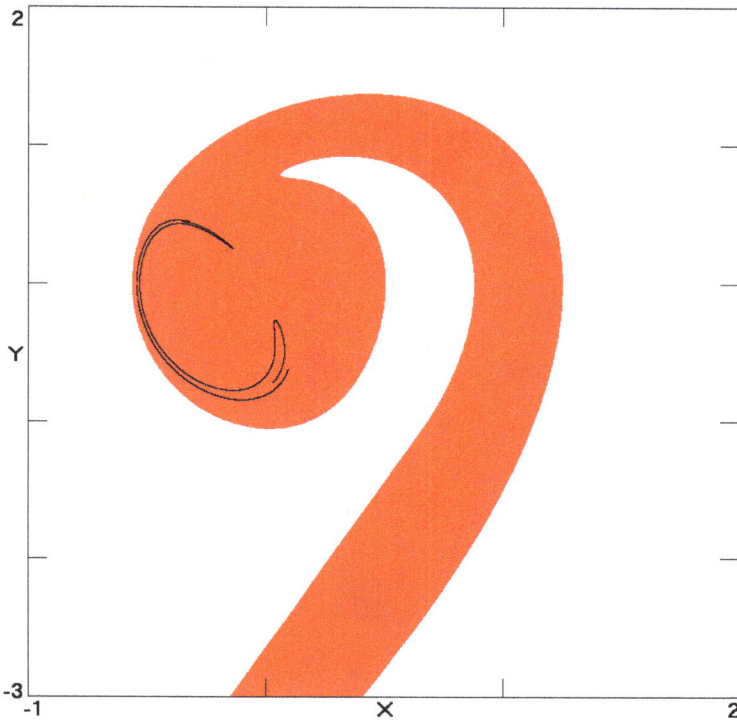

Fig. 24.3   Basin of attraction for Eq. (24.2) with $a = -4$ in the $z = 0$ plane.

24.9   Robustness

One measure of the robustness of a chaotic system is the amount by which the parameters can be changed from their nominal values before the probability of chaos decreases to 50% [Sprott (2022)]. For the system in Eq. (24.2) with $a = -4$ and initial conditions $(0, -1, 0)$, after 1568 trials, it is estimated that the parameter a can be changed by 60% before the chaos is more likely to be lost than not. Thus the system is highly robust. This result is consistent with the data in Fig. 24.4.

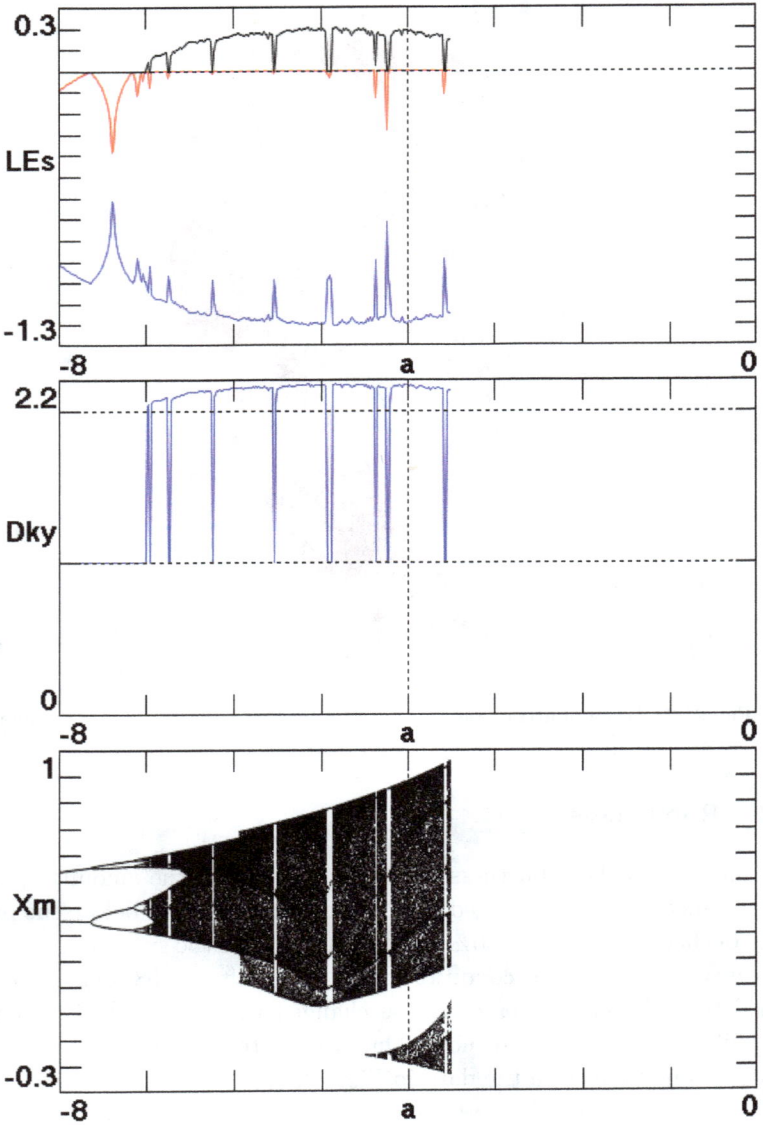

Fig. 24.4 Bifurcation diagram for Eq. (24.2) as a function of the parameter a.

Chapter 25

Simplest Chaotic System

With only five terms and a single quadratic nonlinearity, this is arguably the simplest example of a dissipative autonomous chaotic flow, although it is relatively fragile. It is in the form of a jerk equation for which several additional examples are given in the chapters that follow.

25.1 Introduction

The simplest dissipative autonomous flow [Sprott (1997)] written in its most general form with an adjustable coefficient in each of the three terms is given by

$$\dot{x} = y$$
$$\dot{y} = z \quad \quad (25.1)$$
$$\dot{z} = a_1 z + a_2 y^2 + a_3 x.$$

The usual parameters are $a_1 = -2.017$, $a_2 = 1$, $a_3 = -1$. It is written here with parameters that can be either positive or negative with a preference for as many positive as possible. It can be written in terms of the third derivative of x (the *jerk*) as $\dddot{x} = a_1 \ddot{x} + a_2 \dot{x}^2 + a_3 x$. Zhang and Heidel (1997) proved that no simpler chaotic case can exist.

The following sections were written by the computer program that performed the optimization, carried out the analysis of the resulting system, and produced the corresponding figures, all without human intervention.

25.2 Simplified System

After about 3×10^4 trials, of which 20 were chaotic, simplified parameters for Eq. (25.1) that give chaotic solutions are $a_1 = -2.02$, $a_2 = 1$, $a_3 = -1$.

Thus Eq. (25.1) can be written more compactly as

$$\dot{x} = y$$
$$\dot{y} = z \qquad\qquad (25.2)$$
$$\dot{z} = az + y^2 - x,$$

where $a = -2.02$.

Note that with five terms, Eq. (25.2) should have one independent parameter through a linear rescaling of the three variables plus time, and so the dynamics is completely captured by the single parameter a, which could be put in any of the five terms, albeit with a different numerical value.

25.3   Equilibria

The system in Eq. (25.2) with $a = -2.02$ has an unstable saddle focus at $(0, 0, 0)$ with eigenvalues $(0.1012 - 0.6631i, 0.1012 + 0.6631i, -2.2225)$ and a Poincaré index of 0 in the $z = 0$ plane.

The strange attractor is self-excited, and the system has no symmetry.

25.4   Attractor

Figure 25.1 shows various views of the attractor for Eq. (25.2) with $a = -2.02$ and initial conditions $(-1, 0, 0)$. The rainbow of colors shows the local value of the largest Lyapunov exponent with red indicating the most positive values (regions of worst predictability) and blue indicating the most negative values (regions of best predictability).

25.5   Time Series

Figure 25.2 shows the time series for the three variables along with the local value of the largest Lyapunov exponent (LL) for Eq. (25.2) with $a = -2.02$. Red color in the Lyapunov exponent indicates that the error is growing parallel to the orbit, while blue indicates growth perpendicular to the orbit. Note that the orbit passes through regions where the local Lyapunov exponent is strongly positive and other regions where it is strongly negative as is typical for a chaotic system and is also reflected by the colors in Fig. 25.1.

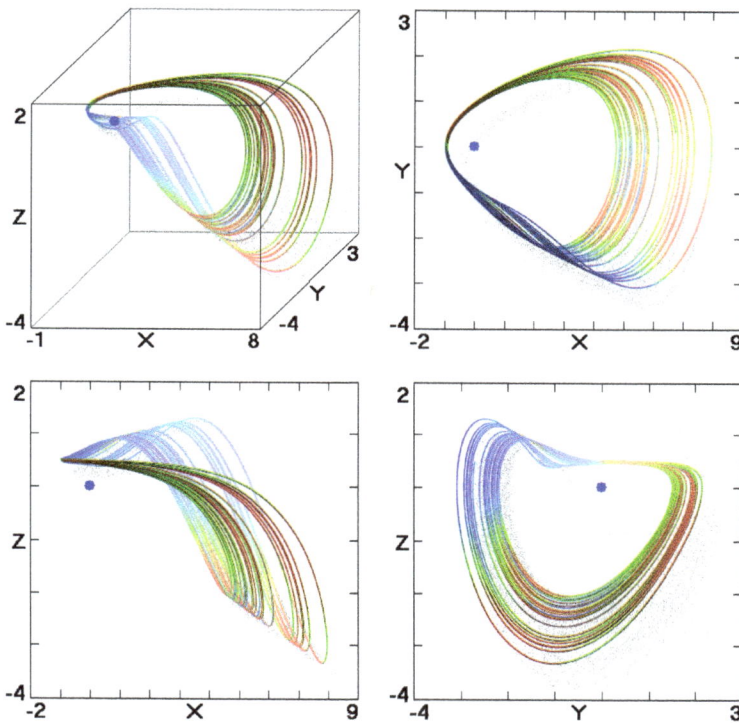

Fig. 25.1   Views of the attractor for Eq. (25.2) with $a = -2.02$ and initial conditions $(-1, 0, 0)$.

25.6   Lyapunov Exponents

The global Lyapunov exponents are determined by averaging the local Lyapunov exponents along the orbit. The values typically converge very slowly because of the large variation along the orbit, and an integration time of order 10^8 is required to obtain 4-digit accuracy.

The results of such a calculation for the system in Eq. (25.2) with $a = -2.02$ after a time of 2×10^6 are LE = $(0.0487, 0, -2.0687)$ with a Kaplan–Yorke dimension of 2.0235, where the last digit in the quoted values is only an approximation. The positive value of the largest Lyapunov exponent indicates that the system is chaotic, and the negative sum of the exponents (-2.0200) indicates that the system is dissipative with a strange attractor.

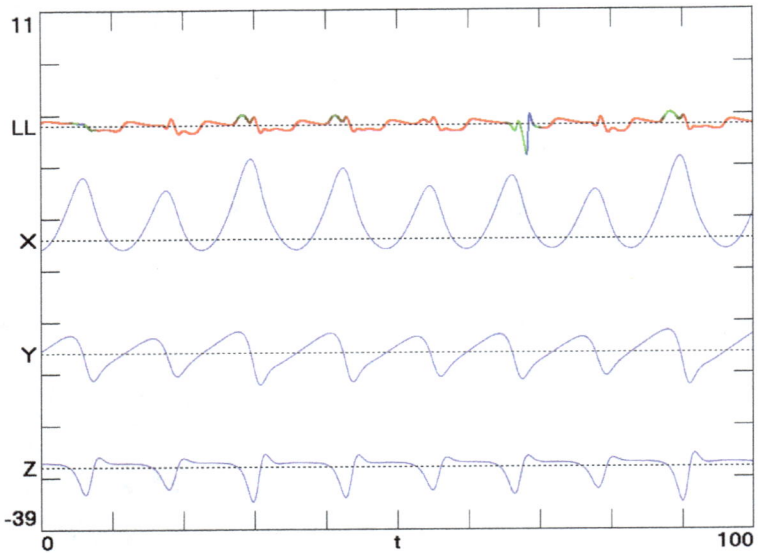

Fig. 25.2   Time series for the variables in Eq. (25.2) with $a = -2.02$ along with the local Lyapunov exponent (LL).

25.7   Basin of Attraction

Figure 25.3 shows (in red) the basin of attraction for Eq. (25.2) with $a = -2.02$ in the $z = 0$ plane. Also shown (in black) is the cross-section of the attractor in the same plane.

25.8   Bifurcations

Figure 25.4 shows the bifurcation diagram for Eq. (25.2) as a function of the parameter a from $a = -4.04$ to $a = 0$. The initial condition was taken as $(-1, 0, 0)$ at $a = -4.04$ and was not changed as a slowly varied toward $a = 0$. Each of the 500 values of a was calculated for a time of about 1×10^4.

The upper plot shows the three Lyapunov exponents. The middle plot shows the Kaplan–Yorke dimension, and the lower plot shows the local maxima of x. The chaotic region is in the vicinity of $a = -2.02$, and the route to chaos is clearly shown.

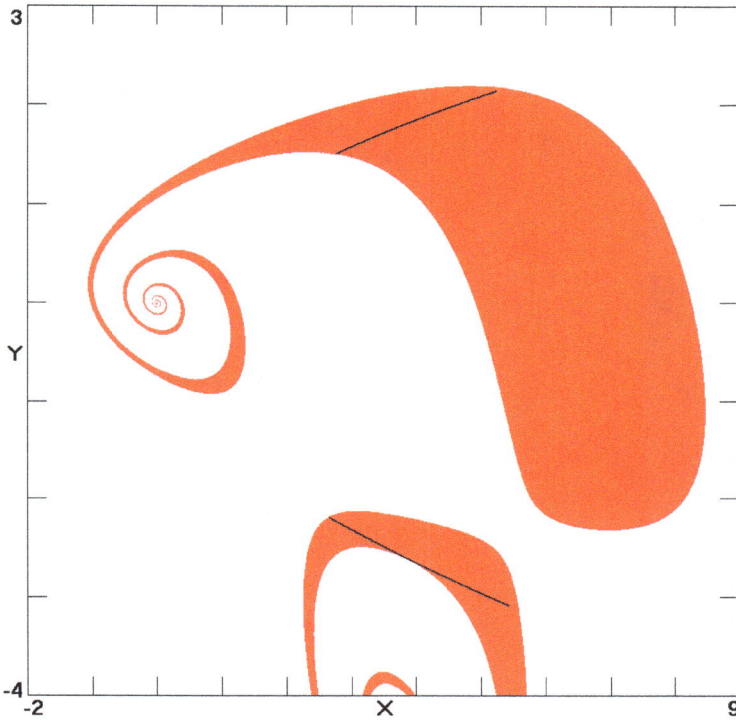

Fig. 25.3   Basin of attraction for Eq. (25.2) with $a = -2.02$ in the $z = 0$ plane.

25.9   Robustness

One measure of the robustness of a chaotic system is the amount by which the parameters can be changed from their nominal values before the probability of chaos decreases to 50% [Sprott (2022)]. For the system in Eq. (25.2) with $a = -2.02$ and initial conditions $(-1, 0, 0)$, after 10588 trials, it is estimated that the parameter a can be changed by 1.8% before the chaos is more likely to be lost than not. Thus the system is highly fragile. This result is consistent with the data in Fig. 25.4.

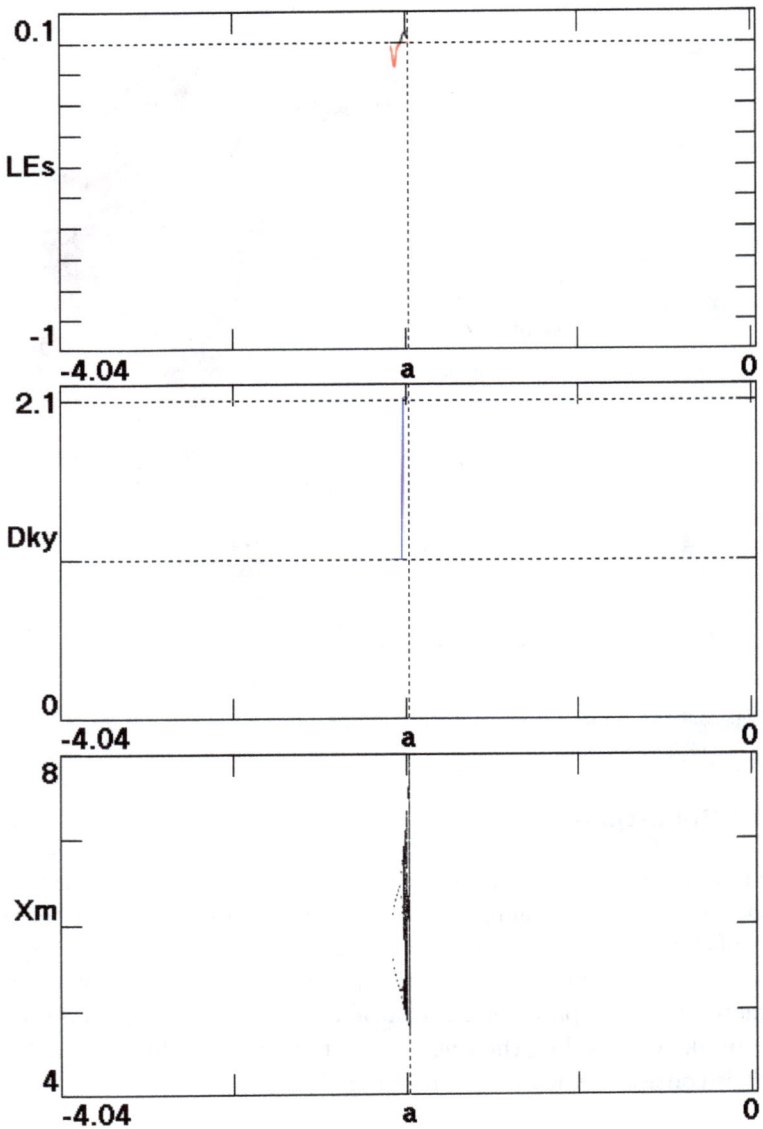

Fig. 25.4   Bifurcation diagram for Eq. (25.2) as a function of the parameter a.

Chapter 26

Malasoma System

A variation of the system in the previous chapter is probably the simplest dissipative autonomous chaotic flow with a cubic rather than a quadratic nonlinearity. It is also a jerk system with five terms and a single nonlinearity but with inversion symmetry, and it is relatively fragile.

26.1 Introduction

The Malasoma (2000) system written in its most general form with an adjustable coefficient in each of the three terms is given by

$$\dot{x} = y$$
$$\dot{y} = z \qquad\qquad (26.1)$$
$$\dot{z} = a_1 z + a_2 xy^2 + a_3 x.$$

The usual parameters are $a_1 = -2.028$, $a_2 = 1$, $a_3 = -1$. It is written here with parameters that can be either positive or negative with a preference for as many positive as possible. It can be written in terms of the third derivative of x as the jerk equation $\dddot{x} = a_1\ddot{x} + a_2 x\dot{x}^2 + a_3 x$.

The following sections were written by the computer program that performed the optimization, carried out the analysis of the resulting system, and produced the corresponding figures, all without human intervention.

26.2 Simplified System

After about 2×10^5 trials, of which eight were chaotic, simplified parameters for Eq. (26.1) that give chaotic solutions are $a_1 = -2.03$, $a_2 = 1$, $a_3 = -1$. Thus Eq. (26.1) can be written more compactly as

$$\dot{x} = y$$
$$\dot{y} = z \qquad\qquad (26.2)$$
$$\dot{z} = az + xy^2 - x,$$

where $a = -2.03$.

Note that with five terms, Eq. (26.2) should have one independent parameter through a linear rescaling of the three variables plus time, and so the dynamics is completely captured by the single parameter a, which could be put in any of the five terms, albeit with a different numerical value.

26.3   Equilibria

The system in Eq. (26.2) with $a = -2.03$ has an unstable saddle focus at $(0, 0, 0)$ with eigenvalues $(0.1005 - 0.6619i, 0.1005 + 0.6619i, -2.2309)$ and a Poincaré index of 0 in the $z = 0$ plane.

The strange attractor is self-excited, and the system is symmetric under the transformation $x \to -x$, $y \to -y$, $z \to -z$.

26.4   Attractor

Figure 26.1 shows various views of the attractor for Eq. (26.2) with $a = -2.03$ and initial conditions $(-4, 0, 3)$. The rainbow of colors shows the local value of the largest Lyapunov exponent with red indicating the most positive values (regions of worst predictability) and blue indicating the most negative values (regions of best predictability).

26.5   Time Series

Figure 26.2 shows the time series for the three variables along with the local value of the largest Lyapunov exponent (LL) for Eq. (26.2) with $a = -2.03$. Red color in the Lyapunov exponent indicates that the error is growing parallel to the orbit, while blue indicates growth perpendicular to the orbit. Note that the orbit passes through regions where the local Lyapunov exponent is strongly positive and other regions where it is strongly negative as is typical for a chaotic system and is also reflected by the colors in Fig. 26.1.

26.6   Lyapunov Exponents

The global Lyapunov exponents are determined by averaging the local Lyapunov exponents along the orbit. The values typically converge very slowly because of the large variation along the orbit, and an integration time of order 10^8 is required to obtain 4-digit accuracy.

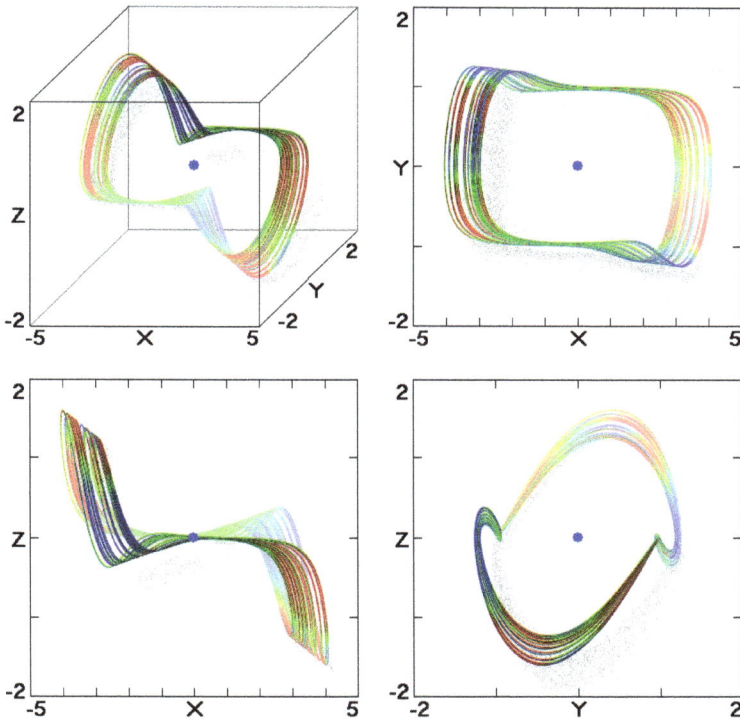

Fig. 26.1   Views of the attractor for Eq. (26.2) with $a = -2.03$ and initial conditions $(-4, 0, 3)$.

The results of such a calculation for the system in Eq. (26.2) with $a = -2.03$ after a time of 2×10^6 are LE = $(0.0768, 0, -2.1068)$ with a Kaplan–Yorke dimension of 2.0364, where the last digit in the quoted values is only an approximation. The positive value of the largest Lyapunov exponent indicates that the system is chaotic, and the negative sum of the exponents (-2.0300) indicates that the system is dissipative with a strange attractor.

26.7   Basin of Attraction

Figure 26.3 shows (in red) the basin of attraction for Eq. (26.2) with $a = -2.03$ in the $z = 0$ plane. Also shown (in black) is the cross-section of the attractor in the same plane.

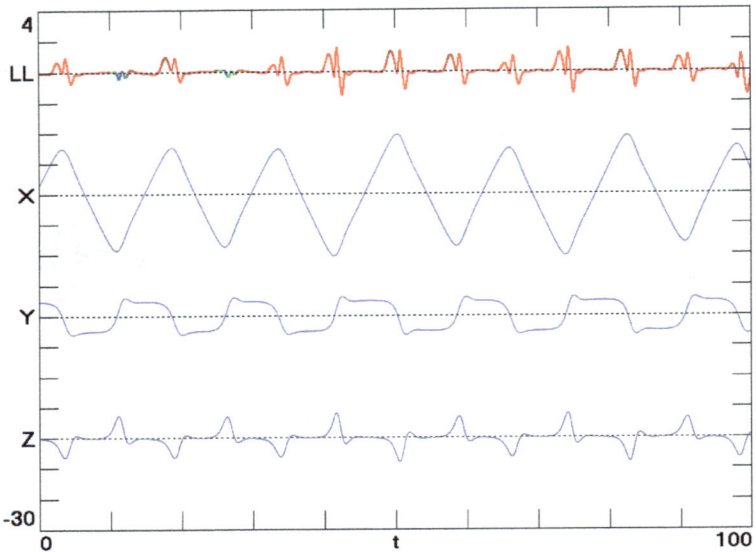

Fig. 26.2   Time series for the variables in Eq. (26.2) with $a = -2.03$ along with the local Lyapunov exponent (LL).

26.8   Bifurcations

Figure 26.4 shows the bifurcation diagram for Eq. (26.2) as a function of the parameter a from $a = -4.06$ to $a = 0$. The initial condition was taken as $(-4, 0, 3)$ at $a = -4.06$ and was not changed as a slowly varied toward $a = 0$. Each of the 500 values of a was calculated for a time of about 1×10^4.

The upper plot shows the three Lyapunov exponents. The middle plot shows the Kaplan–Yorke dimension, and the lower plot shows the local maxima of x. The chaotic region is in the vicinity of $a = -2.03$, and the route to chaos is clearly shown.

26.9   Robustness

One measure of the robustness of a chaotic system is the amount by which the parameters can be changed from their nominal values before the probability of chaos decreases to 50% [Sprott (2022)]. For the system in Eq. (26.2) with $a = -2.03$ and initial conditions $(-4, 0, 3)$, after 9798 trials, it is estimated that the parameter a can be changed by 2.1% before the chaos is

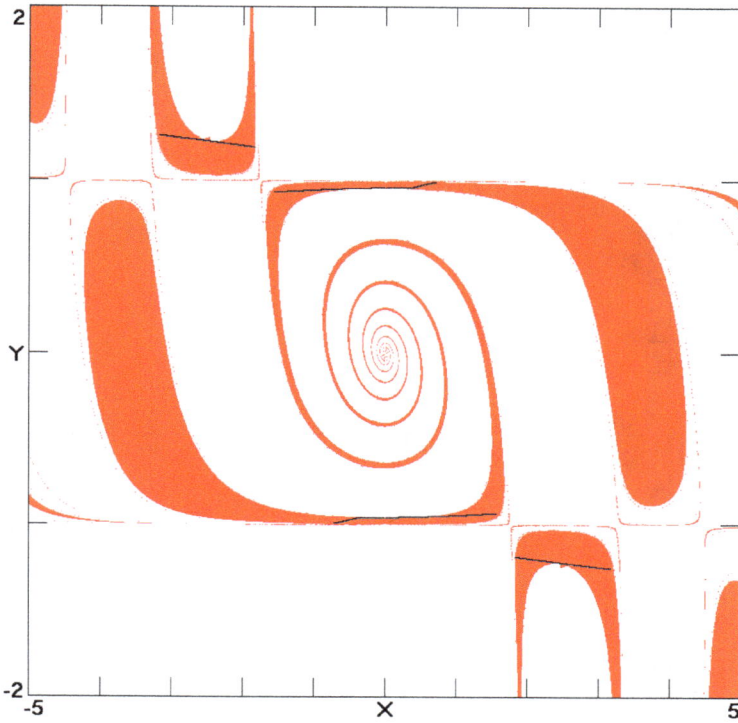

Fig. 26.3　Basin of attraction for Eq. (26.2) with $a = -2.03$ in the $z = 0$ plane.

more likely to be lost than not. Thus the system is somewhat fragile. This result is consistent with the data in Fig. 26.4.

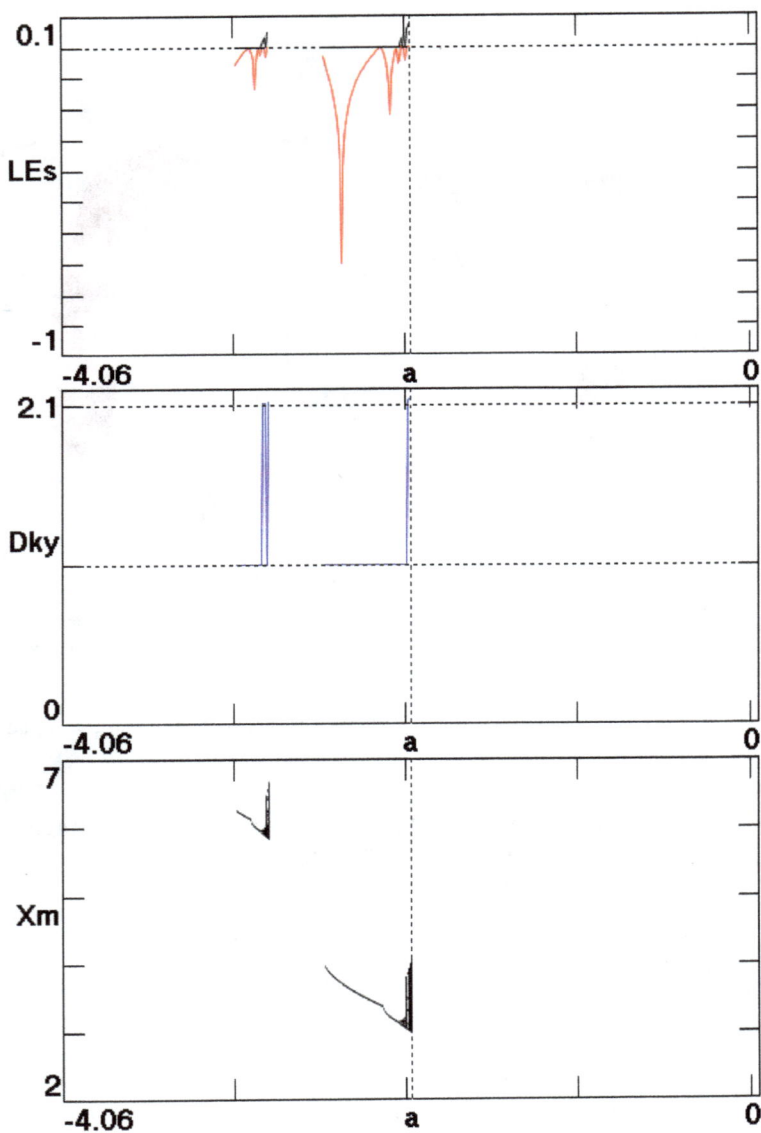

Fig. 26.4   Bifurcation diagram for Eq. (26.2) as a function of the parameter a.

Chapter 27

Moore–Spiegel System

Only slightly more complicated and considerably predating the Malasoma system is another jerk system with a cubic nonlinearity but with six terms. This system was developed to model the irregular variability in the luminosity of stars. The authors noted the aperiodicity and sensitive dependence on initial conditions but were apparently unaware of the similar concurrent work of Lorenz. The system is inversion symmetric with a symmetric pair of strange attractors for the given parameters.

27.1 Introduction

The Moore–Spiegel system [Moore and Spiegel (1966)] written in its most general form with an adjustable coefficient in each of the four terms is given by

$$\dot{x} = y$$
$$\dot{y} = z \qquad\qquad (27.1)$$
$$\dot{z} = a_1 z + a_2 y + a_3 x^2 y + a_4 x.$$

The usual parameters are $a_1 = -1$, $a_2 = 14$, $a_3 = -20$, $a_4 = -6$. It is written here with parameters that can be either positive or negative with a preference for as many positive as possible. It can be written in terms of the third derivative of x as the jerk equation $\dddot{x} = a_1\ddot{x} + a_2\dot{x} + a_3 x^2\dot{x} + a_4 x$.

The following sections were written by the computer program that performed the optimization, carried out the analysis of the resulting system, and produced the corresponding figures, all without human intervention.

27.2 Simplified System

After about 6×10^4 trials, of which 64 were chaotic, simplified parameters for Eq. (27.1) that give chaotic solutions are $a_1 = -0.2$, $a_2 = 9$, $a_3 = -1$,

$a_4 = -1$. Thus Eq. (27.1) can be written more compactly as

$$\dot{x} = y$$
$$\dot{y} = z \tag{27.2}$$
$$\dot{z} = az + by - x^2 y - x,$$

where $a = -0.2$, $b = 9$.

Note that with six terms, Eq. (27.2) should have two independent parameters through a linear rescaling of the three variables plus time, and so the dynamics is completely captured by the given parameters, which could be put in any of the six terms, albeit with different numerical values.

27.3   Equilibria

The system in Eq. (27.2) with $a = -0.2$, $b = 9$ has an unstable saddle node at (0, 0, 0) with eigenvalues (0.1115, 2.8425, −3.154) and a Poincaré index of 0 in the $z = 0$ plane.

The strange attractor is self-excited, and the system is symmetric under the transformation $x \to -x$, $y \to -y$, $z \to -z$.

27.4   Attractor

Figure 27.1 shows various views of one of the attractors for Eq. (27.2) with $a = -0.2$, $b = 9$ and initial conditions (−1, −4, 9). The rainbow of colors shows the local value of the largest Lyapunov exponent with red indicating the most positive values (regions of worst predictability) and blue indicating the most negative values (regions of best predictability).

27.5   Time Series

Figure 27.2 shows the time series for the three variables along with the local value of the largest Lyapunov exponent (LL) for Eq. (27.2) with $a = -0.2$, $b = 9$. Red color in the Lyapunov exponent indicates that the error is growing parallel to the orbit, while blue indicates growth perpendicular to the orbit. Note that the orbit passes through regions where the local Lyapunov exponent is strongly positive and other regions where it is strongly negative as is typical for a chaotic system and is also reflected by the colors in Fig. 27.1.

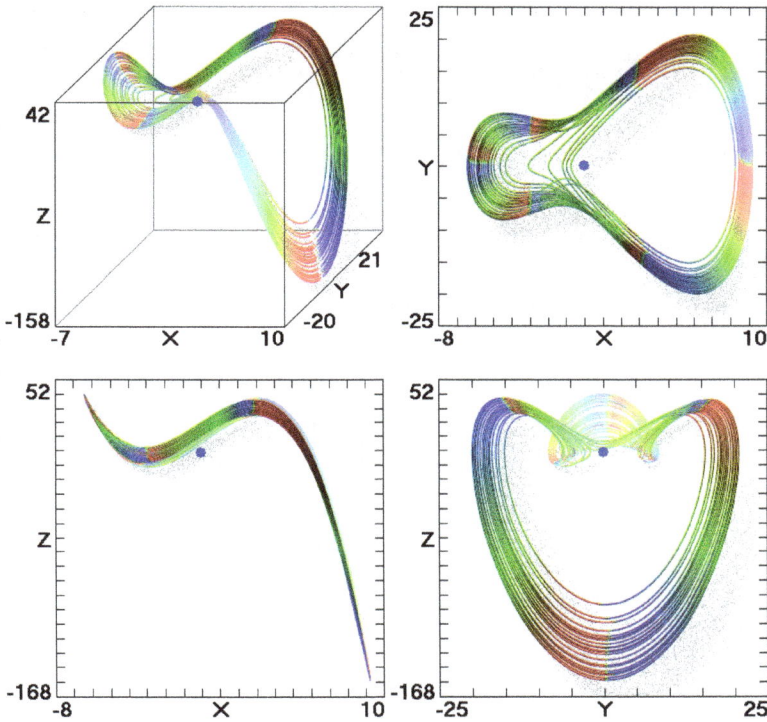

Fig. 27.1   Views of one of the attractors for Eq. (27.2) with $a = -0.2$, $b = 9$ and initial conditions $(-1, -4, 9)$.

27.6   Lyapunov Exponents

The global Lyapunov exponents are determined by averaging the local Lyapunov exponents along the orbit. The values typically converge very slowly because of the large variation along the orbit, and an integration time of order 10^8 is required to obtain 4-digit accuracy.

The results of such a calculation for the system in Eq. (27.2) with $a = -0.2$, $b = 9$ after a time of 4×10^6 are LE $= (0.0583, 0, -0.2583)$ with a Kaplan–Yorke dimension of 2.2258, where the last digit in the quoted values is only an approximation. The positive value of the largest Lyapunov exponent indicates that the system is chaotic, and the negative sum of the exponents (-0.2000) indicates that the system is dissipative with a strange attractor.

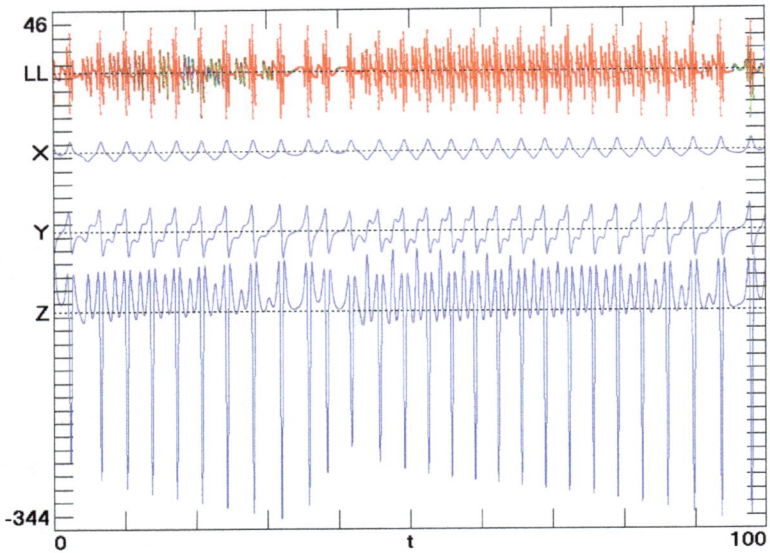

Fig. 27.2　Time series for the variables in Eq. (27.2) with $a = -0.2$, $b = 9$ along with the local Lyapunov exponent (LL).

27.7　Basin of Attraction

Figure 27.3 shows (in red) the basin of attraction for Eq. (27.2) with $a = -0.2$, $b = 9$ in the $z = 0$ plane. Also shown (in black) is the cross-section of the attractors in the same plane.

27.8　Bifurcations

Figure 27.4 shows the bifurcation diagram for Eq. (27.2) as a function of the parameter a from $a = -0.4$ to $a = 0$ for $b = 9$. The initial condition was taken as $(-1, -4, 9)$ at $a = -0.4$ and was not changed as a slowly varied toward $a = 0$. Each of the 500 values of a was calculated for a time of about 2×10^4.

The upper plot shows the three Lyapunov exponents. The middle plot shows the Kaplan–Yorke dimension, and the lower plot shows the local maxima of x. The chaotic region is in the vicinity of $a = -0.2$, $b = 9$, and the route to chaos is clearly shown.

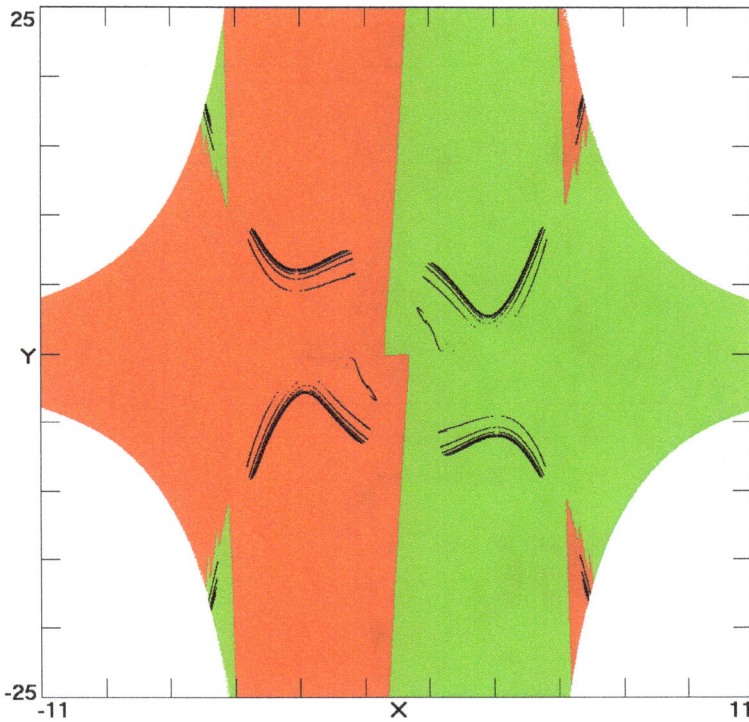

Fig. 27.3   Basin of attraction for Eq. (27.2) with $a = -0.2$, $b = 9$ in the $z = 0$ plane.

27.9   Robustness

One measure of the robustness of a chaotic system is the amount by which the parameters can be changed from their nominal values before the probability of chaos decreases to 50% [Sprott (2022)]. For the system in Eq. (27.2) with $a = -0.2$, $b = 9$ and initial conditions $(-1, -4, 9)$, after 1959 trials, it is estimated that the parameters can be changed by 49% before the chaos is more likely to be lost than not. Thus the system is somewhat robust.

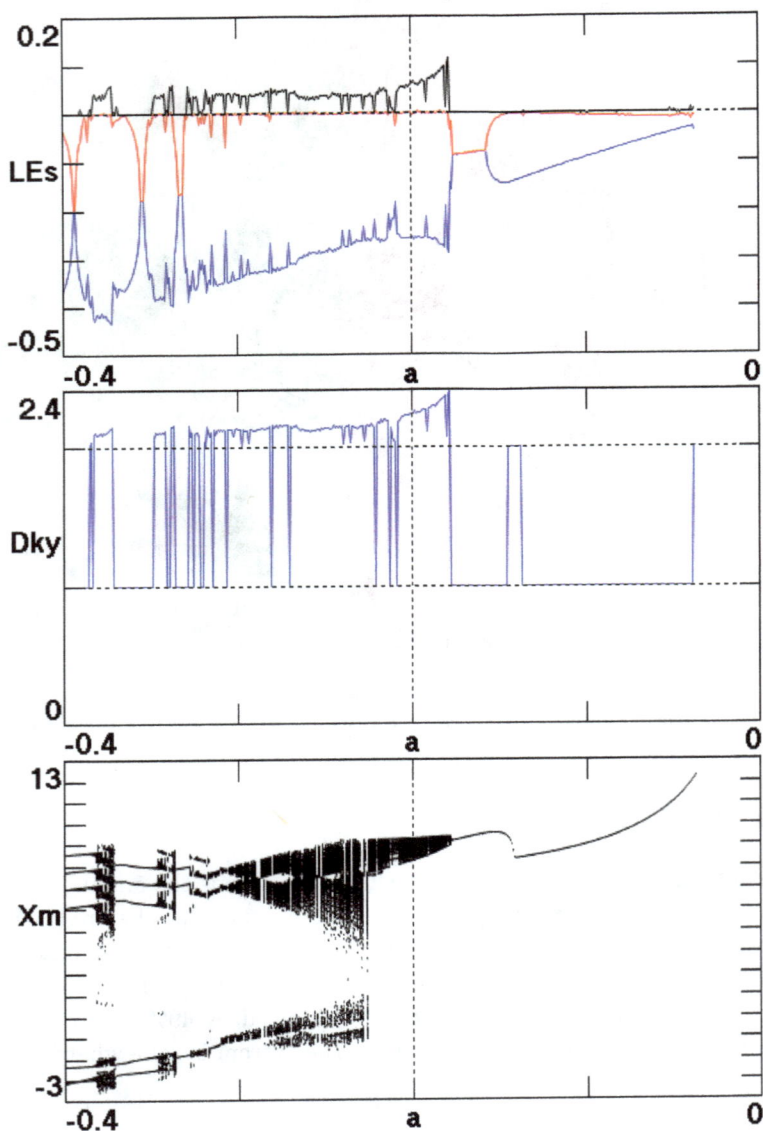

Fig. 27.4 Bifurcation diagram for Eq. (27.2) as a function of the parameter a for $b = 9$.

Chapter 28

Linz–Sprott System

Perhaps the simplest chaotic jerk system with a piecewise-linear nonlinearity is the Linz–Sprott system. It has an absolute value function that makes it especially suitable for electronic implementation using diodes. It also provides an example of a memory oscillator of which there are many other cases.

28.1 Introduction

The Linz–Sprott system [Linz and Sprott (1999)] written in its most general form with an adjustable coefficient in each of the four terms is given by

$$\dot{x} = y$$
$$\dot{y} = z \tag{28.1}$$
$$\dot{z} = a_1 z + a_2 y + a_3 |x| + a_4.$$

The usual parameters are $a_1 = -0.6$, $a_2 = a_4 = -1$, $a_3 = 1$. It is written here with parameters that can be either positive or negative with a preference for as many positive as possible. There is a conservative version of this system with $a_1 = 0$ and a tiny chaotic sea, but we consider here only the dissipative version with $a_1 < 0$. This is an example of a memory oscillator because it can be written as a damped harmonic oscillator driven by a time-dependent term that depends on the past value of x: $\ddot{x} - a_1 \dot{x} - a_2 x = \int_{-\infty}^{t} (a_3 |x(t)| + a_4) dt$.

The following sections were written by the computer program that performed the optimization, carried out the analysis of the resulting system, and produced the corresponding figures, all without human intervention.

28.2 Simplified System

After about 2×10^5 trials, of which 18 were chaotic, simplified parameters for Eq. (28.1) that give chaotic solutions are $a_1 = -0.6$, $a_2 = -1$, $a_3 = -1$, $a_4 = 1$. Thus Eq. (28.1) can be written more compactly as

$$\dot{x} = y$$
$$\dot{y} = z \qquad\qquad (28.2)$$
$$\dot{z} = az - y - |x| + 1,$$

where $a = -0.6$.

Note that with six terms, Eq. (28.2) should have two independent parameters through a linear rescaling of the three variables plus time. However, one of the two parameters has a value of ± 1 and can be placed in any of the remaining five terms.

28.3 Equilibria

The system in Eq. (28.2) with $a = -0.6$ has two equilibrium points:

Equilibrium # 1 is an unstable saddle focus at $(-1, 0, 0)$ with eigenvalues $(0.5885, -0.5942 - 1.1603i, -0.5942 + 1.1603i)$ and a Poincaré index of 0 in the $z = 0$ plane.

Equilibrium # 2 is an unstable saddle focus at $(1, 0, 0)$ with eigenvalues $(-0.8356, 0.1178 - 1.0876i, 0.1178 + 1.0876i)$ and a Poincaré index of 0 in the $z = 0$ plane.

The strange attractor is self-excited, and the system has no symmetry.

28.4 Attractor

Figure 28.1 shows various views of the attractor for Eq. (28.2) with $a = -0.6$ and initial conditions $(-1, 1, 0)$. The rainbow of colors shows the local value of the largest Lyapunov exponent with red indicating the most positive values (regions of worst predictability) and blue indicating the most negative values (regions of best predictability).

28.5 Time Series

Figure 28.2 shows the time series for the three variables along with the local value of the largest Lyapunov exponent (LL) for Eq. (28.2) with $a = -0.6$. Red color in the Lyapunov exponent indicates that the error is growing

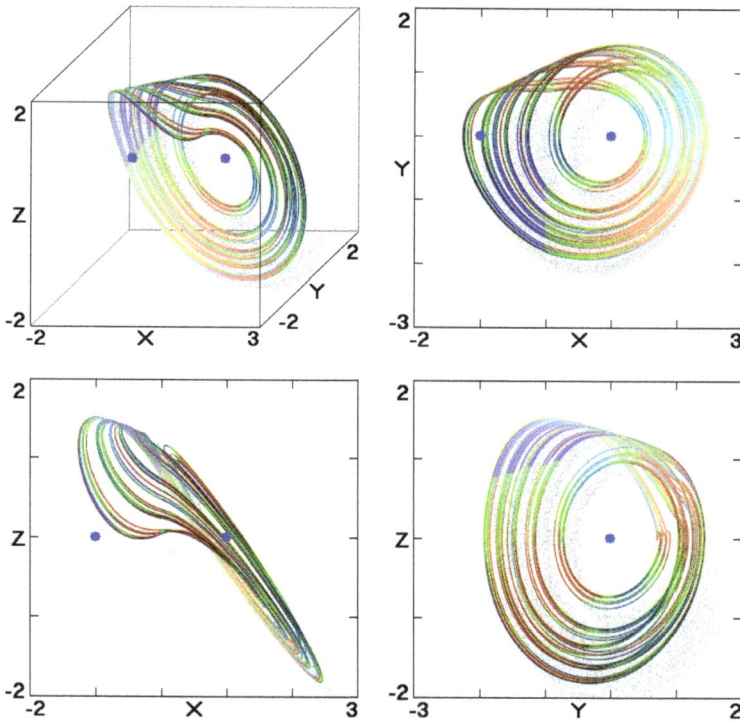

Fig. 28.1 Views of the attractor for Eq. (28.2) with $a = -0.6$ and initial conditions $(-1, 1, 0)$.

parallel to the orbit, while blue indicates growth perpendicular to the orbit. Note that the orbit passes through regions where the local Lyapunov exponent is strongly positive and other regions where it is strongly negative as is typical for a chaotic system and is also reflected by the colors in Fig. 28.1.

28.6 Lyapunov Exponents

The global Lyapunov exponents are determined by averaging the local Lyapunov exponents along the orbit. The values typically converge very slowly because of the large variation along the orbit, and an integration time of order 10^8 is required to obtain 4-digit accuracy.

The results of such a calculation for the system in Eq. (28.2) with $a =$

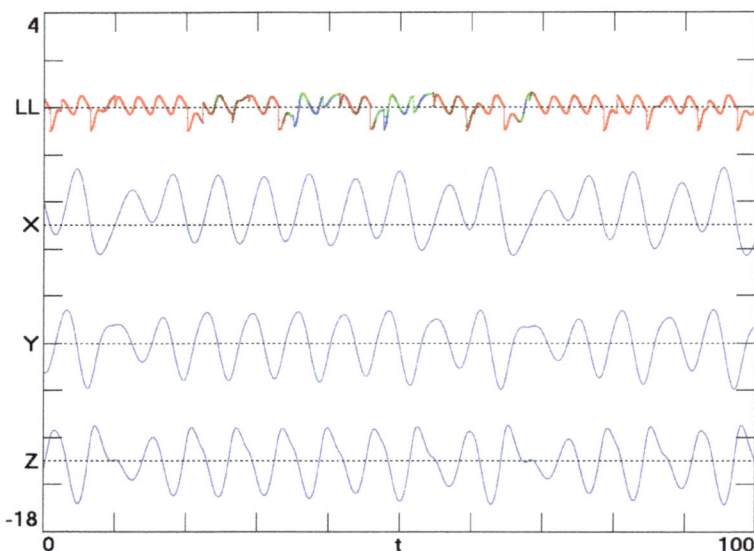

Fig. 28.2   Time series for the variables in Eq. (28.2) with $a = -0.6$ along with the local Lyapunov exponent (LL).

-0.6 after a time of 2×10^6 are LE $= (0.0364, 0, -0.6364)$ with a Kaplan–Yorke dimension of 2.0572, where the last digit in the quoted values is only an approximation. The positive value of the largest Lyapunov exponent indicates that the system is chaotic, and the negative sum of the exponents (-0.6000) indicates that the system is dissipative with a strange attractor.

28.7   Basin of Attraction

Figure 28.3 shows (in red) the basin of attraction for Eq. (28.2) with $a = -0.6$ in the $z = 0$ plane. Also shown (in black) is the cross-section of the attractor in the same plane.

28.8   Bifurcations

Figure 28.4 shows the bifurcation diagram for Eq. (28.2) as a function of the parameter a from $a = -1.2$ to $a = 0$. The initial condition was taken as $(-1, 1, 0)$ at $a = -1.2$ and was not changed as a slowly varied toward $a = 0$. Each of the 500 values of a was calculated for a time of about 1×10^4.

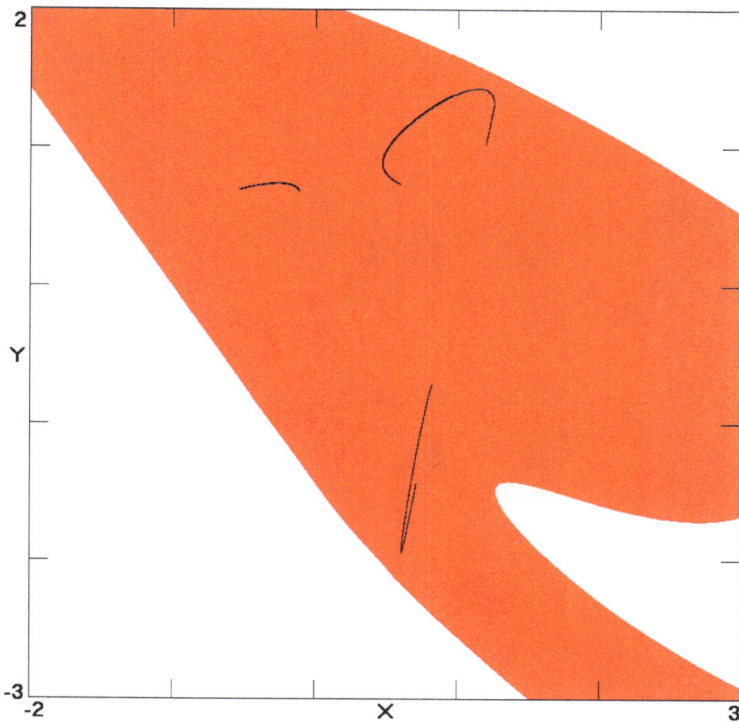

Fig. 28.3   Basin of attraction for Eq. (28.2) with $a = -0.6$ in the $z = 0$ plane.

The upper plot shows the three Lyapunov exponents. The middle plot shows the Kaplan–Yorke dimension, and the lower plot shows the local maxima of x. The chaotic region is in the vicinity of $a = -0.6$, and the route to chaos is clearly shown.

28.9   Robustness

One measure of the robustness of a chaotic system is the amount by which the parameters can be changed from their nominal values before the probability of chaos decreases to 50% [Sprott (2022)]. For the system in Eq. (28.2) with $a = -0.6$ and initial conditions $(-1, 1, 0)$, after 4479 trials, it is estimated that the parameter a can be changed by 10% before the chaos is more likely to be lost than not. Thus the system is somewhat fragile. This result is consistent with the data in Fig. 28.4.

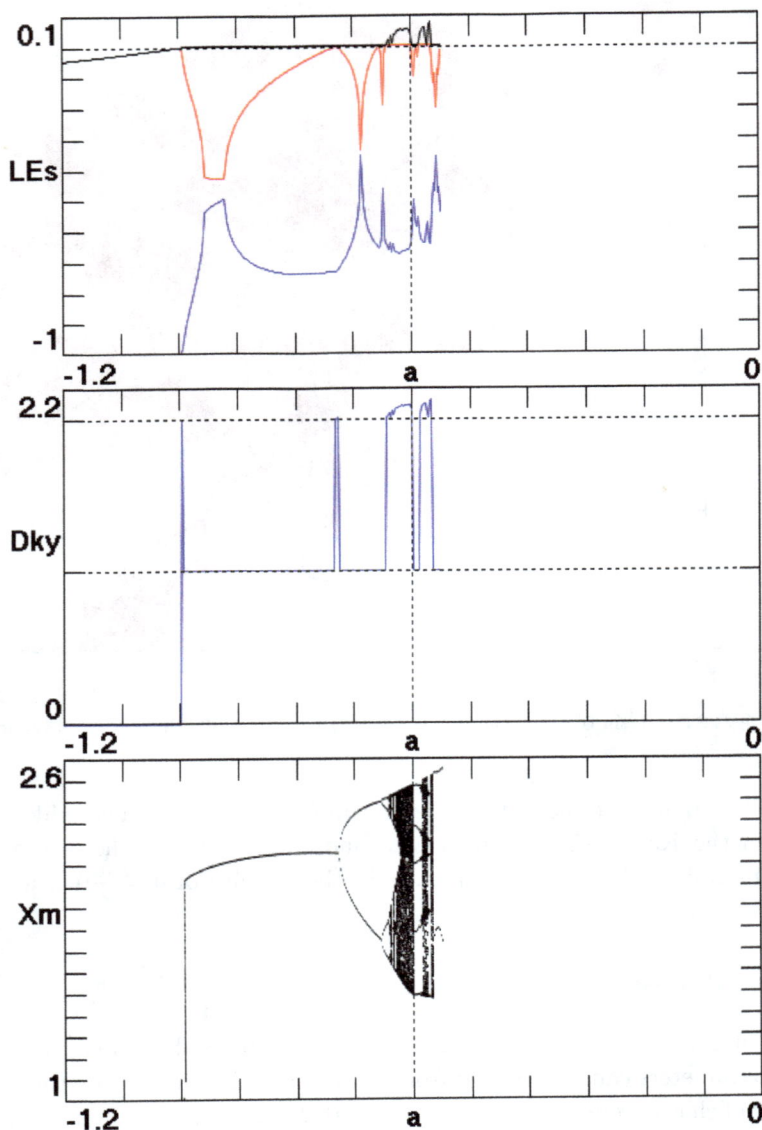

Fig. 28.4   Bifurcation diagram for Eq. (28.2) as a function of the parameter a.

Chapter 29

Elwakil–Kennedy System

Perhaps the simplest chaotic jerk system with a signum nonlinearity is the Elwakil–Kennedy system. It has inversion symmetry, produces double scrolls, and is especially suitable for electronic implementation using saturating amplifiers. It is another example of a memory oscillator of which there are many other cases.

29.1 Introduction

The Elwakil–Kennedy system [Elwakil and Kennedy (2001)] written in its most general form with an adjustable coefficient in each of the four terms is given by

$$
\begin{aligned}
\dot{x} &= y \\
\dot{y} &= z \\
\dot{z} &= a_1 z + a_2 y + a_3 x + a_4 \operatorname{sgn} x.
\end{aligned}
\tag{29.1}
$$

The usual parameters are $a_1 = a_2 = a_3 = -0.8$, $a_4 = 0.8$. It is written here with parameters that can be either positive or negative with a preference for as many positive as possible. There is a conservative version of this system with $a_1 = 0$ and a tiny chaotic sea, but we consider here only the dissipative version with $a_1 < 0$. To avoid the discontinuity in the flow, the $\operatorname{sgn} x$ function is replaced by $\tanh(100x)$ for calculating the Lyapunov exponents. This is an example of a memory oscillator because it can be written as a damped harmonic oscillator driven by a time-dependent term that depends on the past value of x: $\ddot{x} - a_1\dot{x} - a_2 x = \int_{-\infty}^{t} [a_3 x(t) + a_4 \operatorname{sgn} x(t)]dt$.

The following sections were written by the computer program that performed the optimization, carried out the analysis of the resulting system, and produced the corresponding figures, all without human intervention.

29.2   Simplified System

After about 2×10^4 trials, of which 80 were chaotic, simplified parameters for Eq. (29.1) that give chaotic solutions are $a_1 = -1$, $a_2 = -1$, $a_3 = -2$, $a_4 = 1$. Thus Eq. (29.1) can be written more compactly as

$$\begin{aligned} \dot{x} &= y \\ \dot{y} &= z \\ \dot{z} &= -z - y + ax + \operatorname{sgn} x, \end{aligned} \qquad (29.2)$$

where $a = -2$.

Note that with six terms, Eq. (29.2) should have two independent parameters through a linear rescaling of the three variables plus time. However, one of the two parameters has a value of ± 1 and can be placed in any of the remaining five terms.

29.3   Equilibria

The system in Eq. (29.2) with $a = -2$ has three equilibrium points:

Equilibrium # 1 is an unstable saddle focus at $(0.5000, 0, 0)$ with eigenvalues $(0.1766 - 1.2028i, 0.1766 + 1.2028i, -1.3532)$ and a Poincaré index of 0 in the $z = 0$ plane.

Equilibrium # 2 is an unstable saddle focus at $(-0.5000, 0, 0)$ with eigenvalues $(0.1766 - 1.2028i, 0.1766 + 1.2028i, -1.3532)$ and a Poincaré index of 0 in the $z = 0$ plane.

Equilibrium # 3 is an unstable saddle focus at $(0, 0, 0)$ with eigenvalues $(4.233, -2.6165 - 4.038i, -2.6165 + 4.038i)$ and a Poincaré index of 0 in the $z = 0$ plane.

The strange attractor is self-excited, and the system is symmetric under the transformation $x \to -x$, $y \to -y$, $z \to -z$.

29.4   Attractor

Figure 29.1 shows various views of the attractor for Eq. (29.2) with $a = -2$ and initial conditions $(-1, 1, 0)$. The rainbow of colors shows the local value of the largest Lyapunov exponent with red indicating the most positive values (regions of worst predictability) and blue indicating the most negative values (regions of best predictability).

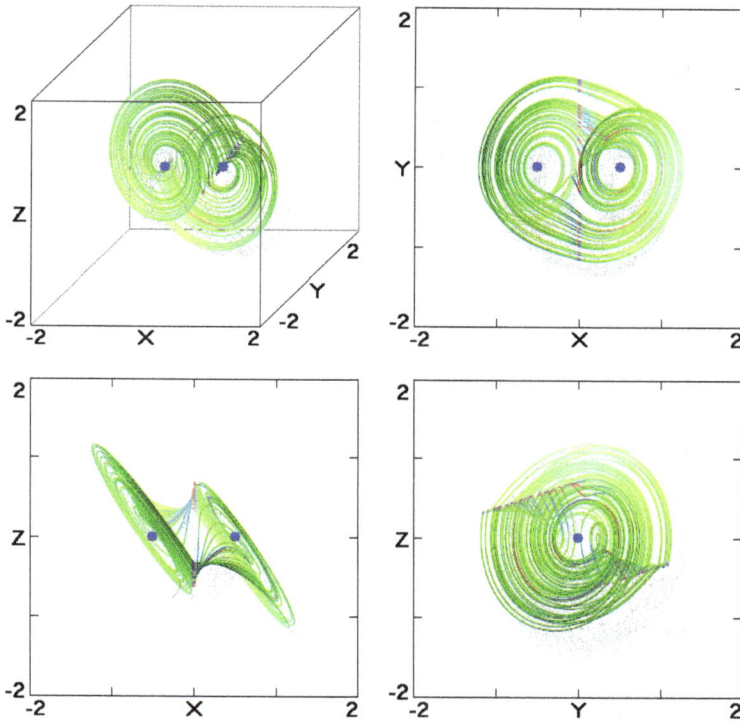

Fig. 29.1 Views of the attractor for Eq. (29.2) with $a = -2$ and initial conditions $(-1, 1, 0)$.

29.5 Time Series

Figure 29.2 shows the time series for the three variables along with the local value of the largest Lyapunov exponent (LL) for Eq. (29.2) with $a = -2$. Red color in the Lyapunov exponent indicates that the error is growing parallel to the orbit, while blue indicates growth perpendicular to the orbit. Note that the orbit passes through regions where the local Lyapunov exponent is strongly positive and other regions where it is strongly negative as is typical for a chaotic system and is also reflected by the colors in Fig. 29.1.

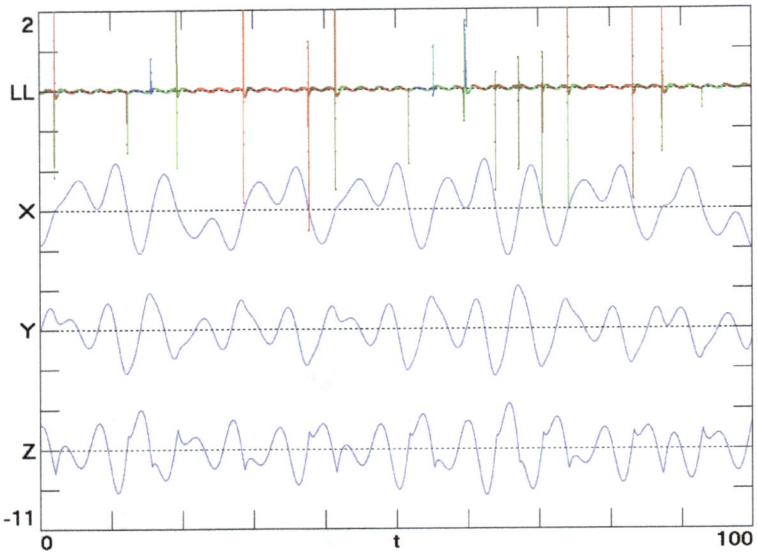

Fig. 29.2   Time series for the variables in Eq. (29.2) with $a = -2$ along with the local Lyapunov exponent (LL).

29.6   Lyapunov Exponents

The global Lyapunov exponents are determined by averaging the local Lyapunov exponents along the orbit. The values typically converge very slowly because of the large variation along the orbit, and an integration time of order 10^8 is required to obtain 4-digit accuracy.

The results of such a calculation for the system in Eq. (29.2) with $a = -2$ after a time of 3×10^6 are LE = $(0.1492, 0, -1.1492)$ with a Kaplan–Yorke dimension of 2.1297, where the last digit in the quoted values is only an approximation. The positive value of the largest Lyapunov exponent indicates that the system is chaotic, and the negative sum of the exponents (-1) indicates that the system is dissipative with a strange attractor.

29.7   Basin of Attraction

Figure 29.3 shows (in red) the basin of attraction for Eq. (29.2) with $a = -2$ in the $z = 0$ plane. Also shown (in black) is the cross-section of the attractor in the same plane.

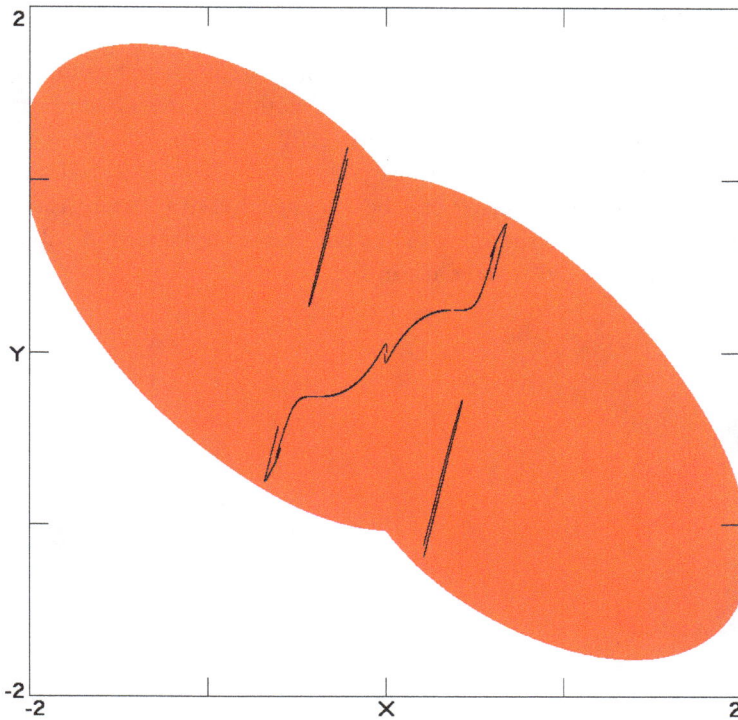

Fig. 29.3   Basin of attraction for Eq. (29.2) with $a = -2$ in the $z = 0$ plane.

29.8   Bifurcations

Figure 29.4 shows the bifurcation diagram for Eq. (29.2) as a function of the parameter a from $a = -4$ to $a = 0$. The initial condition was taken as $(-1, 1, 0)$ at $a = -4$ and was not changed as a slowly varied toward $a = 0$. Each of the 500 values of a was calculated for a time of about 1×10^4.

The upper plot shows the three Lyapunov exponents. The middle plot shows the Kaplan–Yorke dimension, and the lower plot shows the local maxima of x. The chaotic region is in the vicinity of $a = -2$, and the route to chaos is clearly shown.

29.9   Robustness

One measure of the robustness of a chaotic system is the amount by which the parameters can be changed from their nominal values before the probability of chaos decreases to 50% [Sprott (2022)]. For the system in Eq. (29.2) with $a = -2$ and initial conditions $(-1, 1, 0)$, after 2035 trials, it is estimated that the parameter a can be changed by 49% before the chaos is more likely to be lost than not. Thus the system is somewhat robust. This result is consistent with the data in Fig. 29.4.

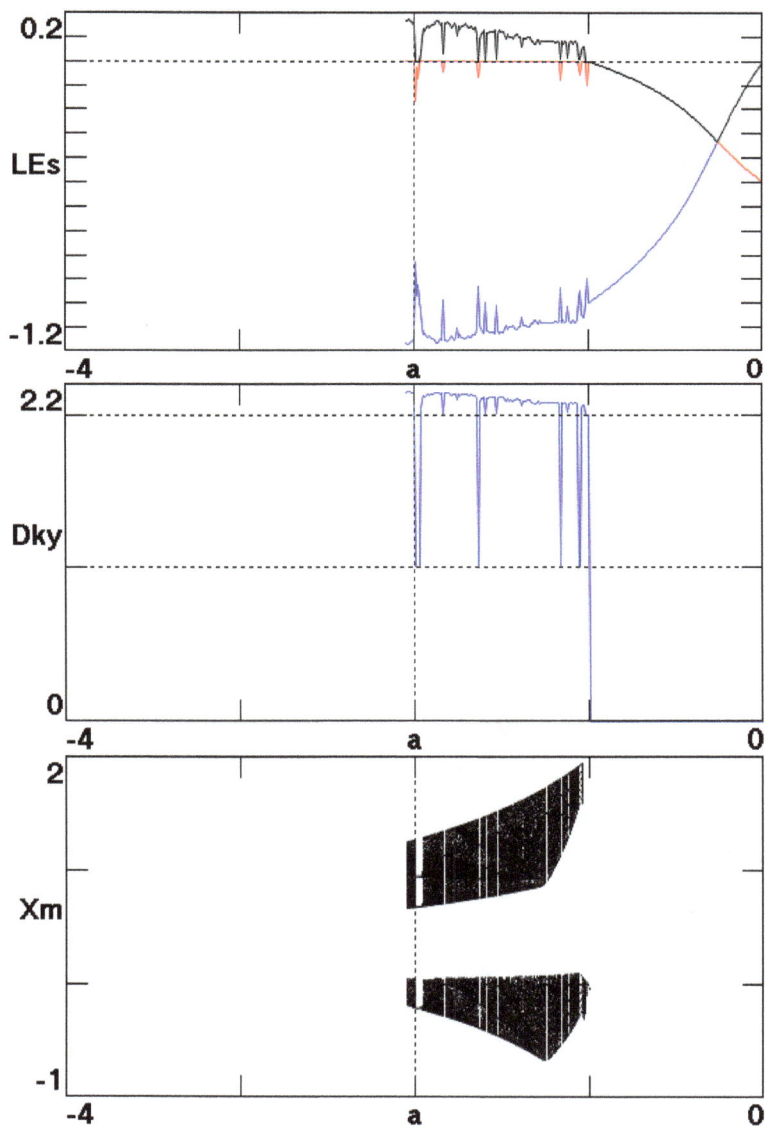

Fig. 29.4   Bifurcation diagram for Eq. (29.2) as a function of the parameter a.

Chapter 30

Chua System

Chua's circuit is one of the oldest and most widely studied chaotic electronic circuits. It can be modeled in a number of ways, one of the simplest of which uses a piecewise-linear nonlinearity to model Chua's diode. It can produce a wide variety of strange attractors including some that are hidden. The system is inversion symmetric with a symmetric double-scroll strange attractor for some values of the parameters and a symmetric pair of single-scroll strange attractors coexisting with a symmetric pair of limit cycles for the case shown here.

30.1   Introduction

The Chua system [Matsumoto *et al.* (1985)] written in its most general form with an adjustable coefficient in each of the seven terms is given by

$$
\begin{aligned}
\dot{x} &= a_1 y - a_2 x + a_3(|x+1| - |x-1|) \\
\dot{y} &= a_4 x - a_5 y + a_6 z \\
\dot{z} &= -a_7 y.
\end{aligned}
\tag{30.1}
$$

The usual parameters are $a_1 = 9$, $a_2 = 18/7$, $a_3 = 9/7$, $a_4 = a_5 = a_6 = 1$, $a_7 = 100/7$. It is written here with the signs constrained and non-negative parameters since it is a model for a particular circuit.

   The following sections were written by the computer program that performed the optimization, carried out the analysis of the resulting system, and produced the corresponding figures, all without human intervention.

30.2   Simplified System

After about 6×10^4 trials, of which 90 were chaotic, simplified parameters for Eq. (30.1) that give chaotic solutions are $a_1 = 1$, $a_2 = 1$, $a_3 = 1$, $a_4 = 1$,

$a_5 = 0$, $a_6 = 1$, $a_7 = 3.5$. Thus Eq. (30.1) can be written more compactly as

$$\dot{x} = y - x + (|x+1| - |x-1|)$$
$$\dot{y} = x + z \tag{30.2}$$
$$\dot{z} = -ay,$$

where $a = 3.5$.

Note that with six terms, Eq. (30.2) should have two independent parameters through a linear rescaling of the three variables plus time. However, one of the two parameters has a value of ± 1 and can be placed in any of the remaining five terms.

30.3 Equilibria

The system in Eq. (30.2) with $a = 3.5$ has three equilibrium points:

Equilibrium # 1 is an unstable saddle focus at $(-2, 0, 2)$ with eigenvalues $(-1.2467, 0.1233 - 1.671i, 0.1233 + 1.671i)$ and a Poincaré index of -1 in the $z = 2$ plane.

Equilibrium # 2 is an unstable saddle focus at $(2, 0, -2)$ with eigenvalues $(-1.2467, 0.1233 - 1.671i, 0.1233 + 1.671i)$ and a Poincaré index of -1 in the $z = -2$ plane.

Equilibrium # 3 is an unstable saddle focus at $(0, 0, 0)$ with eigenvalues $(1.2467, -0.1233 - 1.671i, -0.1233 + 1.671i)$ and a Poincaré index of -1 in the $z = 0$ plane.

The strange attractor is self-excited, and the system is symmetric under the transformation $x \to -x$, $y \to -y$, $z \to -z$.

30.4 Attractor

Figure 30.1 shows various views of one of the attractors for Eq. (30.2) with $a = 3.5$ and initial conditions $(3, 1, -7)$. The rainbow of colors shows the local value of the largest Lyapunov exponent with red indicating the most positive values (regions of worst predictability) and blue indicating the most negative values (regions of best predictability).

30.5 Time Series

Figure 30.2 shows the time series for the three variables along with the local value of the largest Lyapunov exponent (LL) for Eq. (30.2) with $a = 3.5$. Red color in the Lyapunov exponent indicates that the error is growing

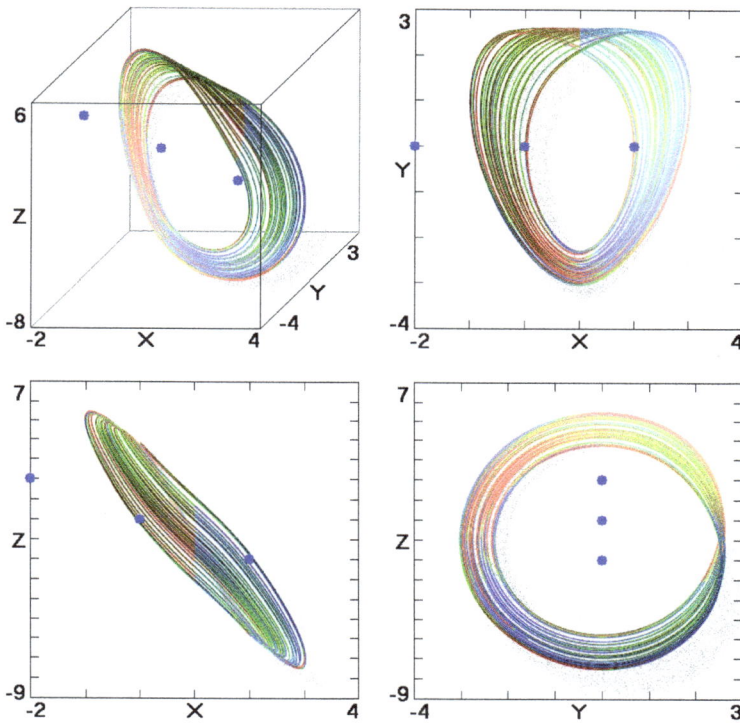

Fig. 30.1   Views of one of the attractors for Eq. (30.2) with $a = 3.5$ and initial conditions $(3, 1, -7)$.

parallel to the orbit, while blue indicates growth perpendicular to the orbit. Note that the orbit passes through regions where the local Lyapunov exponent is strongly positive and other regions where it is strongly negative as is typical for a chaotic system and is also reflected by the colors in Fig. 30.1.

30.6   Lyapunov Exponents

The global Lyapunov exponents are determined by averaging the local Lyapunov exponents along the orbit. The values typically converge very slowly because of the large variation along the orbit, and an integration time of order 10^8 is required to obtain 4-digit accuracy.

The results of such a calculation for the system in Eq. (30.2) with $a = 3.5$

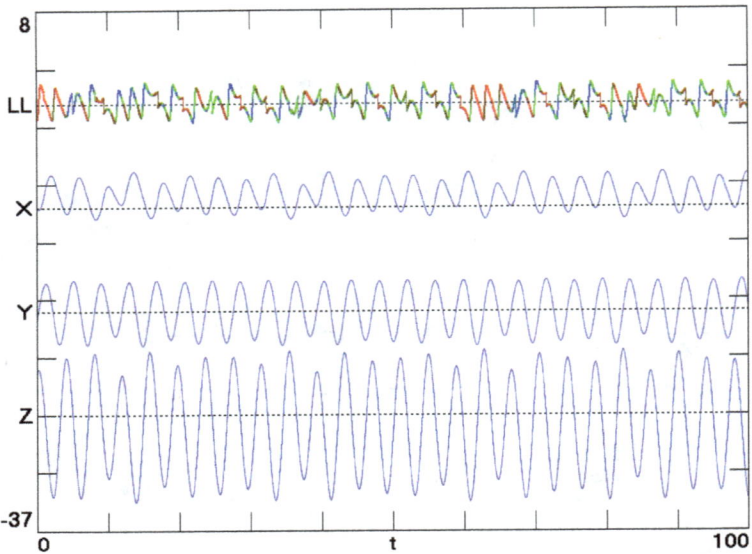

Fig. 30.2   Time series for the variables in Eq. (30.2) with $a = 3.5$ along with the local Lyapunov exponent (LL).

after a time of 2×10^6 are LE $= (0.0395,\ 0,\ -0.0552)$ with a Kaplan–Yorke dimension of 2.7154, where the last digit in the quoted values is only an approximation. The positive value of the largest Lyapunov exponent indicates that the system is chaotic, and the negative sum of the exponents (-0.0157) indicates that the system is dissipative with a strange attractor.

30.7   Basin of Attraction

Figure 30.3 shows (in red) the basin of attraction for Eq. (30.2) with $a = 3.5$ in the $z = 0$ plane. Also shown (in black) is the cross-section of the attractors in the same plane.

30.8   Bifurcations

Figure 30.4 shows the bifurcation diagram for Eq. (30.2) as a function of the parameter a from 0 to 7. The initial condition was taken as $(3, 1, -7)$ for each value of a. Each of the 500 values of a was calculated for a time of about 1×10^4.

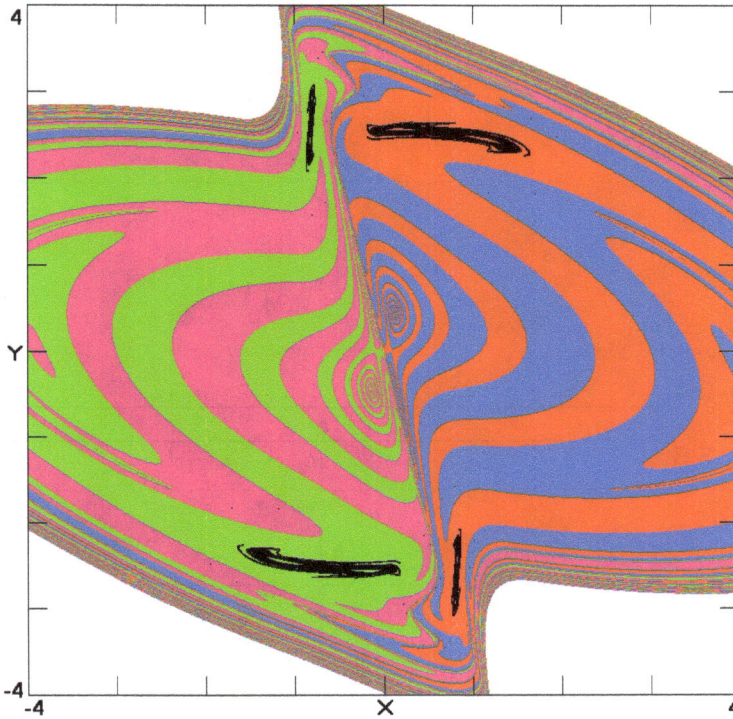

Fig. 30.3   Basin of attraction for Eq. (30.2) with $a = 3.5$ in the $z = 0$ plane.

The upper plot shows the three Lyapunov exponents. The middle plot shows the Kaplan–Yorke dimension, and the lower plot shows the local maxima of x. The chaotic region is in the vicinity of $a = 3.5$, and the route to chaos is clearly shown.

30.9   Robustness

One measure of the robustness of a chaotic system is the amount by which the parameters can be changed from their nominal values before the probability of chaos decreases to 50% [Sprott (2022)]. For the system in Eq. (30.2) with $a = 3.5$ and initial conditions $(3, 1, -7)$, after 5107 trials, it is estimated that the parameter a can be changed by 7.7% before the chaos is more likely to be lost than not. Thus the system is somewhat fragile. This result is consistent with the data in Fig. 30.4.

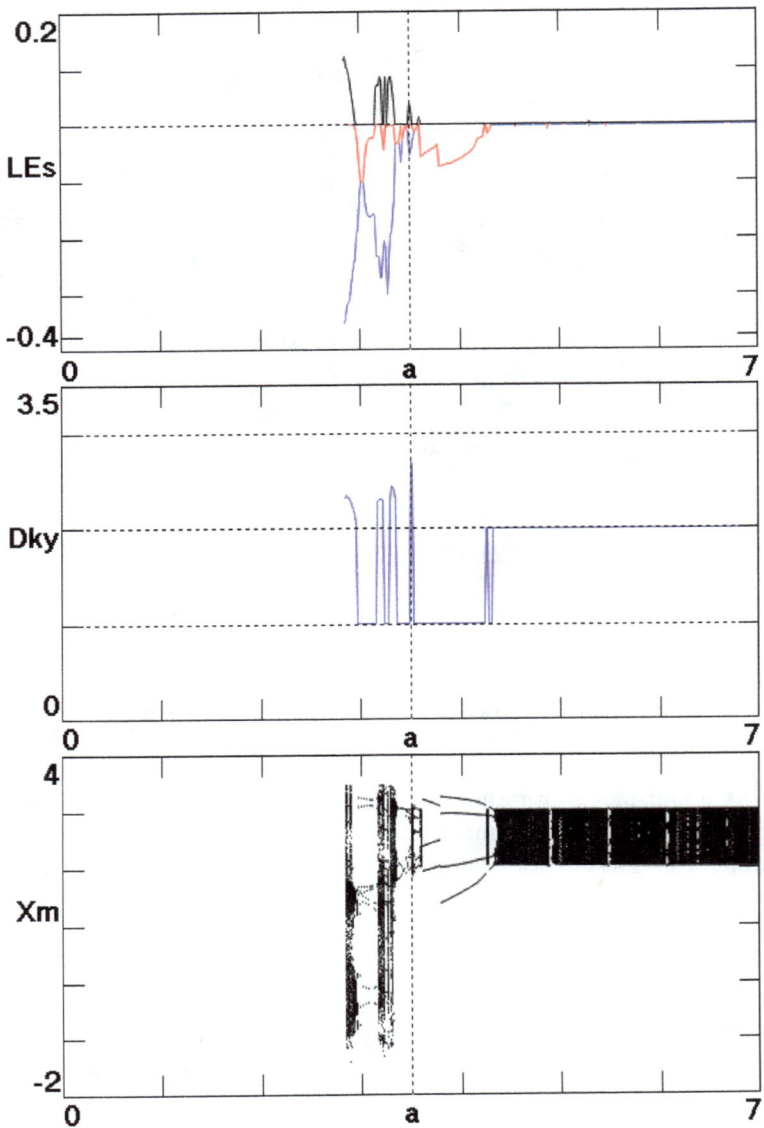

Fig. 30.4   Bifurcation diagram for Eq. (30.2) as a function of the parameter a.

Chapter 31

Chen System

The Chen system is another relatively old and widely studied chaotic system. It is a time-reversed version of the Lorenz system with a particular constraint on the parameters. It is symmetric with respect to a 180° rotation about the z axis.

31.1 Introduction

The Chen system [Chen and Ueta (1999)] written in its most general form with an adjustable coefficient in each of the seven terms is given by

$$
\begin{aligned}
\dot{x} &= a_1 y - a_2 x \\
\dot{y} &= -a_3 x - a_4 xz + a_5 y \\
\dot{z} &= a_6 xy - a_7 z.
\end{aligned}
\tag{31.1}
$$

However, the parameters are constrained by $a_2 = a_1$ and $a_3 = a_1 - a_5$. The usual parameters are $a_1 = a_2 = 35$, $a_3 = 7$, $a_4 = a_6 = 1$, $a_5 = 28$, $a_7 = 3$. It is written here with the signs constrained and non-negative parameters to preserve the form of the system [Algaba $et\ al.$ (2013)].

The following sections were written by the computer program that performed the optimization, carried out the analysis of the resulting system, and produced the corresponding figures, all without human intervention.

31.2 Simplified System

After about 3×10^4 trials, of which ten were chaotic, simplified parameters for Eq. (31.1) that give chaotic solutions are $a_1 = 1$, $a_2 = 1$, $a_3 = 0.3$, $a_4 = 1$, $a_5 = 0.7$, $a_6 = 1$, $a_7 = 0.1$. Thus Eq. (31.1) can be written more compactly as

$$\dot{x} = y - x$$
$$\dot{y} = -ax - xz + by \qquad (31.2)$$
$$\dot{z} = xy - cz,$$

where $a = 0.3$, $b = 0.7$, $c = 0.1$.

Note that with seven terms, Eq. (31.2) should have three independent parameters through a linear rescaling of the three variables plus time, and so the dynamics is completely captured by the given parameters, which could be put in any of the seven terms, albeit with different numerical values.

31.3   Equilibria

The system in Eq. (31.2) with $a = 0.3$, $b = 0.7$, $c = 0.1$ has three equilibrium points:

Equilibrium # 1 is an unstable saddle focus at $(-0.2000, -0.2000, 0.4000)$ with eigenvalues $(-0.5427, 0.0713 - 0.3773i, 0.0713 + 0.3773i)$ and a Poincaré index of -1 in the $z = 0.4$ plane.

Equilibrium # 2 is an unstable saddle focus at $(0.2000, 0.2000, 0.4000)$ with eigenvalues $(-0.5427, 0.0713 - 0.3773i, 0.0713 + 0.3773i)$ and a Poincaré index of 0 in the $z = 0.4$ plane.

Equilibrium # 3 is an unstable saddle node at $(0, 0, 0)$ with eigenvalues $(-0.8, 0.5, -0.1)$ and a Poincaré index of -1 in the $z = 0$ plane.

The strange attractor is self-excited, and the system is symmetric under the transformation $x \rightarrow -x$, $y \rightarrow -y$.

31.4   Attractor

Figure 31.1 shows various views of the attractor for Eq. (31.2) with $a = 0.3$, $b = 0.7$, $c = 0.1$ and initial conditions $(-2, -3, 0)$. The rainbow of colors shows the local value of the largest Lyapunov exponent with red indicating the most positive values (regions of worst predictability) and blue indicating the most negative values (regions of best predictability).

31.5   Time Series

Figure 31.2 shows the time series for the three variables along with the local value of the largest Lyapunov exponent (LL) for Eq. (31.2) with $a = 0.3$,

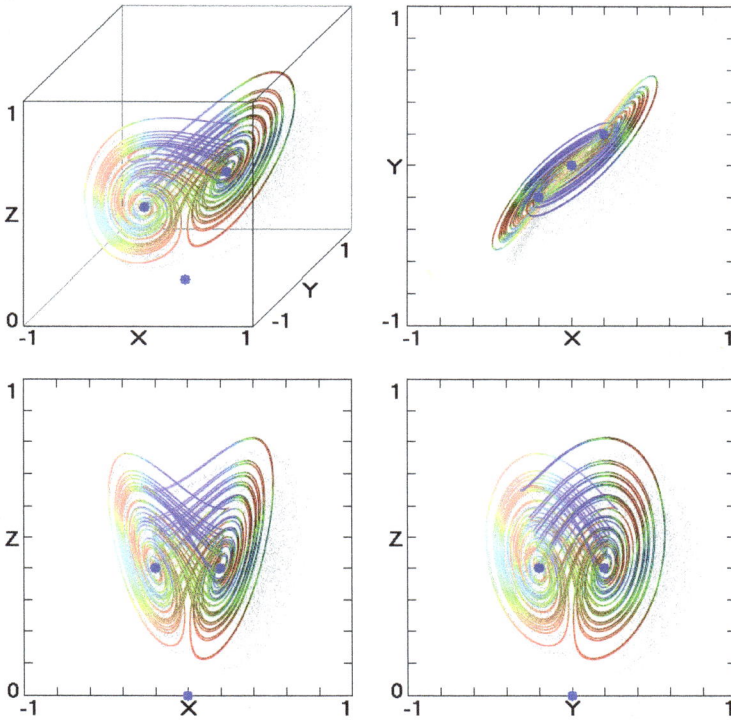

Fig. 31.1   Views of the attractor for Eq. (31.2) with $a = 0.3$, $b = 0.7$, $c = 0.1$ and initial conditions $(-2, -3, 0)$.

$b = 0.7$, $c = 0.1$. Red color in the Lyapunov exponent indicates that the error is growing parallel to the orbit, while blue indicates growth perpendicular to the orbit. Note that the orbit passes through regions where the local Lyapunov exponent is strongly positive and other regions where it is strongly negative as is typical for a chaotic system and is also reflected by the colors in Fig. 31.1.

31.6   Lyapunov Exponents

The global Lyapunov exponents are determined by averaging the local Lyapunov exponents along the orbit. The values typically converge very slowly because of the large variation along the orbit, and an integration time of order 10^8 is required to obtain 4-digit accuracy.

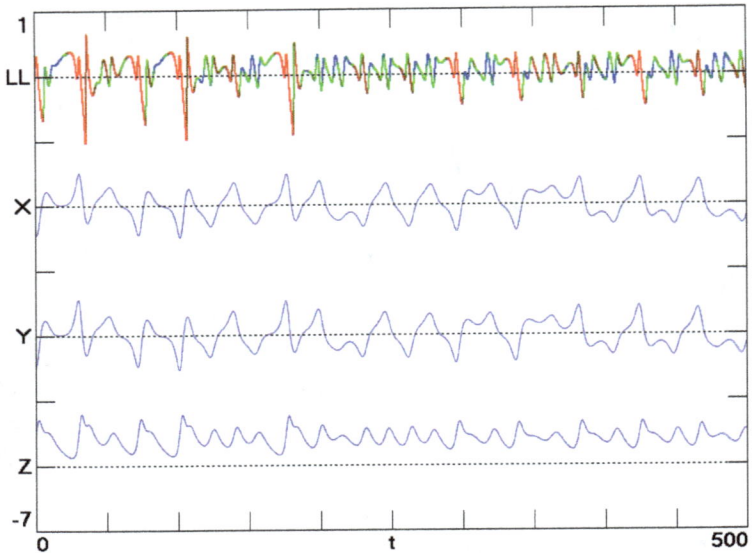

Fig. 31.2 Time series for the variables in Eq. (31.2) with $a = 0.3$, $b = 0.7$, $c = 0.1$ along with the local Lyapunov exponent (LL).

The results of such a calculation for the system in Eq. (31.2) with $a = 0.3$, $b = 0.7$, $c = 0.1$ after a time of 2×10^6 are LE $= (0.0601, 0, -0.4601)$ with a Kaplan–Yorke dimension of 2.1307, where the last digit in the quoted values is only an approximation. The positive value of the largest Lyapunov exponent indicates that the system is chaotic, and the negative sum of the exponents (-0.4000) indicates that the system is dissipative with a strange attractor.

31.7 Basin of Attraction

Figure 31.3 shows (in red) the basin of attraction for Eq. (31.2) with $a = 0.3$, $b = 0.7$, $c = 0.1$ in the $z = 0.4$ plane. Also shown (in black) is the cross-section of the attractor in the same plane.

31.8 Bifurcations

Figure 31.4 shows the bifurcation diagram for Eq. (31.2) as a function of the parameter a from 0 to 0.6 for $b = 0.7$, $c = 0.1$. The initial condition

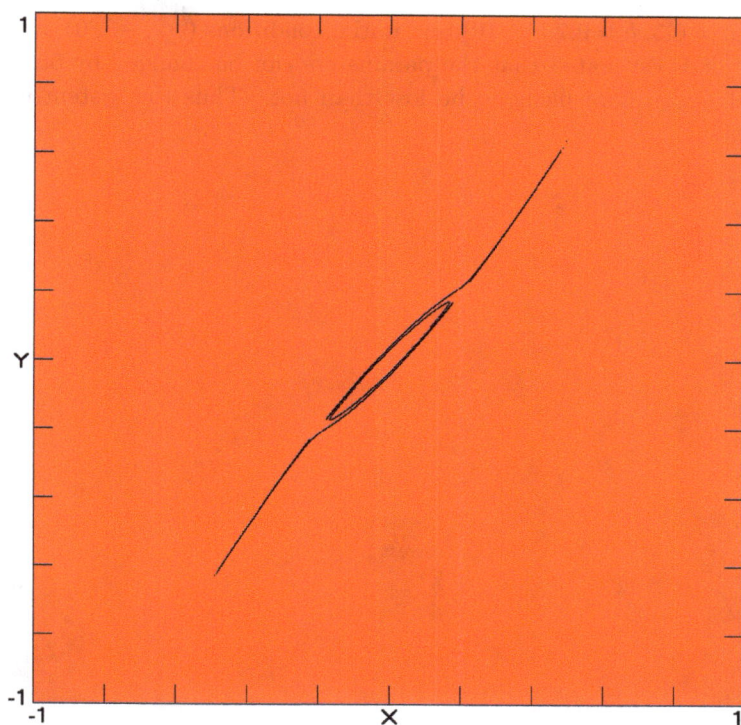

Fig. 31.3 Basin of attraction for Eq. (31.2) with $a = 0.3$, $b = 0.7$, $c = 0.1$ in the $z = 0.4$ plane.

was taken as $(-2, -3, 0)$ at $a = 0.6$ and was not changed as a slowly varied toward $a = 0$. Each of the 500 values of a was calculated for a time of about 1×10^4.

The upper plot shows the three Lyapunov exponents. The middle plot shows the Kaplan–Yorke dimension, and the lower plot shows the local maxima of x. The chaotic region is in the vicinity of $a = 0.3$, $b = 0.7$, $c = 0.1$, and the route to chaos is clearly shown.

31.9 Robustness

One measure of the robustness of a chaotic system is the amount by which the parameters can be changed from their nominal values before the probability of chaos decreases to 50% [Sprott (2022)]. For the system in Eq. (31.2) with $a = 0.3$, $b = 0.7$, $c = 0.1$ and initial conditions $(-2, -3, 0)$, after 2491 trials, it is estimated that the parameters can be changed by 56% before the chaos is more likely to be lost than not. Thus the system is highly robust.

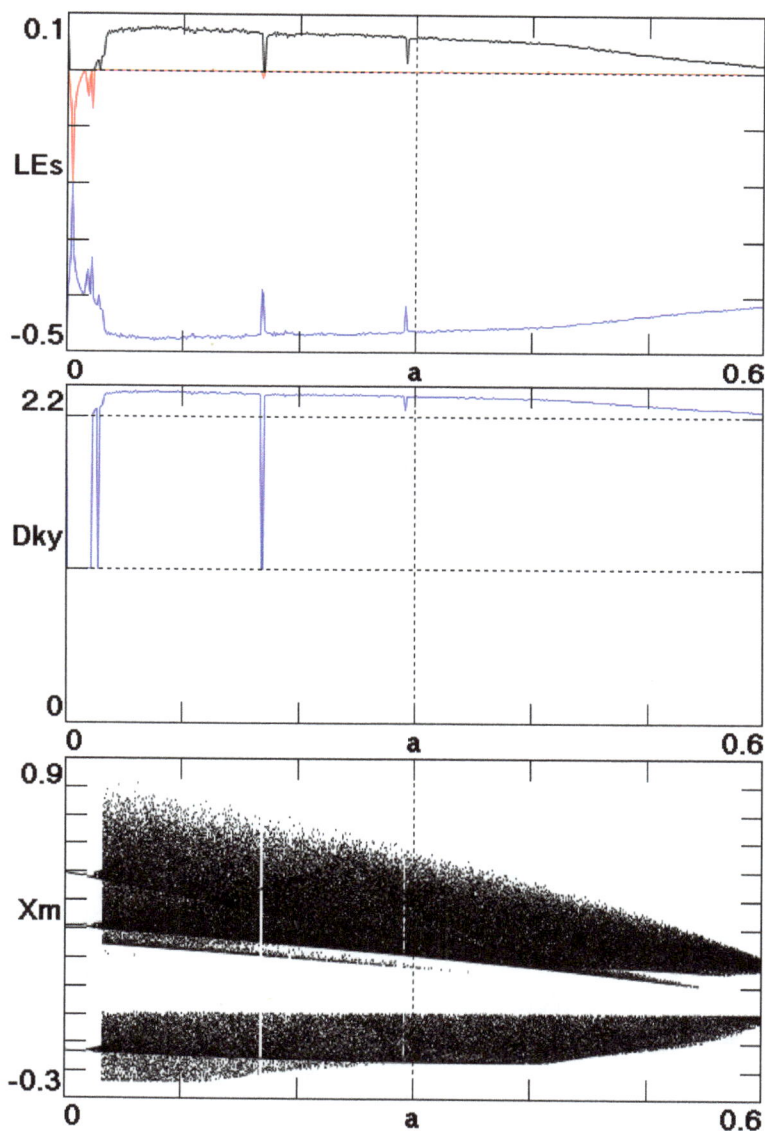

Fig. 31.4   Bifurcation diagram for Eq. (31.2) as a function of the parameter a for $b = 0.7$, $c = 0.1$.

Chapter 32

Halvorsen System

This is an unpublished chaotic system proposed by Arne Dehli Halvorsen with the property that the equations are cyclically symmetric so that the strange attractor is symmetric with respect to a 90° rotation about any of the three axes. The resulting attractor has a special elegance and beauty.

32.1 Introduction

The Halvorsen system [Sprott (2003)] written in its most general form with an adjustable coefficient in each of the twelve terms is given by

$$\dot{x} = -a_1 x - a_2 y - a_3 z - a_4 y^2$$
$$\dot{y} = -a_1 y - a_2 z - a_3 x - a_4 z^2 \qquad (32.1)$$
$$\dot{z} = -a_1 z - a_2 x - a_3 y - a_4 x^2.$$

The usual parameters are $a_1 = 1.27$, $a_2 = a_3 = 4$, $a_4 = 1$. Unlike the previous cases, the four parameters are repeated in the equations to ensure the desired symmetry. The system has a conservative variant ($a_1 = 0$), but only the dissipative case ($a_1 > 0$) is considered here.

The following sections were written by the computer program that performed the optimization, carried out the analysis of the resulting system, and produced the corresponding figures, all without human intervention.

32.2 Simplified System

After about 2×10^4 trials, of which 328 were chaotic, simplified parameters for Eq. (32.1) that give chaotic solutions are $a_1 = 0.32$, $a_2 = 1$, $a_3 = 1$, $a_4 = 1$. Thus Eq. (32.1) can be written more compactly as

$$\dot{x} = -ax - y - z - y^2$$
$$\dot{y} = -ay - z - x - z^2 \qquad (32.2)$$
$$\dot{z} = -az - x - y - x^2,$$

where $a = 0.32$.

Note that with twelve terms, Eq. (32.2) should have eight independent parameters through a linear rescaling of the three variables plus time. However, seven of the eight parameters have values of ± 1 and can be placed in any of the remaining eleven terms.

32.3 Equilibria

The system in Eq. (32.2) with $a = 0.32$ has two equilibrium points:

Equilibrium # 1 is an unstable saddle focus at $(-2.3200, -2.3200, -2.3200)$ with eigenvalues $(-1.64 - 4.0184i, -1.64 + 4.0184i, 2.32)$ and a Poincaré index of 1 in the $z = -2.32$ plane.

Equilibrium # 2 is an unstable saddle node at $(0, 0, 0)$ with eigenvalues $(0.68, -2.32, 0.68)$ and a Poincaré index of -1 in the $z = 0$ plane.

The strange attractor is self-excited, and the system has no symmetry.

32.4 Attractor

Figure 32.1 shows various views of the attractor for Eq. (32.2) with $a = 0.32$ and initial conditions $(-2, 0, 0)$. The rainbow of colors shows the local value of the largest Lyapunov exponent with red indicating the most positive values (regions of worst predictability) and blue indicating the most negative values (regions of best predictability).

32.5 Time Series

Figure 32.2 shows the time series for the three variables along with the local value of the largest Lyapunov exponent (LL) for Eq. (32.2) with $a = 0.32$. Red color in the Lyapunov exponent indicates that the error is growing parallel to the orbit, while blue indicates growth perpendicular to the orbit. Note that the orbit passes through regions where the local Lyapunov exponent is strongly positive and other regions where it is strongly negative as is typical for a chaotic system and is also reflected by the colors in Fig. 32.1.

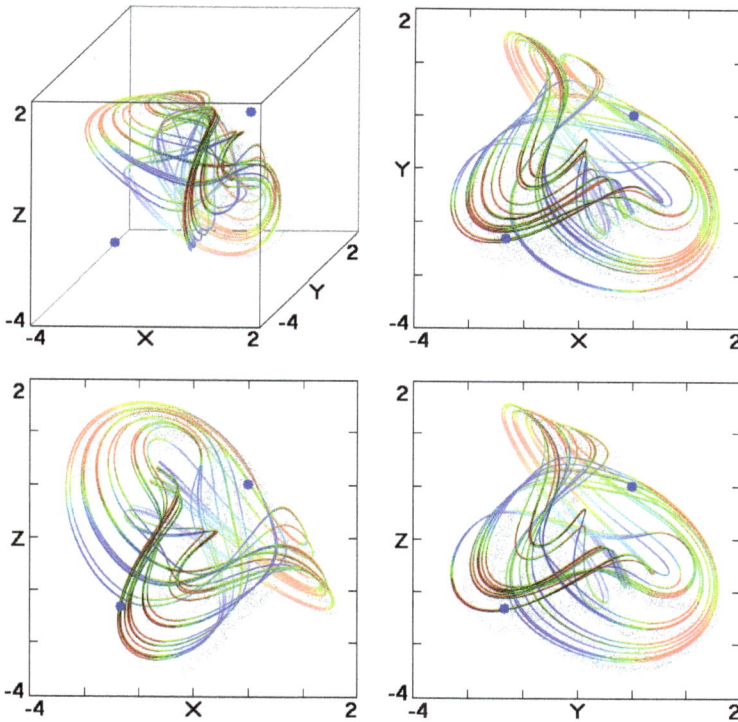

Fig. 32.1   Views of the attractor for Eq. (32.2) with $a = 0.32$ and initial conditions $(-2, 0, 0)$.

32.6   Lyapunov Exponents

The global Lyapunov exponents are determined by averaging the local Lyapunov exponents along the orbit. The values typically converge very slowly because of the large variation along the orbit, and an integration time of order 10^8 is required to obtain 4-digit accuracy.

The results of such a calculation for the system in Eq. (32.2) with $a = 0.32$ after a time of 3×10^6 are LE = $(0.1926, 0, -1.1526)$ with a Kaplan–Yorke dimension of 2.1671, where the last digit in the quoted values is only an approximation. The positive value of the largest Lyapunov exponent indicates that the system is chaotic, and the negative sum of the exponents (-0.9600) indicates that the system is dissipative with a strange attractor.

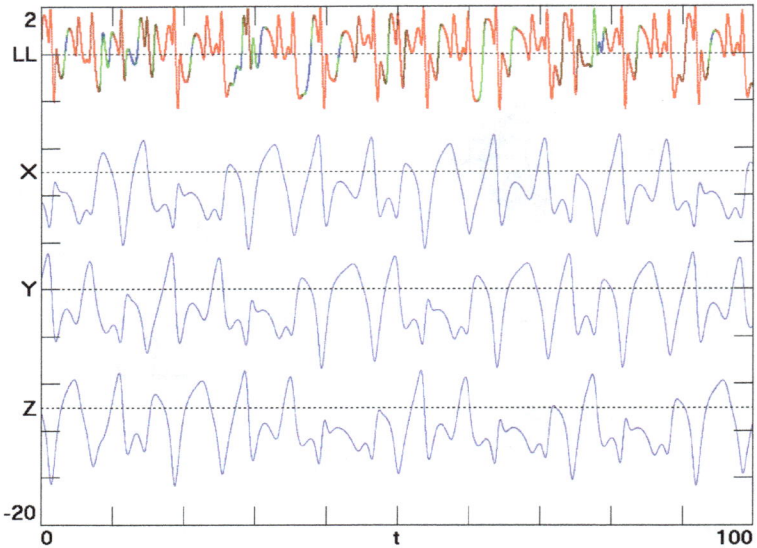

Fig. 32.2   Time series for the variables in Eq. (32.2) with $a = 0.32$ along with the local Lyapunov exponent (LL).

32.7   Basin of Attraction

Figure 32.3 shows (in red) the basin of attraction for Eq. (32.2) with $a = 0.32$ in the $z = 0$ plane. Also shown (in black) is the cross-section of the attractor in the same plane.

32.8   Bifurcations

Figure 32.4 shows the bifurcation diagram for Eq. (32.2) as a function of the parameter a from 0 to 0.64. The initial condition was taken as $(-2, 0, 0)$ at $a = 0.64$ and was not changed as a slowly varied toward $a = 0$. Each of the 500 values of a was calculated for a time of about 1×10^4.

The upper plot shows the three Lyapunov exponents. The middle plot shows the Kaplan–Yorke dimension, and the lower plot shows the local maxima of x. The chaotic region is in the vicinity of $a = 0.32$, and the route to chaos is clearly shown.

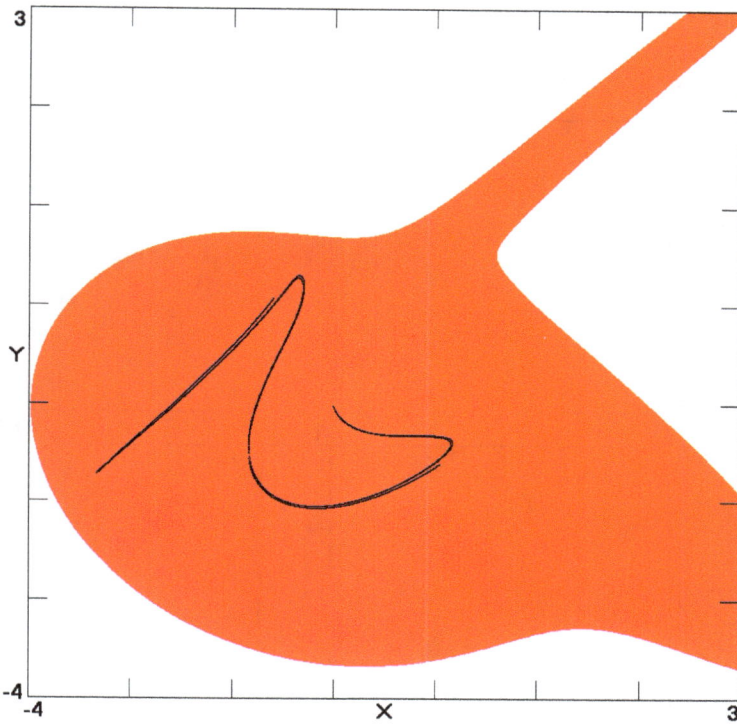

Fig. 32.3   Basin of attraction for Eq. (32.2) with $a = 0.32$ in the $z = 0$ plane.

32.9   Robustness

One measure of the robustness of a chaotic system is the amount by which the parameters can be changed from their nominal values before the probability of chaos decreases to 50% [Sprott (2022)]. For the system in Eq. (32.2) with $a = 0.32$ and initial conditions $(-2, 0, 0)$, after 2827 trials, it is estimated that the parameter a can be changed by 24% before the chaos is more likely to be lost than not. Thus the system is somewhat robust. This result is consistent with the data in Fig. 32.4.

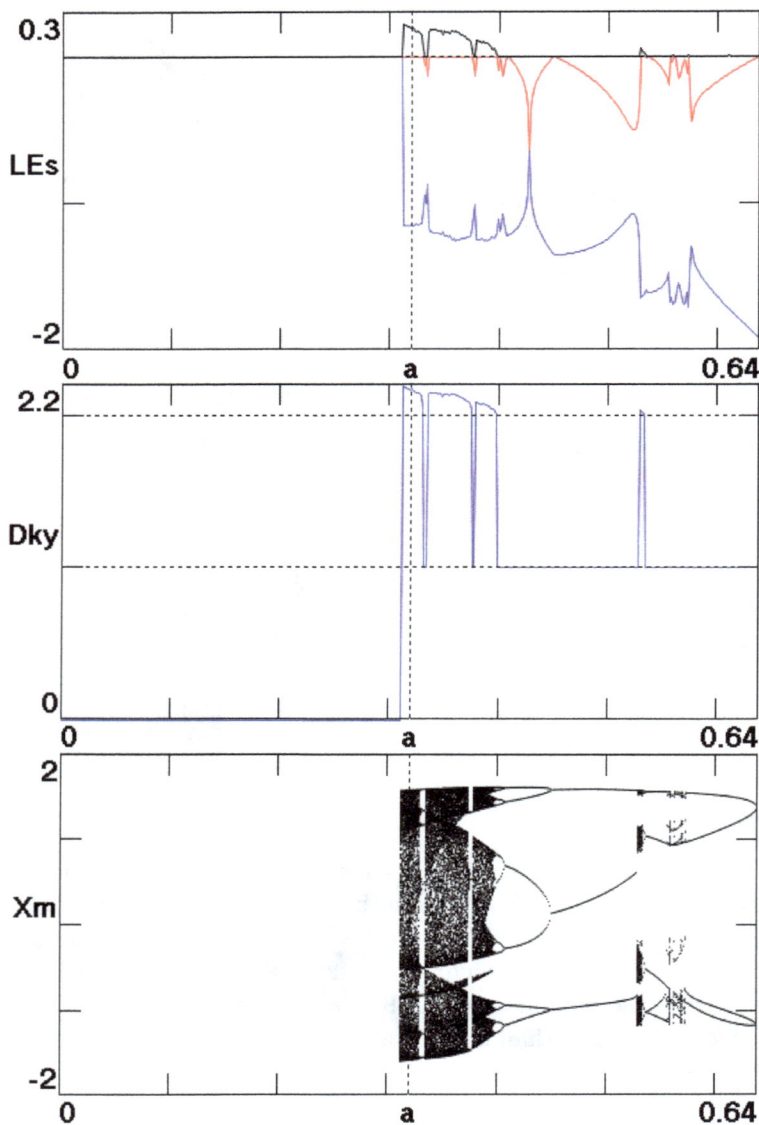

Fig. 32.4   Bifurcation diagram for Eq. (32.2) as a function of the parameter *a*.

Chapter 33

Thomas System

René Thomas proposed an even simpler cyclically symmetric system, variants of which have been widely studied. Its strange attractor is symmetric with respect to a 90° rotation about any of the three axes as well as inversion symmetric, and it has a special elegance and beauty. It has an infinite line of equilibrium points, only a sample of which is shown.

33.1 Introduction

The Thomas (1999) system written in its most general form with an adjustable coefficient in each of the six terms is given by

$$
\begin{aligned}
\dot{x} &= -a_1 x + a_2 \sin y \\
\dot{y} &= -a_1 y + a_2 \sin z \\
\dot{z} &= -a_1 z + a_2 \sin x.
\end{aligned}
\tag{33.1}
$$

The usual parameters are $a_1 = 0.18$, $a_2 = 1$. The two parameters are repeated in the equations to ensure the desired symmetry. The system has a conservative variant ($a_1 = 0$) giving rise to so-called 'labyrinth chaos', but only the dissipative case ($a_1 > 0$) is considered here.

The following sections were written by the computer program that performed the optimization, carried out the analysis of the resulting system, and produced the corresponding figures, all without human intervention.

33.2 Simplified System

After about 2×10^2 trials, of which 43 were chaotic, simplified parameters for Eq. (33.1) that give chaotic solutions are $a_1 = 0.1$, $a_2 = 1$. Thus

Eq. (33.1) can be written more compactly as

$$\dot{x} = -ax + \sin y$$
$$\dot{y} = -ay + \sin z \qquad (33.2)$$
$$\dot{z} = -az + \sin x,$$

where $a = 0.1$.

Note that with six terms, Eq. (33.2) should have two independent parameters through a linear rescaling of the three variables plus time. However, one of the two parameters has a value of ± 1 and can be placed in any of the remaining five terms.

33.3 Equilibria

The system in Eq. (33.2) with $a = 0.1$ has many equilibrium points, six of which are as follows:

Equilibrium # 1 is an unstable saddle focus at $(0.0319, 3.1384, 0.3192)$ with eigenvalues $(-1.0827, 0.3913 - 0.851i, 0.3913 + 0.851i)$ and a Poincaré index of 1 in the $z = 0.3192$ plane.

Equilibrium # 2 is an unstable saddle focus at $(3.1129, 2.8250, 0.2864)$ with eigenvalues $(-0.5847 - 0.8396i, -0.5847 + 0.8396i, 0.8695)$ and a Poincaré index of 1 in the $z = 0.2864$ plane.

Equilibrium # 3 is an unstable saddle focus at $(-2.8523, -2.8523, -2.8523)$ with eigenvalues $(-1.0585, 0.3792 - 0.83i, 0.3792 + 0.83i)$ and a Poincaré index of 1 in the $z = -2.8523$ plane.

Equilibrium # 4 is an unstable saddle focus at $(0.3192, 0.0319, 3.1384)$ with eigenvalues $(-1.0827, 0.3913 - 0.851i, 0.3913 + 0.851i)$ and a Poincaré index of 1 in the $z = 3.1384$ plane.

Equilibrium # 5 is an unstable saddle focus at $(3.1384, 0.3192, 0.0319)$ with eigenvalues $(-1.0827, 0.3913 - 0.851i, 0.3913 + 0.851i)$ and a Poincaré index of 1 in the $z = 0.0319$ plane.

Equilibrium # 6 is an unstable saddle focus at $(-3.1384, -0.3192, -0.0319)$ with eigenvalues $(-1.0827, 0.3913 - 0.851i, 0.3913 + 0.851i)$ and a Poincaré index of 1 in the $z = -0.0319$ plane.

The strange attractor is self-excited, and the system is symmetric under the transformation $x \rightarrow -x$, $y \rightarrow -y$, $z \rightarrow -z$.

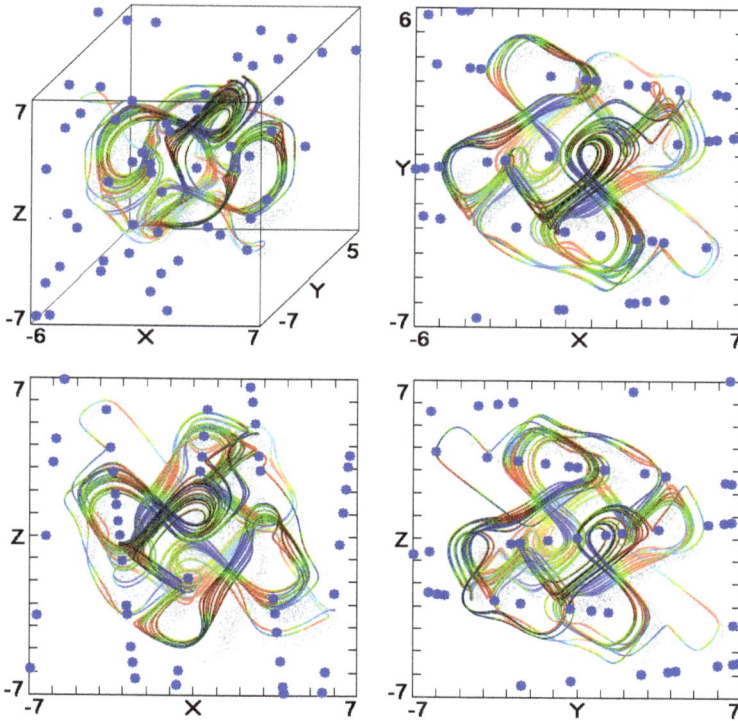

Fig. 33.1   Views of the attractor for Eq. (33.2) with $a = 0.1$ and initial conditions $(-1, -2, 2)$.

33.4   Attractor

Figure 33.1 shows various views of the attractor for Eq. (33.2) with $a = 0.1$ and initial conditions $(-1, -2, 2)$. The rainbow of colors shows the local value of the largest Lyapunov exponent with red indicating the most positive values (regions of worst predictability) and blue indicating the most negative values (regions of best predictability).

33.5   Time Series

Figure 33.2 shows the time series for the three variables along with the local value of the largest Lyapunov exponent (LL) for Eq. (33.2) with $a = 0.1$. Red color in the Lyapunov exponent indicates that the error is growing parallel to the orbit, while blue indicates growth perpendicular to the

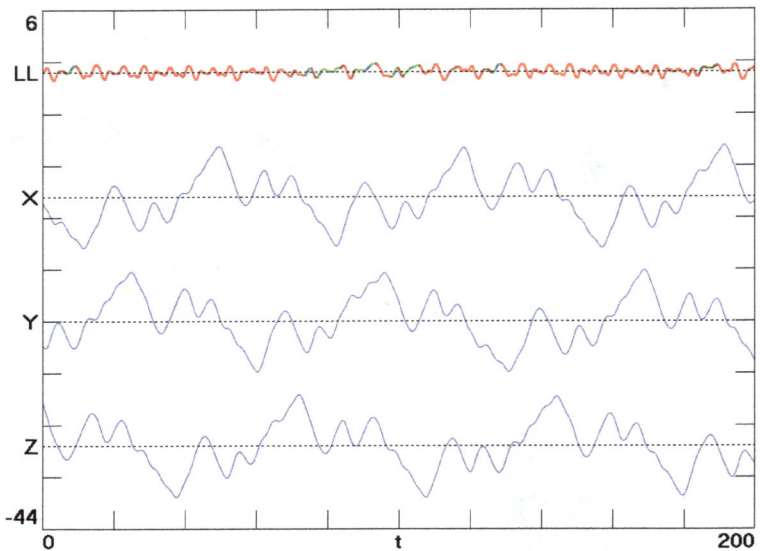

Fig. 33.2 Time series for the variables in Eq. (33.2) with $a = 0.1$ along with the local Lyapunov exponent (LL).

orbit. Note that the orbit passes through regions where the local Lyapunov exponent is strongly positive and other regions where it is strongly negative as is typical for a chaotic system and is also reflected by the colors in Fig. 33.1.

33.6 Lyapunov Exponents

The global Lyapunov exponents are determined by averaging the local Lyapunov exponents along the orbit. The values typically converge very slowly because of the large variation along the orbit, and an integration time of order 10^8 is required to obtain 4-digit accuracy.

The results of such a calculation for the system in Eq. (33.2) with $a = 0.1$ after a time of 3×10^6 are LE = (0.0552, 0, −0.3552) with a Kaplan–Yorke dimension of 2.1553, where the last digit in the quoted values is only an approximation. The positive value of the largest Lyapunov exponent indicates that the system is chaotic, and the negative sum of the exponents (−0.3000) indicates that the system is dissipative with a strange attractor.

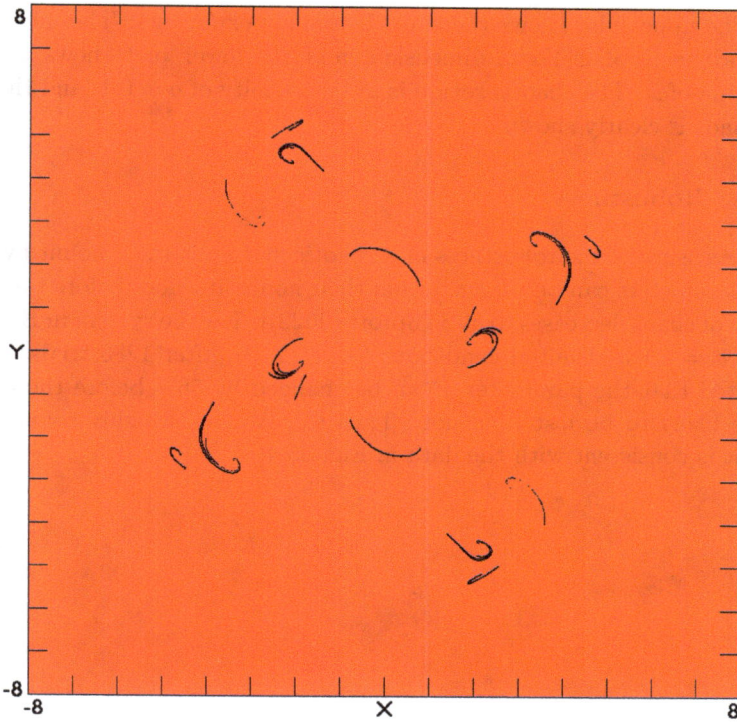

Fig. 33.3   Basin of attraction for Eq. (33.2) with $a = 0.1$ in the $z = 0$ plane.

33.7   Basin of Attraction

Figure 33.3 shows (in red) the basin of attraction for Eq. (33.2) with $a = 0.1$ in the $z = 0$ plane. Also shown (in black) is the cross-section of the attractor in the same plane.

33.8   Bifurcations

Figure 33.4 shows the bifurcation diagram for Eq. (33.2) as a function of the parameter a from 0 to 0.2. The initial condition was taken as $(-1, -2, 2)$ at $a = 0.2$ and was not changed as a slowly varied toward $a = 0$. Each of the 500 values of a was calculated for a time of about 2×10^4.

The upper plot shows the three Lyapunov exponents. The middle plot shows the Kaplan–Yorke dimension, and the lower plot shows the local maxima of x. The chaotic region is in the vicinity of $a = 0.1$, and the route to chaos is clearly shown.

33.9 Robustness

One measure of the robustness of a chaotic system is the amount by which the parameters can be changed from their nominal values before the probability of chaos decreases to 50% [Sprott (2022)]. For the system in Eq. (33.2) with $a = 0.1$ and initial conditions $(-1, -2, 2)$, after 1286 trials, it is estimated that the parameter a can be changed by 56% before the chaos is more likely to be lost than not. Thus the system is highly robust. This result is consistent with the data in Fig. 33.4.

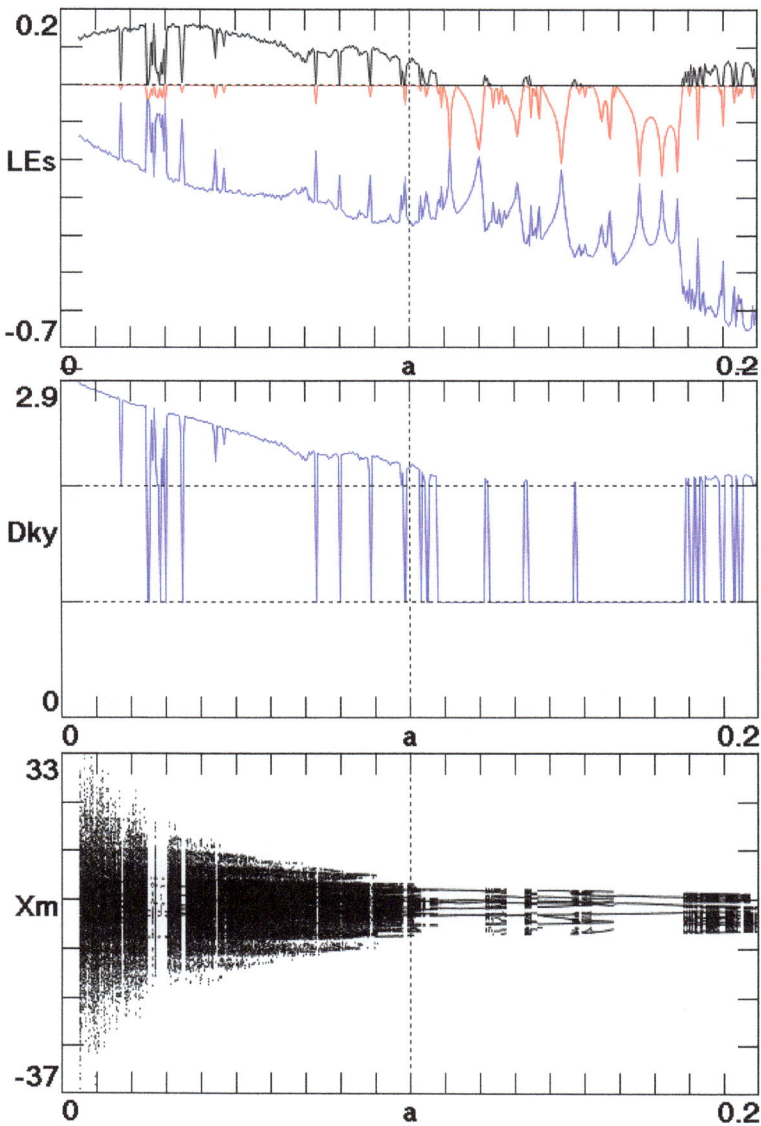

Fig. 33.4   Bifurcation diagram for Eq. (33.2) as a function of the parameter a.

Chapter 34

Rabinovich–Fabrikant System

This is an old system originally devised to model the onset of turbulence in a fluid. With ten terms and five nonlinearities, it is more complicated than most of the previous systems, but it is included here for historical interest. The equations are symmetric under a 180° rotation about the z axis with a symmetric pair of strange attractors for the chosen parameters.

34.1   Introduction

The Rabinovich–Fabrikant system [Rabinovich and Fabrikant (1979)] written in its most general form with an adjustable coefficient in each of the ten terms is given by

$$\dot{x} = a_1 yz - a_2 y + a_3 x^2 y + a_4 x$$
$$\dot{y} = a_5 xz + a_6 x - a_7 x^3 + a_8 y \qquad (34.1)$$
$$\dot{z} = -a_9 z - a_{10} xyz.$$

The usual parameters are $a_1 = a_2 = a_3 = a_6 = a_7 = 1$, $a_4 = a_8 = 0.87$, $a_5 = 3$, $a_9 = 2.2$, $a_{10} = 2$. The signs are preserved, and the parameters are chosen greater than zero to prevent the system from collapsing to one of the simpler systems described earlier.

The following sections were written by the computer program that performed the optimization, carried out the analysis of the resulting system, and produced the corresponding figures, all without human intervention.

34.2   Simplified System

After about 7×10^4 trials, of which 39 were chaotic, simplified parameters for Eq. (34.1) that give chaotic solutions are $a_1 = 1$, $a_2 = 1$, $a_3 = 0.6$,

$a_4 = 1$, $a_5 = 1$, $a_6 = 1$, $a_7 = 1$, $a_8 = 0.07$, $a_9 = 1$, $a_{10} = 1$. Thus Eq. (34.1) can be written more compactly as

$$\dot{x} = yz - y + ax^2y + x$$
$$\dot{y} = xz + x - x^3 + by \qquad (34.2)$$
$$\dot{z} = -z - xyz,$$

where $a = 0.6$, $b = 0.07$.

Note that with ten terms, Eq. (34.2) should have six independent parameters through a linear rescaling of the three variables plus time. However, four of the six parameters have values of ± 1 and can be placed in any of the remaining eight terms.

34.3   Equilibria

The system in Eq. (34.2) with $a = 0.6$, $b = 0.07$ has five equilibrium points:

Equilibrium # 1 is an unstable saddle focus at $(0, 0, 0)$ with eigenvalues $(0.535 - 0.8853i, 0.535 + 0.8853i, -1)$ and a Poincaré index of 1 in the $z = 0$ plane.

Equilibrium # 2 is an unstable saddle focus at $(-1.8160, 0.5507, 2.3192)$ with eigenvalues $(0.086 - 5.4832i, 0.086 + 5.4832i, -0.3019)$ and a Poincaré index of 1 in the $z = 2.3192$ plane.

Equilibrium # 3 is an unstable saddle focus at $(1.8160, -0.5507, 2.3192)$ with eigenvalues $(0.086 - 5.4832i, 0.086 + 5.4832i, -0.3019)$ and a Poincaré index of 1 in the $z = 2.3192$ plane.

Equilibrium # 4 is an unstable saddle focus at $(-0.1881, 5.3168, 1.0142)$ with eigenvalues $(-0.1343 - 5.3518i, -0.1343 + 5.3518i, 0.1385)$ and a Poincaré index of -1 in the $z = 1.0142$ plane.

Equilibrium # 5 is an unstable saddle focus at $(0.1881, -5.3168, 1.0142)$ with eigenvalues $(-0.1343 - 5.3518i, -0.1343 + 5.3518i, 0.1385)$ and a Poincaré index of -1 in the $z = 1.0142$ plane.

The strange attractor is self-excited, and the system is symmetric under the transformation $x \to -x$, $y \to -y$.

34.4   Attractor

Figure 34.1 shows various views of the two attractors for Eq. (34.2) with $a = 0.6$, $b = 0.07$ and initial conditions $(2, -1, 1)$.

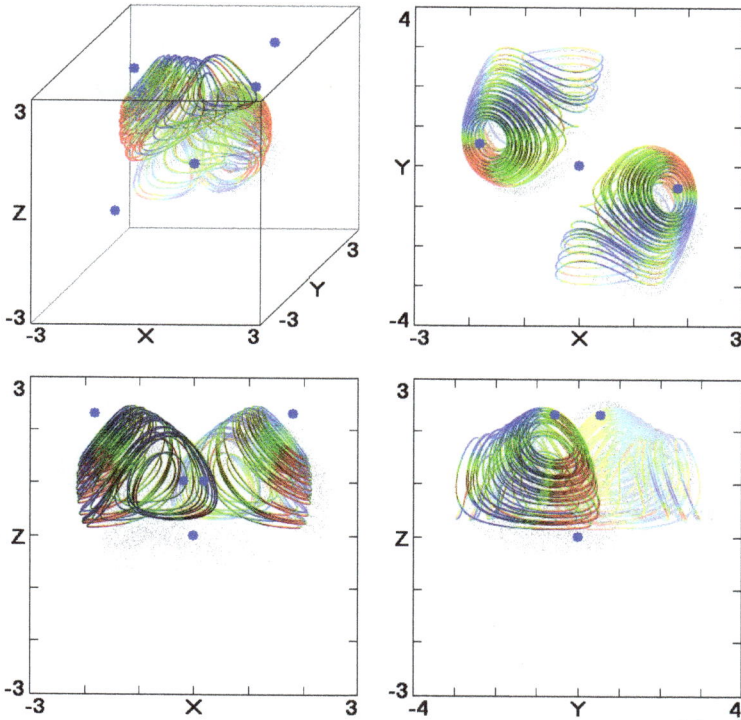

Fig. 34.1   Views of the two attractors for Eq. (34.2) with $a = 0.6$, $b = 0.07$.

34.5   Time Series

Figure 34.2 shows the time series for the three variables along with the local value of the largest Lyapunov exponent (LL) for Eq. (34.2) with $a = 0.6$, $b = 0.07$. Red color in the Lyapunov exponent indicates that the error is growing parallel to the orbit, while blue indicates growth perpendicular to the orbit. Note that the orbit passes through regions where the local Lyapunov exponent is strongly positive and other regions where it is strongly negative as is typical for a chaotic system and is also reflected by the colors in Fig. 34.1.

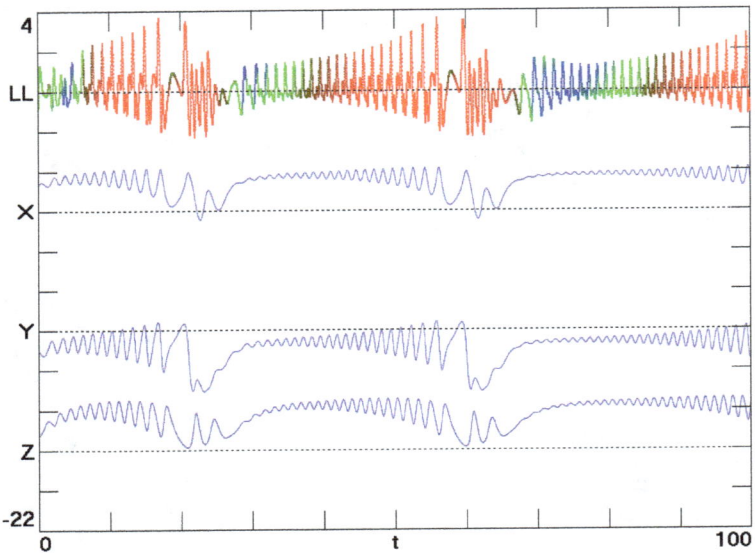

Fig. 34.2   Time series for the variables in Eq. (34.2) with $a = 0.6$, $b = 0.07$ along with the local Lyapunov exponent (LL).

34.6   Lyapunov Exponents

The global Lyapunov exponents are determined by averaging the local Lyapunov exponents along the orbit. The values typically converge very slowly because of the large variation along the orbit, and an integration time of order 10^8 is required to obtain 4-digit accuracy.

The results of such a calculation for the system in Eq. (34.2) with $a = 0.6$, $b = 0.07$ after a time of 4×10^6 are LE = (0.2136, 0, −0.3436) with a Kaplan–Yorke dimension of 2.6216, where the last digit in the quoted values is only an approximation. The positive value of the largest Lyapunov exponent indicates that the system is chaotic, and the negative sum of the exponents (−0.1300) indicates that the system is dissipative with a strange attractor.

34.7   Basin of Attraction

Figure 34.3 shows (in red) the basin of attraction for Eq. (34.2) with $a = 0.6$, $b = 0.07$ in the $z = 1.1042$ plane. Also shown (in black) is the cross-section of the attractor in the same plane.

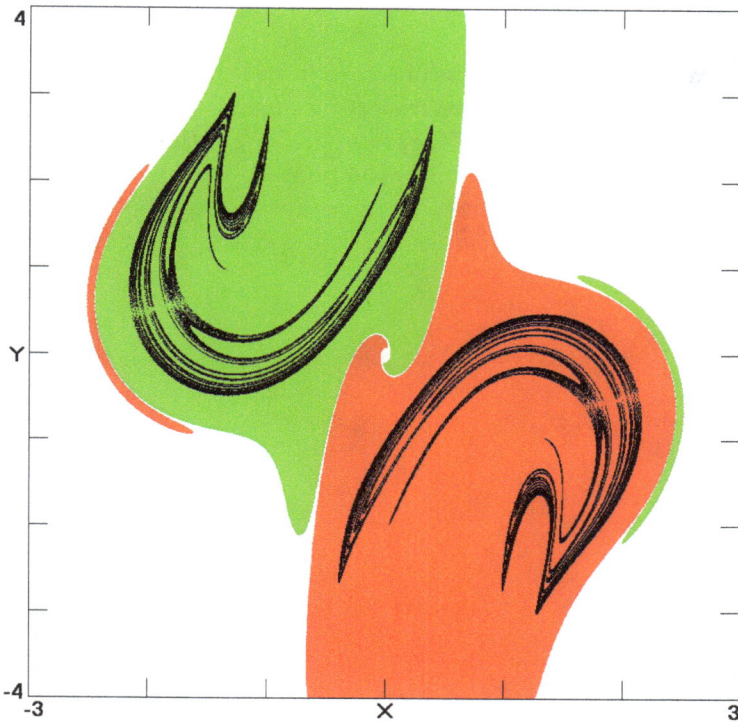

Fig. 34.3   Basin of attraction for Eq. (34.2) with $a = 0.6$, $b = 0.07$ in the $z = 1.1042$ plane.

34.8   Bifurcations

Figure 34.4 shows the bifurcation diagram for Eq. (34.2) as a function of the parameter a from 0 to 1.2 for $b = 0.07$. The initial condition was taken as $(2, -1, 1)$ at $a = 1.2$ and was not changed as a slowly varied toward $a = 0$. Each of the 500 values of a was calculated for a time of about 1×10^4.

The upper plot shows the three Lyapunov exponents. The middle plot shows the Kaplan–Yorke dimension, and the lower plot shows the local maxima of x. The chaotic region is in the vicinity of $a = 0.6$, $b = 0.07$, and the route to chaos is clearly shown.

34.9   Robustness

One measure of the robustness of a chaotic system is the amount by which the parameters can be changed from their nominal values before the probability of chaos decreases to 50% [Sprott (2022)]. For the system in Eq. (34.2) with $a = 0.6$, $b = 0.07$ and initial conditions $(2, -1, 1)$, after 3573 trials, it is estimated that the parameters can be changed by 20% before the chaos is more likely to be lost than not. Thus the system is somewhat fragile.

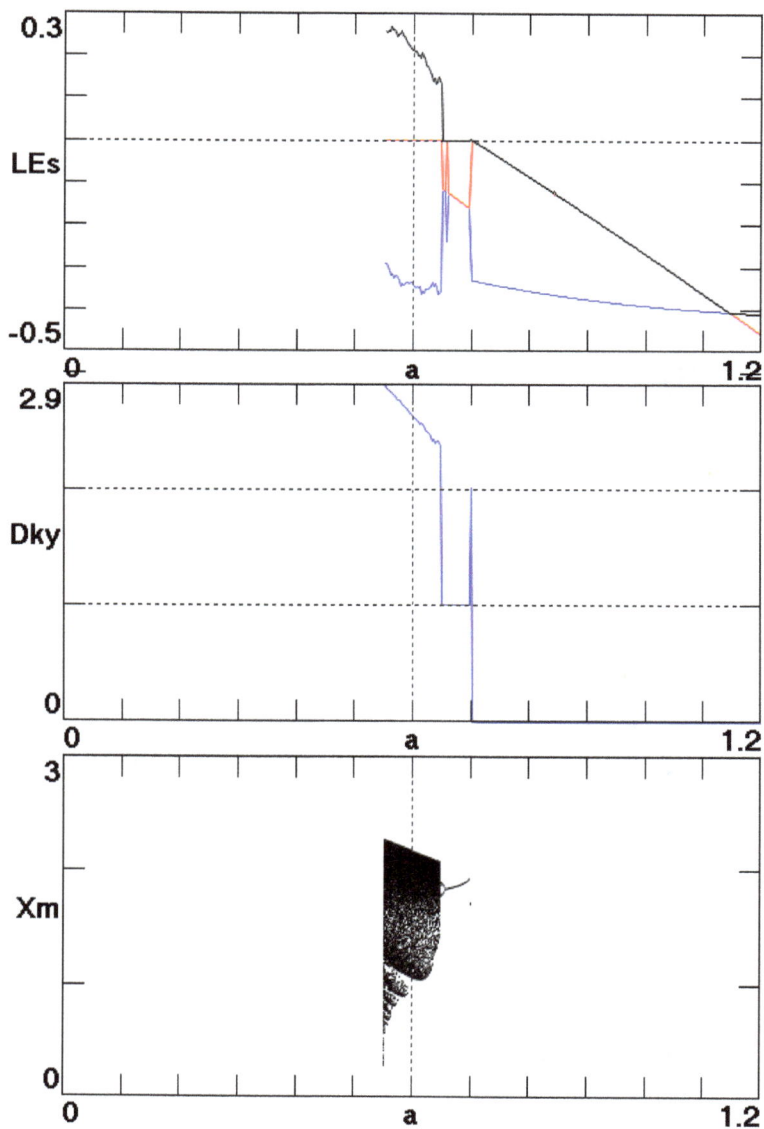

Fig. 34.4 Bifurcation diagram for Eq. (34.2) as a function of the parameter a for $b = 0.07$.

Chapter 35

Leipnik–Newton System

This is another old mathematical chaotic system describing Euler's rigid body rotation with linear feedback control. It is one of the first systems studied in which there is a pair of strange attractors, although the simplified case shown here has a single attractor.

35.1 Introduction

The Leipnik–Newton system [Leipnik and Newton (1981)] written in its most general form with an adjustable coefficient in each of the eight terms is given by

$$\begin{aligned}
\dot{x} &= -a_1 x + a_2 y + a_3 yz \\
\dot{y} &= -a_4 x - a_5 y + a_6 xy \\
\dot{z} &= a_7 z - a_8 xy.
\end{aligned} \tag{35.1}$$

The usual parameters are $a_1 = a_5 = 0.4$, $a_2 = a_4 = 1$, $a_3 = 10$, $a_6 = a_8 = 5$, $a_7 = 0.175$. The signs are preserved, and the parameters are chosen greater than zero to prevent the system from collapsing to one of the simpler systems described earlier.

The following sections were written by the computer program that performed the optimization, carried out the analysis of the resulting system, and produced the corresponding figures, all without human intervention.

35.2 Simplified System

After about 2×10^5 trials, of which 22 were chaotic, simplified parameters for Eq. (35.1) that give chaotic solutions are $a_1 = 1$, $a_2 = 1$, $a_3 = 1$, $a_4 = 32$, $a_5 = 1$, $a_6 = 1$, $a_7 = 1$, $a_8 = 1$. Thus Eq. (35.1) can be written

more compactly as

$$\dot{x} = -x + y + yz$$
$$\dot{y} = -ax - y + xy \qquad\qquad (35.2)$$
$$\dot{z} = z - xy,$$

where $a = 32$.

Note that with eight terms, Eq. (35.2) should have four independent parameters through a linear rescaling of the three variables plus time. However, three of the four parameters have values of ± 1 and can be placed in any of the remaining seven terms.

35.3 Equilibria

The system in Eq. (35.2) with $a = 32$ has three equilibrium points:

Equilibrium # 1 is an unstable saddle focus at $(0.1638, -6.2663, -1.0261)$ with eigenvalues $(0.5052 - 6.2351i, 0.5052 + 6.2351i, -1.8467)$ and a Poincaré index of -1 in the $z = -1.0261$ plane.

Equilibrium # 2 is an unstable saddle focus at $(-0.1970, 5.2663, -1.03744)$ with eigenvalues $(0.491 - 5.2566i, 0.491 + 5.2566i, -2.1789)$ and a Poincaré index of 1 in the $z = -1.0374$ plane.

Equilibrium # 3 is an unstable saddle focus at $(0, 0, 0)$ with eigenvalues $(-1 - 5.6569i, -1 + 5.6569i, 1)$ and a Poincaré index of 1 in the $z = 0$ plane.

The strange attractor is self-excited, and the system has no symmetry.

35.4 Attractor

Figure 35.1 shows various views of the attractor for Eq. (35.2) with $a = 32$ and initial conditions $(-1, 5, -2)$. The rainbow of colors shows the local value of the largest Lyapunov exponent with red indicating the most positive values (regions of worst predictability) and blue indicating the most negative values (regions of best predictability).

35.5 Time Series

Figure 35.2 shows the time series for the three variables along with the local value of the largest Lyapunov exponent (LL) for Eq. (35.2) with $a = 32$. Red color in the Lyapunov exponent indicates that the error is growing parallel to the orbit, while blue indicates growth perpendicular to the orbit. Note that the orbit passes through regions where the local Lyapunov

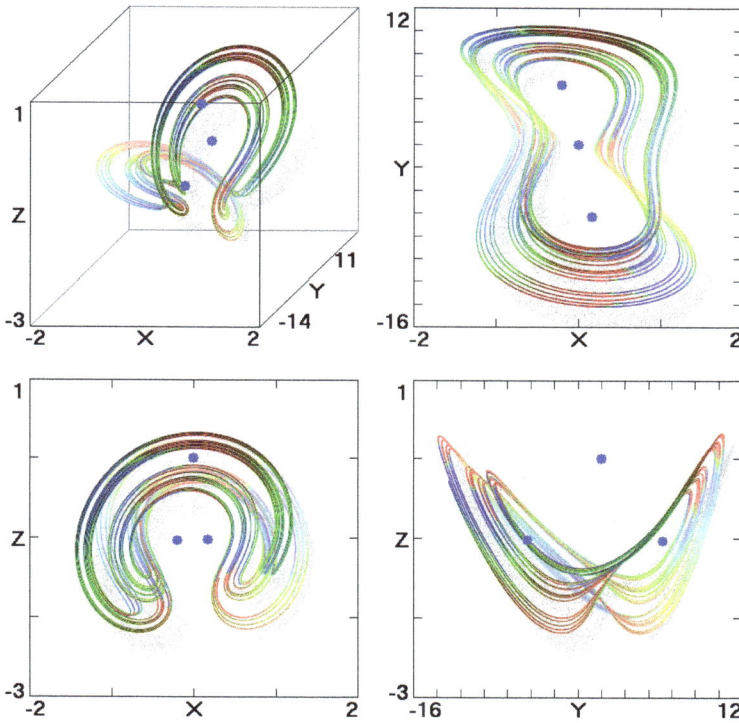

Fig. 35.1 Views of the attractor for Eq. (35.2) with $a = 32$ and initial conditions $(-1, 5, -2)$.

exponent is strongly positive and other regions where it is strongly negative as is typical for a chaotic system and is also reflected by the colors in Fig. 35.1.

35.6 Lyapunov Exponents

The global Lyapunov exponents are determined by averaging the local Lyapunov exponents along the orbit. The values typically converge very slowly because of the large variation along the orbit, and an integration time of order 10^8 is required to obtain 4-digit accuracy.

The results of such a calculation for the system in Eq. (35.2) with $a = 32$ after a time of 4×10^6 are LE $= (0.1174, 0, -1.1563)$ with a Kaplan–Yorke dimension of 2.1015, where the last digit in the quoted values is only

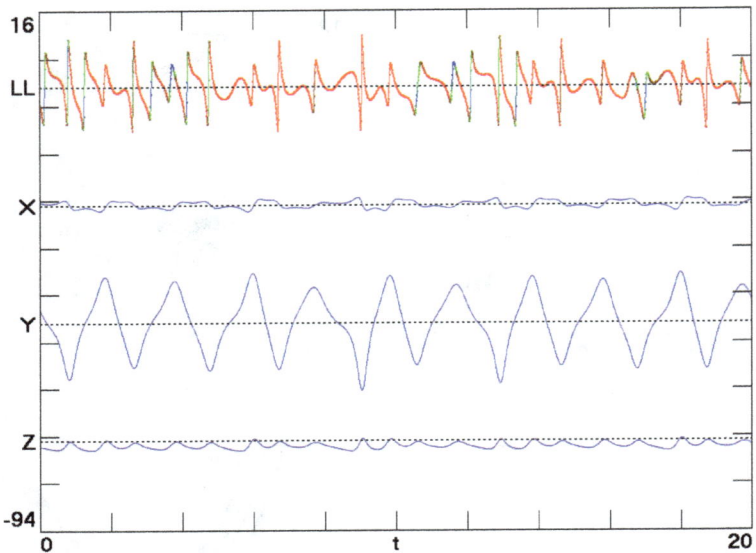

Fig. 35.2   Time series for the variables in Eq. (35.2) with $a = 32$ along with the local Lyapunov exponent (LL).

an approximation. The positive value of the largest Lyapunov exponent indicates that the system is chaotic, and the negative sum of the exponents (-1.0388) indicates that the system is dissipative with a strange attractor.

35.7   Basin of Attraction

Figure 35.3 shows (in red) the basin of attraction for Eq. (35.2) with $a = 32$ in the $z = -1.0261$ plane. Also shown (in black) is the cross-section of the attractor in the same plane.

35.8   Bifurcations

Figure 35.4 shows the bifurcation diagram for Eq. (35.2) as a function of the parameter a from 0 to 64. The initial condition was taken as (-1, 5, -2) at $a = 64$ and was not changed as a slowly varied toward $a = 0$. Each of the 500 values of a was calculated for a time of about 1×10^4.

    The upper plot shows the three Lyapunov exponents. The middle plot shows the Kaplan–Yorke dimension, and the lower plot shows the local

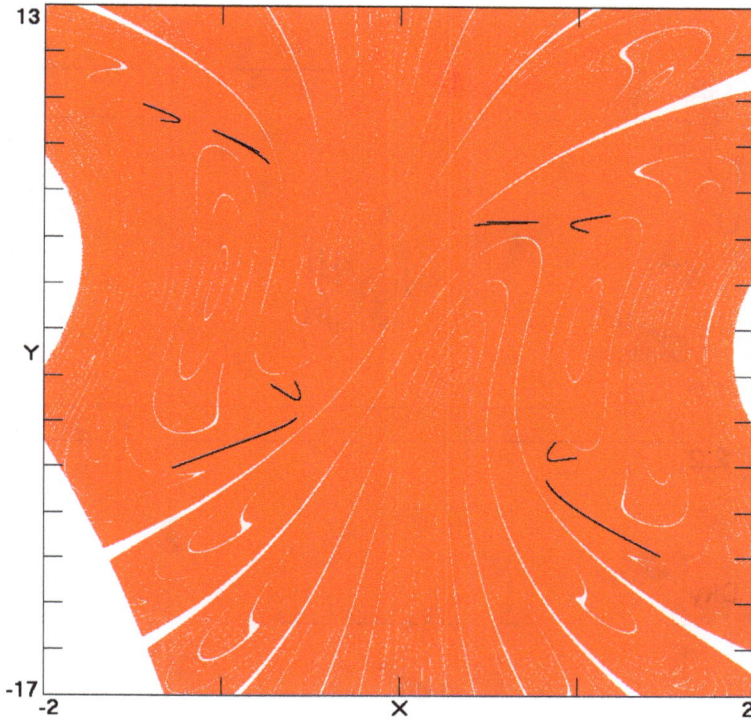

Fig. 35.3 Basin of attraction for Eq. (35.2) with $a = 32$ in the $z = -1.0261$ plane.

maxima of x. The chaotic region is in the vicinity of $a = 32$, and the route to chaos is clearly shown.

35.9 Robustness

One measure of the robustness of a chaotic system is the amount by which the parameters can be changed from their nominal values before the probability of chaos decreases to 50% [Sprott (2022)]. For the system in Eq. (35.2) with $a = 32$ and initial conditions $(-1, 5, -2)$, after 10793 trials, it is estimated that the parameter a can be changed by 1.8% before the chaos is more likely to be lost than not. Thus the system is highly fragile. This result is consistent with the data in Fig. 35.4.

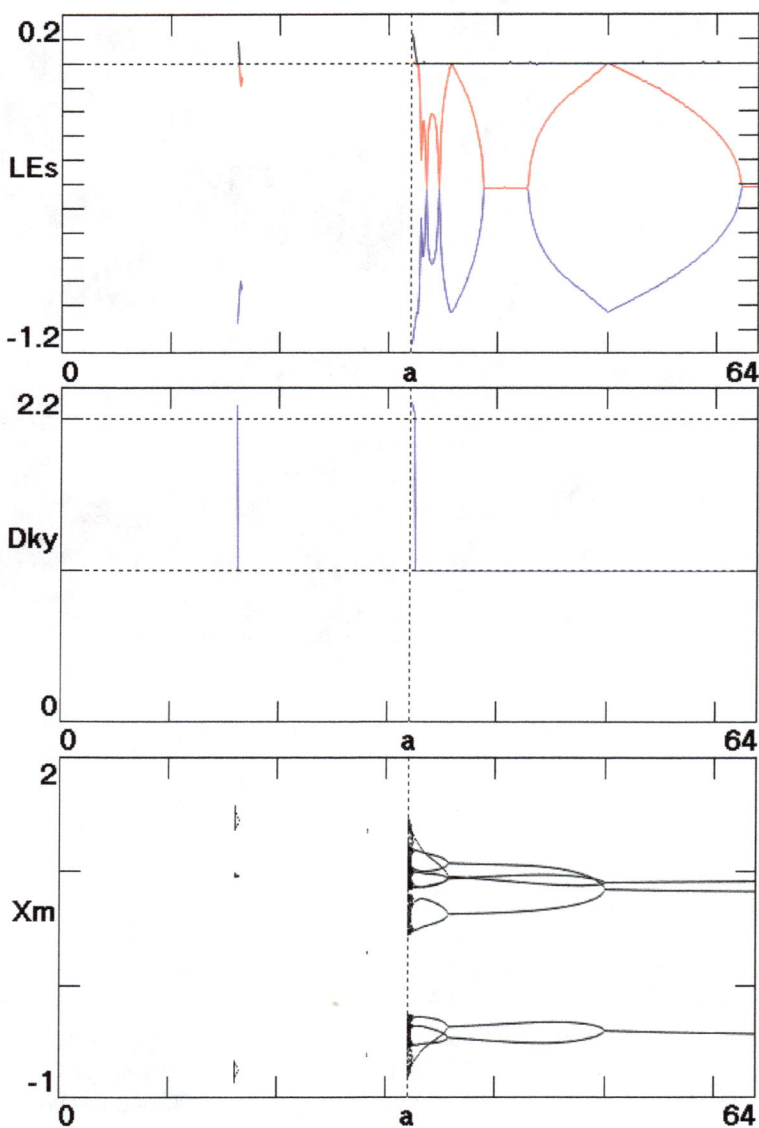

Fig. 35.4   Bifurcation diagram for Eq. (35.2) as a function of the parameter a.

Chapter 36

Arnéodo–Coullet–Tresser System

This is another old chaotic system that was proposed to investigate the onset of turbulence. Like the Lorenz system, this system is invariant under a $180°$ rotation about the z axis, but with a cubic nonlinearity.

36.1 Introduction

The Arnéodo–Coullet–Tresser system [Arnéodo *et al.* (1981)] written in its most general form with an adjustable coefficient in each of the eight terms is given by

$$
\begin{aligned}
\dot{x} &= a_1 x - a_2 y \\
\dot{y} &= -a_3 y + a_4 xz + a_5 x^3 \\
\dot{z} &= -a_6 z + a_7 xy - a_8 z^2.
\end{aligned}
\tag{36.1}
$$

The usual parameters are $a_1 = a_2 = 1.8$, $a_3 = 7.2$, $a_4 = a_7 = 1$, $a_5 = 0.02$, $a_6 = 2.7$, $a_8 = 0.07$. The signs are preserved, and the parameters are chosen greater than zero to prevent the system from collapsing to one of the simpler systems described earlier.

The following sections were written by the computer program that performed the optimization, carried out the analysis of the resulting system, and produced the corresponding figures, all without human intervention.

36.2 Simplified System

After about 1×10^5 trials, of which eight were chaotic, simplified parameters for Eq. (36.1) that give chaotic solutions are $a_1 = 0.58$, $a_2 = 1$, $a_3 = 1$, $a_4 = 1$, $a_5 = 1$, $a_6 = 0.2$, $a_7 = 1$, $a_8 = 1$. Thus Eq. (36.1) can be written

more compactly as

$$\dot{x} = ax - y$$
$$\dot{y} = -y + xz + x^3 \qquad (36.2)$$
$$\dot{z} = -bz + xy - z^2,$$

where $a = 0.58$, $b = 0.2$.

Note that with eight terms, Eq. (36.2) should have four independent parameters through a linear rescaling of the three variables plus time. However, two of the four parameters have values of ± 1 and can be placed in any of the remaining six terms.

36.3 Equilibria

The system in Eq. (36.2) with $a = 0.58$, $b = 0.2$ has six equilibrium points:

Equilibrium # 1 is a stable focus at $(0.5206, 0.3020, 0.3089)$ with eigenvalues $(-0.0007 - 0.7829i, -0.0007 + 0.7829i, -1.2365)$ and a Poincaré index of 1 in the $z = 0.3089$ plane.

Equilibrium # 2 is an unstable saddle focus at $(-1.2919, -0.7493, -1.0889)$ with eigenvalues $(-0.3039 - 1.436i, -0.3039 + 1.436i, 2.1656)$ and a Poincaré index of 1 in the $z = -1.0889$ plane.

Equilibrium # 3 is an unstable saddle node at $(0, 0, -0.2000)$ with eigenvalues $(0.6978, -1.1178, 0.2)$ and a Poincaré index of -1 in the $z = -0.2$ plane.

Equilibrium # 4 is a stable focus at $(-0.5206, -0.3020, 0.3089)$ with eigenvalues $(-0.0007 - 0.7829i, -0.0007 + 0.7829i, -1.2365)$ and a Poincaré index of 1 in the $z = 0.3089$ plane.

Equilibrium # 5 is an unstable saddle focus at $(1.2919, 0.7493, -1.0889)$ with eigenvalues $(-0.3039 - 1.436i, -0.3039 + 1.436i, 2.1656)$ and a Poincaré index of 1 in the $z = -1.0889$ plane.

Equilibrium # 6 is an unstable saddle node at $(0, 0, 0)$ with eigenvalues $(0.58, -1, -0.2)$ and a Poincaré index of -1 in the $z = 0$ plane.

The strange attractor is self-excited, and the system is symmetric under the transformation $x \to -x$, $y \to -y$.

36.4 Attractor

Figure 36.1 shows various views of the attractor for Eq. (36.2) with $a = 0.58$, $b = 0.2$ and initial conditions $(-2, -1, 0)$. The rainbow of colors shows the local value of the largest Lyapunov exponent with red indicating the

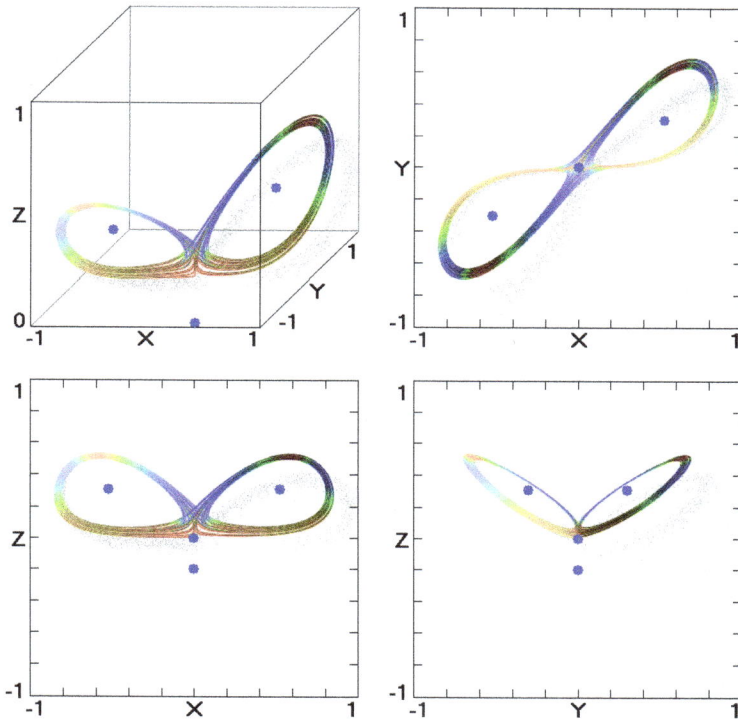

Fig. 36.1 Views of the attractor for Eq. (36.2) with $a = 0.58$, $b = 0.2$ and initial conditions $(-2, -1, 0)$.

most positive values (regions of worst predictability) and blue indicating the most negative values (regions of best predictability).

36.5 Time Series

Figure 36.2 shows the time series for the three variables along with the local value of the largest Lyapunov exponent (LL) for Eq. (36.2) with $a = 0.58$, $b = 0.2$. Red color in the Lyapunov exponent indicates that the error is growing parallel to the orbit, while blue indicates growth perpendicular to the orbit. Note that the orbit passes through regions where the local Lyapunov exponent is strongly positive and other regions where it is strongly negative as is typical for a chaotic system and is also reflected by the colors in Fig. 36.1.

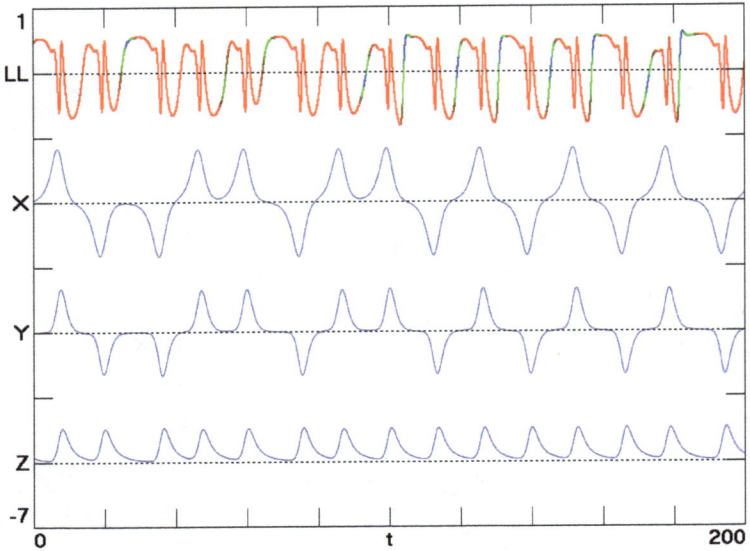

Fig. 36.2   Time series for the variables in Eq. (36.2) with $a = 0.58$, $b = 0.2$ along with the local Lyapunov exponent (LL).

36.6   Lyapunov Exponents

The global Lyapunov exponents are determined by averaging the local Lyapunov exponents along the orbit. The values typically converge very slowly because of the large variation along the orbit, and an integration time of order 10^8 is required to obtain 4-digit accuracy.

The results of such a calculation for the system in Eq. (36.2) with $a = 0.58$, $b = 0.2$ after a time of 2×10^6 are LE $= (0.0441, 0, -1.0302)$ with a Kaplan–Yorke dimension of 2.0427, where the last digit in the quoted values is only an approximation. The positive value of the largest Lyapunov exponent indicates that the system is chaotic, and the negative sum of the exponents (-0.9861) indicates that the system is dissipative with a strange attractor.

36.7   Basin of Attraction

Figure 36.3 shows (in red) the basin of attraction for Eq. (36.2) with $a = 0.58$, $b = 0.2$ in the $z = 0.3089$ plane. Also shown (in black) is the cross-section of the attractor in the same plane.

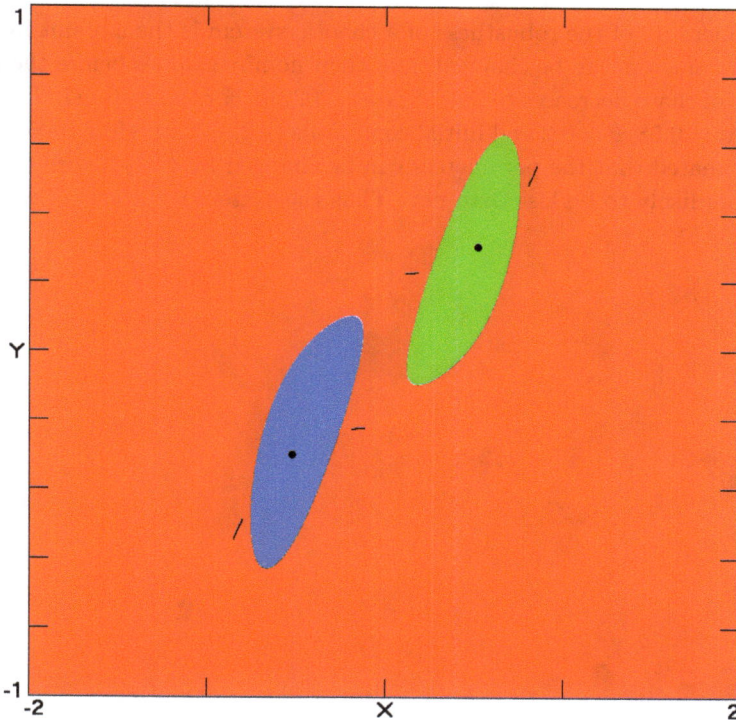

Fig. 36.3   Basin of attraction for Eq. (36.2) with $a = 0.58$, $b = 0.2$ in the $z = 0.3089$ plane.

36.8   Bifurcations

Figure 36.4 shows the bifurcation diagram for Eq. (36.2) as a function of the parameter a from 0 to 1.16 for $b = 0.2$. The initial condition was taken as $(-2, -1, 0)$ for each value of a. Each of the 500 values of a was calculated for a time of about 1×10^4.

The upper plot shows the three Lyapunov exponents. The middle plot shows the Kaplan–Yorke dimension, and the lower plot shows the local maxima of x. The chaotic region is in the vicinity of $a = 0.58$, $b = 0.2$, and the route to chaos is clearly shown.

36.9   Robustness

One measure of the robustness of a chaotic system is the amount by which the parameters can be changed from their nominal values before the probability of chaos decreases to 50% [Sprott (2022)]. For the system in Eq. (36.2) with $a = 0.58$, $b = 0.2$ and initial conditions $(-2, -1, 0)$, after 7900 trials, it is estimated that the parameters can be changed by 4.2% before the chaos is more likely to be lost than not. Thus the system is somewhat fragile.

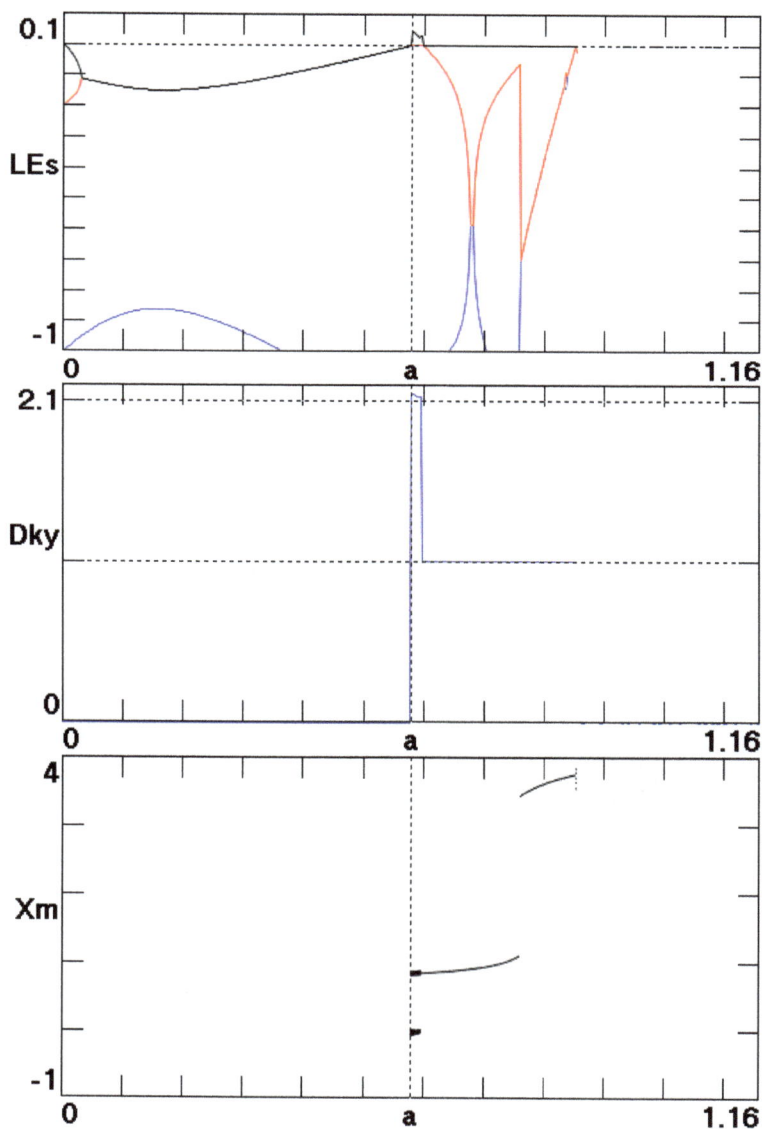

Fig. 36.4   Bifurcation diagram for Eq. (36.2) as a function of the parameter a for $b = 0.2$.

Chapter 37

Lorenz-84 System

This system was proposed by Lorenz as a simplification of the equations used by Hadley to describe the global circulation of winds around the Earth. With eleven terms, five quadratic nonlinearites, and two constant forcing terms, it is one of the most complicated systems in this book, but it is included here for historical reasons.

37.1   Introduction

The Lorenz-84 system [Lorenz (1984)] written in its most general form with an adjustable coefficient in each of the eleven terms is given by

$$
\begin{aligned}
\dot{x} &= -a_1 y^2 - a_2 z^2 - a_3 x + a_4 \\
\dot{y} &= a_5 y - a_6 xz - a_7 y + a_8 \\
\dot{z} &= a_9 xy + a_{10} xz - a_{11} z.
\end{aligned}
\tag{37.1}
$$

The usual parameters are $a_1 = a_2 = a_5 = a_7 = a_8 = a_{10} = a_{11} = 1$, $a_3 = 0.25$, $a_4 = 2$, $a_6 = a_9 = 4$. The signs are preserved, and the parameters are chosen greater than zero to prevent the system from collapsing to one of the simpler systems described earlier.

The following sections were written by the computer program that performed the optimization, carried out the analysis of the resulting system, and produced the corresponding figures, all without human intervention.

37.2   Simplified System

After about 3×10^2 trials, of which 90 were chaotic, simplified parameters for Eq. (37.1) that give chaotic solutions are $a_1 = 1$, $a_2 = 1$, $a_3 = 1$, $a_4 = 1$, $a_5 = 1$, $a_6 = 1$, $a_7 = 0.1$, $a_8 = 1$, $a_9 = 1$, $a_{10} = 1$, $a_{11} = 1$. Thus Eq. (37.1) can be written more compactly as

$$\dot{x} = -y^2 - z^2 - x + 1$$
$$\dot{y} = y - xz - ay + 1 \qquad\qquad (37.2)$$
$$\dot{z} = xy + xz - z,$$

where $a = 0.1$.

Note that with eleven terms, Eq. (37.2) should have seven independent parameters through a linear rescaling of the three variables plus time. However, six of the seven parameters have values of ± 1 and can be placed in any of the remaining ten terms.

37.3   Equilibria

The system in Eq. (37.2) with $a = 0.1$ has an unstable saddle focus at $(-0.9423, 0.3111, -1.3585)$ with eigenvalues $(-3.8663, 0.912 - 3.2301i, 0.912 + 3.2301i)$ and a Poincaré index of -1 in the $z = -1.3585$ plane.

The strange attractor is self-excited, and the system has no symmetry.

37.4   Attractor

Figure 37.1 shows various views of the attractor for Eq. (37.2) with $a = 0.1$ and initial conditions $(-5, -1, -2)$. The rainbow of colors shows the local value of the largest Lyapunov exponent with red indicating the most positive values (regions of worst predictability) and blue indicating the most negative values (regions of best predictability).

37.5   Time Series

Figure 37.2 shows the time series for the three variables along with the local value of the largest Lyapunov exponent (LL) for Eq. (37.2) with $a = 0.1$. Red color in the Lyapunov exponent indicates that the error is growing parallel to the orbit, while blue indicates growth perpendicular to the orbit. Note that the orbit passes through regions where the local Lyapunov exponent is strongly positive and other regions where it is strongly negative as is typical for a chaotic system and is also reflected by the colors in Fig. 37.1.

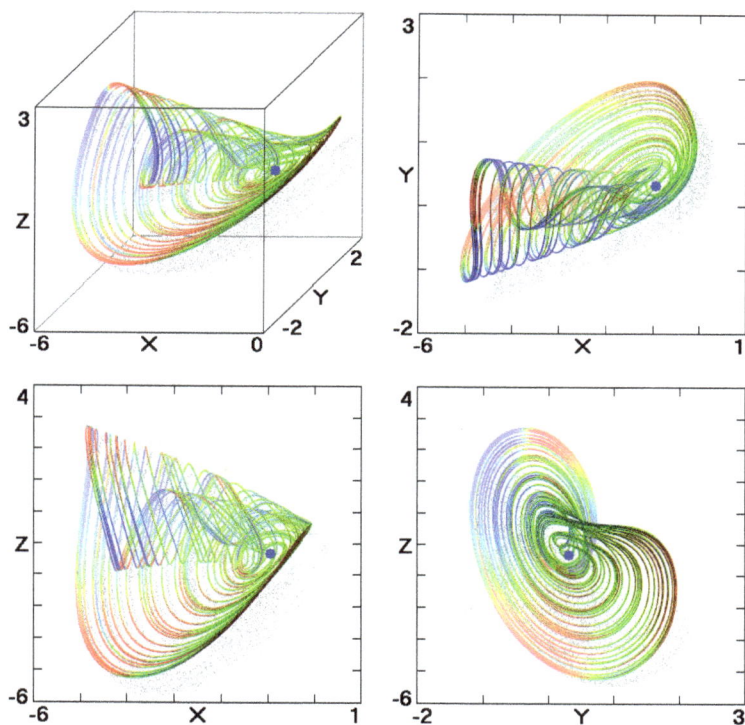

Fig. 37.1   Views of the attractor for Eq. (37.2) with $a = 0.1$ and initial conditions $(-5, -1, -2)$.

37.6   Lyapunov Exponents

The global Lyapunov exponents are determined by averaging the local Lyapunov exponents along the orbit. The values typically converge very slowly because of the large variation along the orbit, and an integration time of order 10^8 is required to obtain 4-digit accuracy.

The results of such a calculation for the system in Eq. (37.2) with $a = 0.1$ after a time of 5×10^6 are LE $= (0.3598, 0, -2.9894)$ with a Kaplan–Yorke dimension of 2.1203, where the last digit in the quoted values is only an approximation. The positive value of the largest Lyapunov exponent indicates that the system is chaotic, and the negative sum of the exponents (-2.6296) indicates that the system is dissipative with a strange attractor.

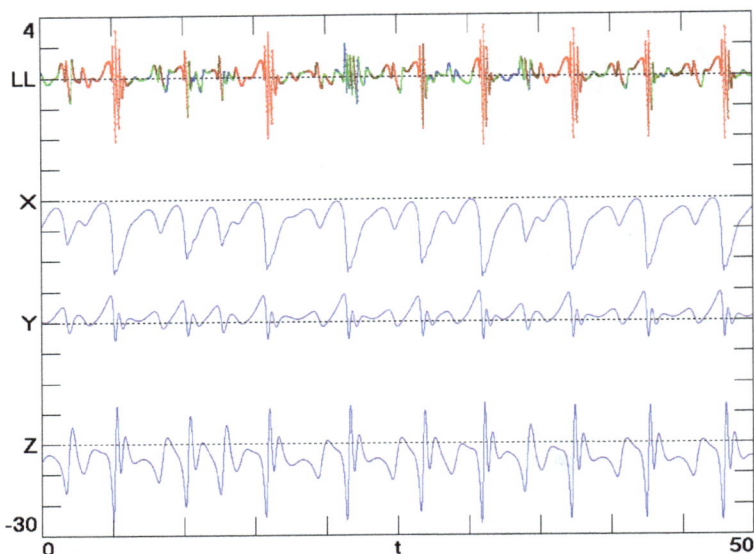

Fig. 37.2   Time series for the variables in Eq. (37.2) with $a = 0.1$ along with the local Lyapunov exponent (LL).

37.7   Basin of Attraction

Figure 37.3 shows (in red) the basin of attraction for Eq. (37.2) with $a = 0.1$ in the $z = -1.3585$ plane. Also shown (in black) is the cross-section of the attractor in the same plane.

37.8   Bifurcations

Figure 37.4 shows the bifurcation diagram for Eq. (37.2) as a function of the parameter a from 0 to 0.2. The initial condition was taken as $(-5, -1, -2)$ at $a = 0.2$ and was not changed as a slowly varied toward $a = 0$. Each of the 500 values of a was calculated for a time of about 2×10^4.

The upper plot shows the three Lyapunov exponents. The middle plot shows the Kaplan–Yorke dimension, and the lower plot shows the local maxima of x. The chaotic region is in the vicinity of $a = 0.1$, and the route to chaos is clearly shown.

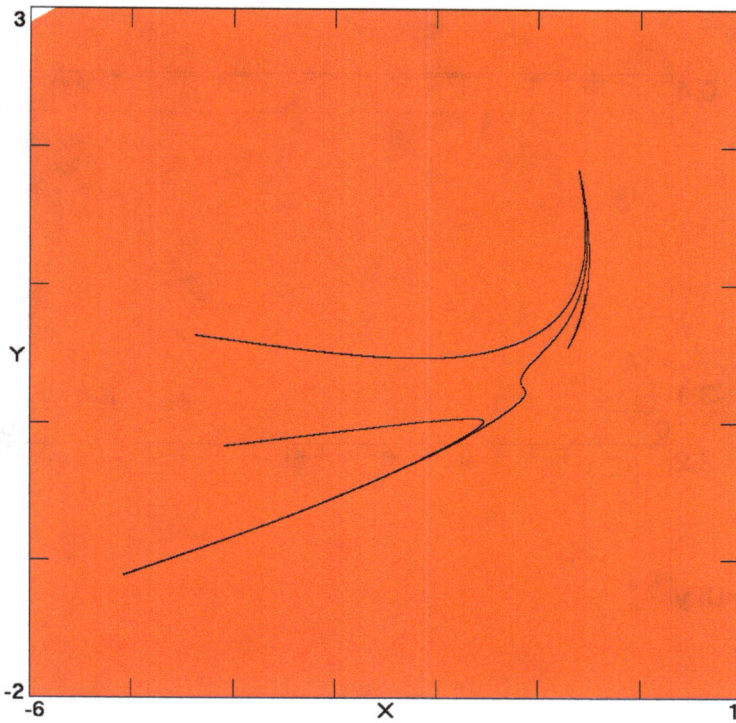

Fig. 37.3   Basin of attraction for Eq. (37.2) with $a = 0.1$ in the $z = -1.3585$ plane.

37.9   Robustness

One measure of the robustness of a chaotic system is the amount by which the parameters can be changed from their nominal values before the probability of chaos decreases to 50% [Sprott (2022)]. For the system in Eq. (37.2) with $a = 0.1$ and initial conditions $(-5, -1, -2)$, after 3400 trials, it is estimated that the parameter a can be changed by 4.4% before the chaos is more likely to be lost than not. Thus the system is somewhat fragile. This result is consistent with the data in Fig. 37.4.

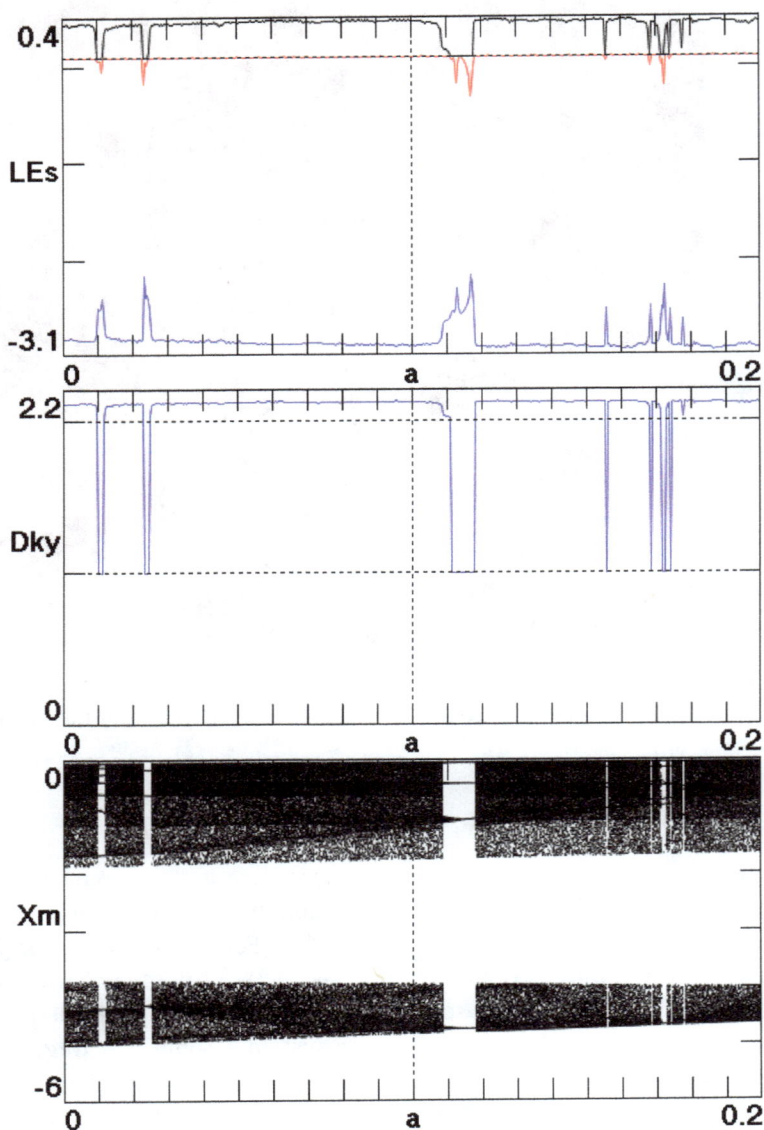

Fig. 37.4   Bifurcation diagram for Eq. (37.2) as a function of the parameter a.

Chapter 38

Wei System

This system is a modification of the Sprott D system with an additional constant term that removes the equilibrium. It is one of the earliest and simplest cases with a hidden strange attractor. Many other such examples are now known. It is also unusual because it is dissipative but time reversible with a symmetric attractor–repellor pair.

38.1 Introduction

The Wei (2011) system written in its most general form with an adjustable coefficient in each of the six terms is given by

$$
\begin{aligned}
\dot{x} &= -a_1 y \\
\dot{y} &= a_2 x + a_3 z \\
\dot{z} &= a_4 xz + a_5 y^2 - a_6.
\end{aligned}
\tag{38.1}
$$

The usual parameters are $a_1 = a_2 = a_3 = a_5 = 1$, $a_4 = 2$, $a_6 = 0.35$. The signs are preserved, and the parameters are chosen greater than zero to prevent the system from collapsing to one of the simpler systems described earlier. With all positive parameters, no equilibrium points can exist.

The following sections were written by the computer program that performed the optimization, carried out the analysis of the resulting system, and produced the corresponding figures, all without human intervention.

38.2 Simplified System

After about 2×10^4 trials, of which 21 were chaotic, simplified parameters for Eq. (38.1) that give chaotic solutions are $a_1 = 1$, $a_2 = 1$, $a_3 = 1$, $a_4 = 0.6$, $a_5 = 1$, $a_6 = 1$. Thus Eq. (38.1) can be written more compactly

as

$$\dot{x} = -y$$
$$\dot{y} = x + z \qquad\qquad (38.2)$$
$$\dot{z} = axz + y^2 - 1,$$

where $a = 0.6$.

Note that with six terms, Eq. (38.2) should have two independent parameters through a linear rescaling of the three variables plus time. However, one of the two parameters has a value of ± 1 and can be placed in any of the remaining five terms.

38.3   Equilibria

The system in Eq. (38.2) with $a = 0.6$ has no equilibrium points. The strange attractor is apparently hidden, and the system is symmetric under the transformation $x \to -x$, $z \to -z$, $t \to -t$.

38.4   Attractor

Figure 38.1 shows various views of the attractor for Eq. (38.2) with $a = 0.6$ and initial conditions $(-3, -1, 0)$. The rainbow of colors shows the local value of the largest Lyapunov exponent with red indicating the most positive values (regions of worst predictability) and blue indicating the most negative values (regions of best predictability).

38.5   Time Series

Figure 38.2 shows the time series for the three variables along with the local value of the largest Lyapunov exponent (LL) for Eq. (38.2) with $a = 0.6$. Red color in the Lyapunov exponent indicates that the error is growing parallel to the orbit, while blue indicates growth perpendicular to the orbit. Note that the orbit passes through regions where the local Lyapunov exponent is strongly positive and other regions where it is strongly negative as is typical for a chaotic system and is also reflected by the colors in Fig. 38.1.

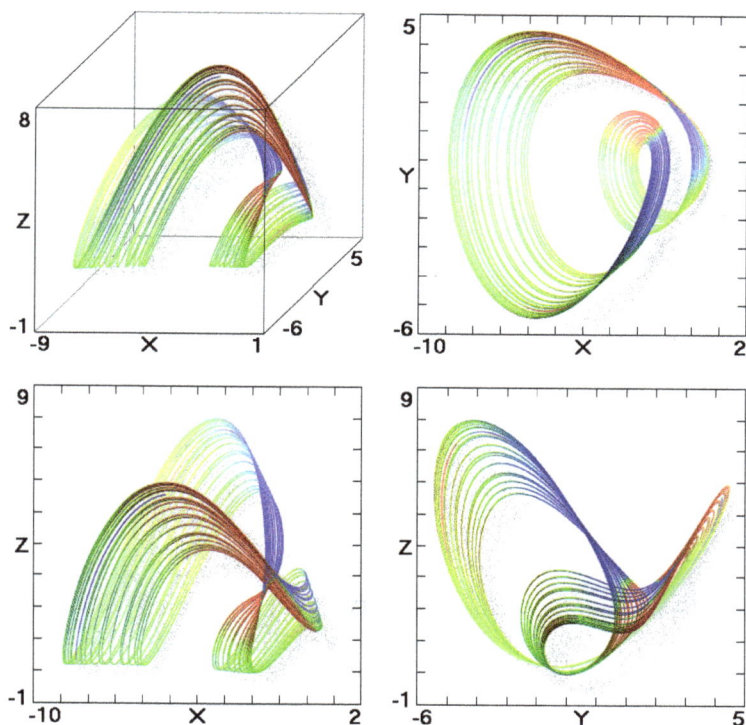

Fig. 38.1   Views of the attractor for Eq. (38.2) with $a = 0.6$ and initial conditions $(-3, -1, 0)$.

38.6   Lyapunov Exponents

The global Lyapunov exponents are determined by averaging the local Lyapunov exponents along the orbit. The values typically converge very slowly because of the large variation along the orbit, and an integration time of order 10^8 is required to obtain 4-digit accuracy.

The results of such a calculation for the system in Eq. (38.2) with $a = 0.6$ after a time of 3×10^3 are LE = $(0.0403, 0, -1.6350)$ with a Kaplan–Yorke dimension of 2.0246, where the last digit in the quoted values is only an approximation. The positive value of the largest Lyapunov exponent indicates that the system is chaotic, and the negative sum of the exponents (-1.5947) indicates that the system is dissipative with a strange attractor.

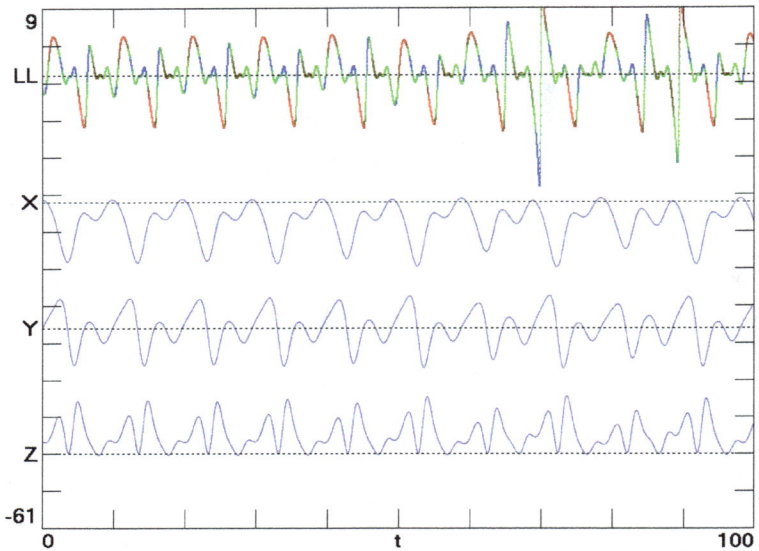

Fig. 38.2   Time series for the variables in Eq. (38.2) with $a = 0.6$ along with the local Lyapunov exponent (LL).

38.7   Basin of Attraction

Figure 38.3 shows (in red) the basin of attraction for Eq. (38.2) with $a = 0.6$ in the $z = 1$ plane. Also shown (in black) is the cross-section of the attractor in the same plane.

38.8   Bifurcations

Figure 38.4 shows the bifurcation diagram for Eq. (38.2) as a function of the parameter a from 0 to 1.2. The initial condition was taken as $(-3, -1, 0)$ at $a = 1.2$ and was not changed as a slowly varied toward $a = 0$. Each of the 500 values of a was calculated for a time of about 1×10^4.

The upper plot shows the three Lyapunov exponents. The middle plot shows the Kaplan–Yorke dimension, and the lower plot shows the local maxima of x. The chaotic region is in the vicinity of $a = 0.6$, and the route to chaos is clearly shown.

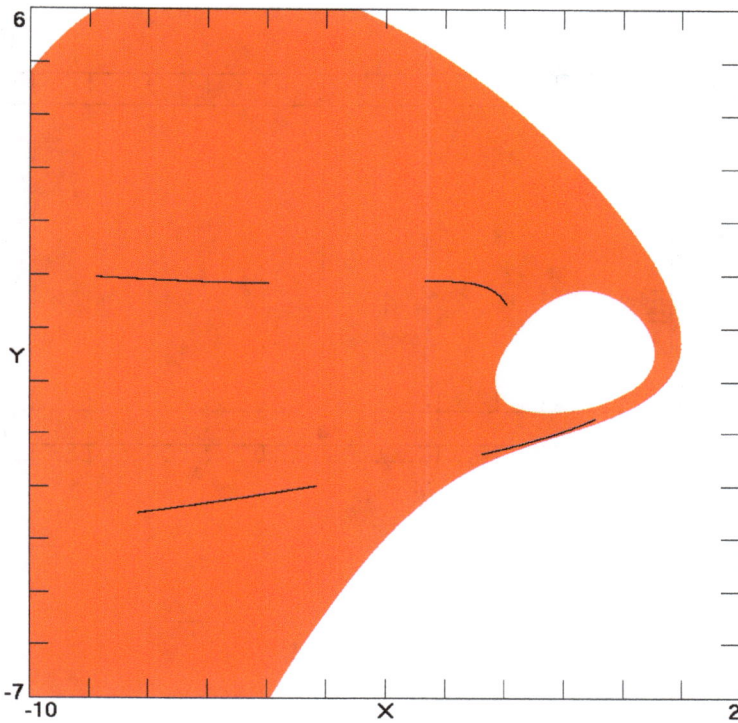

Fig. 38.3   Basin of attraction for Eq. (38.2) with $a = 0.6$ in the $z = 1$ plane.

38.9   Robustness

One measure of the robustness of a chaotic system is the amount by which the parameters can be changed from their nominal values before the probability of chaos decreases to 50% [Sprott (2022)]. For the system in Eq. (38.2) with $a = 0.6$ and initial conditions $(-3, -1, 0)$, after 3916 trials, it is estimated that the parameter a can be changed by 13% before the chaos is more likely to be lost than not. Thus the system is somewhat fragile. This result is consistent with the data in Fig. 38.4.

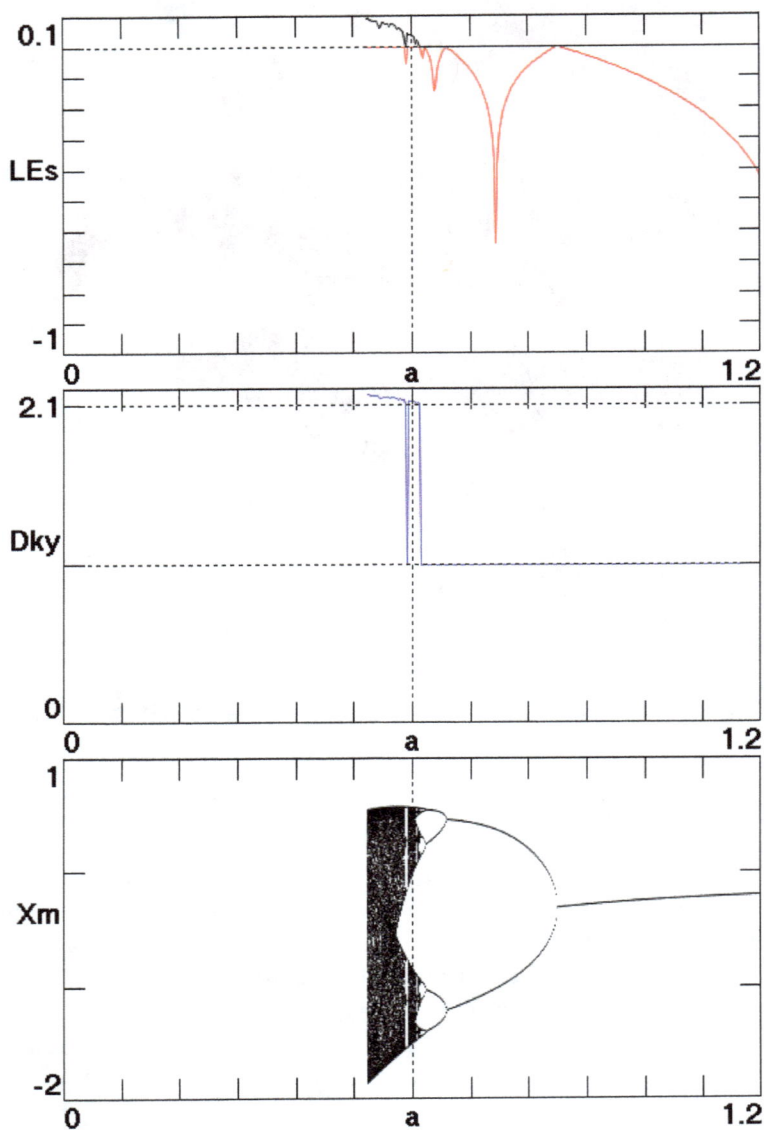

Fig. 38.4 Bifurcation diagram for Eq. (38.2) as a function of the parameter a.

Chapter 39

Wang–Chen System

This system is a modification of the Sprott E system with an additional constant term that gives a single stable equilibrium in addition to the strange attractor. Thus the strange attractor is hidden, and with only six terms and two quadratic nonlinearities, this might be the simplest such example.

39.1   Introduction

The Wang–Chen system [Wang and Chen (2012)] written in its most general form with an adjustable coefficient in each of the six terms is given by

$$\dot{x} = a_1 yz + a_2$$
$$\dot{y} = a_3 x^2 - a_4 y \qquad\qquad (39.1)$$
$$\dot{z} = a_5 - a_6 x.$$

The usual parameters are $a_1 = a_3 = a_4 = a_5 = 1$, $a_2 = 0.01$, $a_6 = 4$. The signs are preserved, and the parameters are chosen greater than zero to prevent the system from collapsing to one of the simpler systems described earlier. For some choices of the parameters such as those given above, the strange attractor coexists with a stable equilibrium and a limit cycle [Sprott et al. (2013)].

The following sections were written by the computer program that performed the optimization, carried out the analysis of the resulting system, and produced the corresponding figures, all without human intervention.

39.2   Simplified System

After about 2×10^4 trials, of which 52 were chaotic, simplified parameters for Eq. (39.1) that give chaotic solutions are $a_1 = 1$, $a_2 = 0.1$, $a_3 = 1$,

$a_4 = 2$, $a_5 = 1$, $a_6 = 1$. Thus Eq. (39.1) can be written more compactly as

$$\dot{x} = yz + a$$
$$\dot{y} = x^2 - by \qquad\qquad (39.2)$$
$$\dot{z} = 1 - x,$$

where $a = 0.1$, $b = 2$.

Note that with six terms, Eq. (39.2) should have two independent parameters through a linear rescaling of the three variables plus time, and so the dynamics is completely captured by the given parameters, which could be put in any of the six terms, albeit with different numerical values.

39.3 Equilibria

The system in Eq. (39.2) with $a = 0.1$, $b = 2$ has a stable focus at (1, 0.5000, −0.2000) with eigenvalues $(-0.0959 - 0.7375i, -0.0959 + 0.7375i, -1.8081)$ and a Poincaré index of 1 in the $z = -0.2$ plane.

The strange attractor is apparently hidden, and the system has no symmetry.

39.4 Attractor

Figure 39.1 shows various views of the attractor for Eq. (39.2) with $a = 0.1$, $b = 2$ and initial conditions $(-3, 8, 1)$. The rainbow of colors shows the local value of the largest Lyapunov exponent with red indicating the most positive values (regions of worst predictability) and blue indicating the most negative values (regions of best predictability).

39.5 Time Series

Figure 39.2 shows the time series for the three variables along with the local value of the largest Lyapunov exponent (LL) for Eq. (39.2) with $a = 0.1$, $b = 2$. Red color in the Lyapunov exponent indicates that the error is growing parallel to the orbit, while blue indicates growth perpendicular to the orbit. Note that the orbit passes through regions where the local Lyapunov exponent is strongly positive and other regions where it is strongly negative as is typical for a chaotic system and is also reflected by the colors in Fig. 39.1.

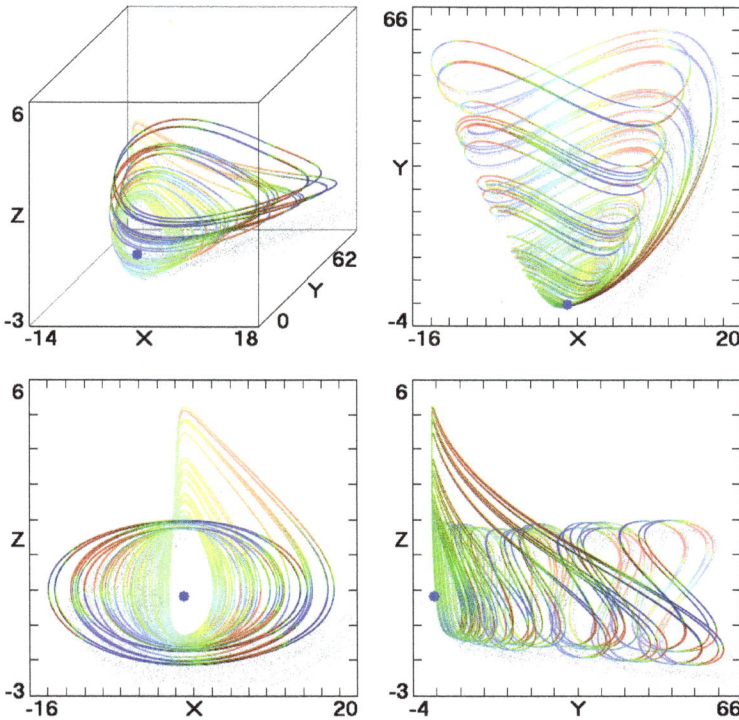

Fig. 39.1   Views of the attractor for Eq. (39.2) with $a = 0.1$, $b = 2$ and initial conditions $(-3, 8, 1)$.

39.6   Lyapunov Exponents

The global Lyapunov exponents are determined by averaging the local Lyapunov exponents along the orbit. The values typically converge very slowly because of the large variation along the orbit, and an integration time of order 10^8 is required to obtain 4-digit accuracy.

The results of such a calculation for the system in Eq. (39.2) with $a = 0.1$, $b = 2$ after a time of 4×10^6 are LE $= (0.1446, 0, -2.1446)$ with a Kaplan–Yorke dimension of 2.0674, where the last digit in the quoted values is only an approximation. The positive value of the largest Lyapunov exponent indicates that the system is chaotic, and the negative sum of the exponents (-2) indicates that the system is dissipative with a strange attractor.

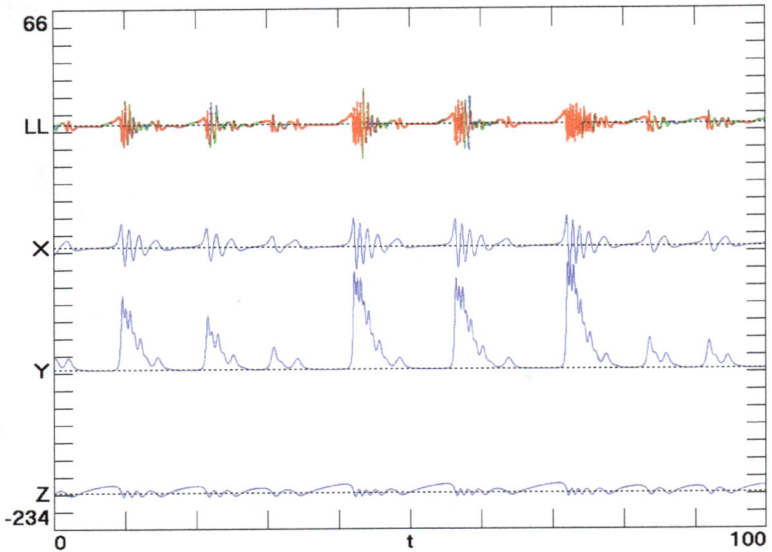

Fig. 39.2   Time series for the variables in Eq. (39.2) with $a = 0.1$, $b = 2$ along with the local Lyapunov exponent (LL).

39.7   Basin of Attraction

Figure 39.3 shows (in red) the basin of attraction for Eq. (39.2) with $a = 0.1$, $b = 2$ in the $z = -0.2$ plane. Also shown (in black) is the cross-section of the attractor in the same plane.

39.8   Bifurcations

Figure 39.4 shows the bifurcation diagram for Eq. (39.2) as a function of the parameter a from 0 to 0.2 for $b = 2$. The initial condition was taken as $(-3, 8, 1)$ at $a = 0.2$ and was not changed as a slowly varied toward $a = 0$. Each of the 500 values of a was calculated for a time of about 1×10^4.

The upper plot shows the three Lyapunov exponents. The middle plot shows the Kaplan–Yorke dimension, and the lower plot shows the local maxima of x. The chaotic region is in the vicinity of $a = 0.1$, $b = 2$, and the route to chaos is clearly shown.

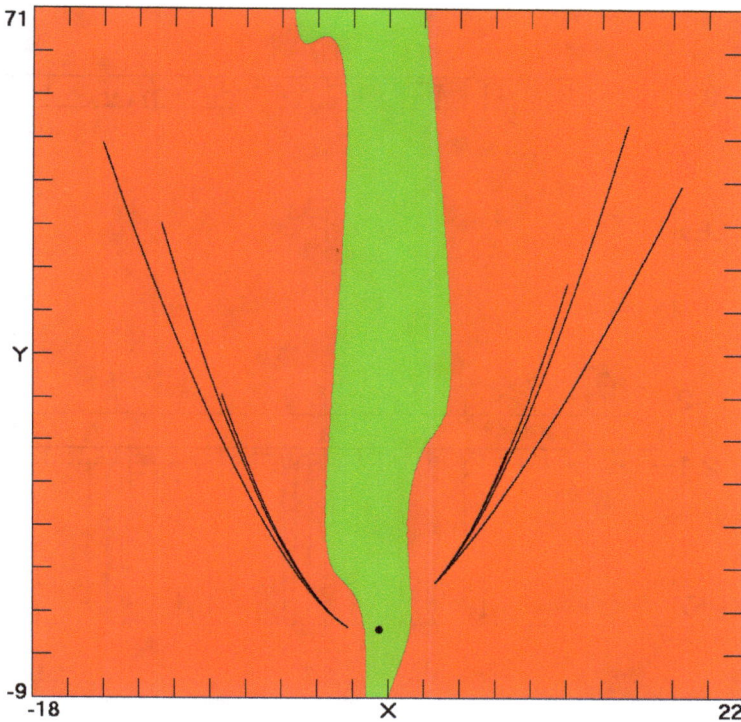

Fig. 39.3   Basin of attraction for Eq. (39.2) with $a = 0.1$, $b = 2$ in the $z = -0.2$ plane.

39.9   Robustness

One measure of the robustness of a chaotic system is the amount by which the parameters can be changed from their nominal values before the probability of chaos decreases to 50% [Sprott (2022)]. For the system in Eq. (39.2) with $a = 0.1$, $b = 2$ and initial conditions $(-3, 8, 1)$, after 2341 trials, it is estimated that the parameters can be changed by 56% before the chaos is more likely to be lost than not. Thus the system is highly robust.

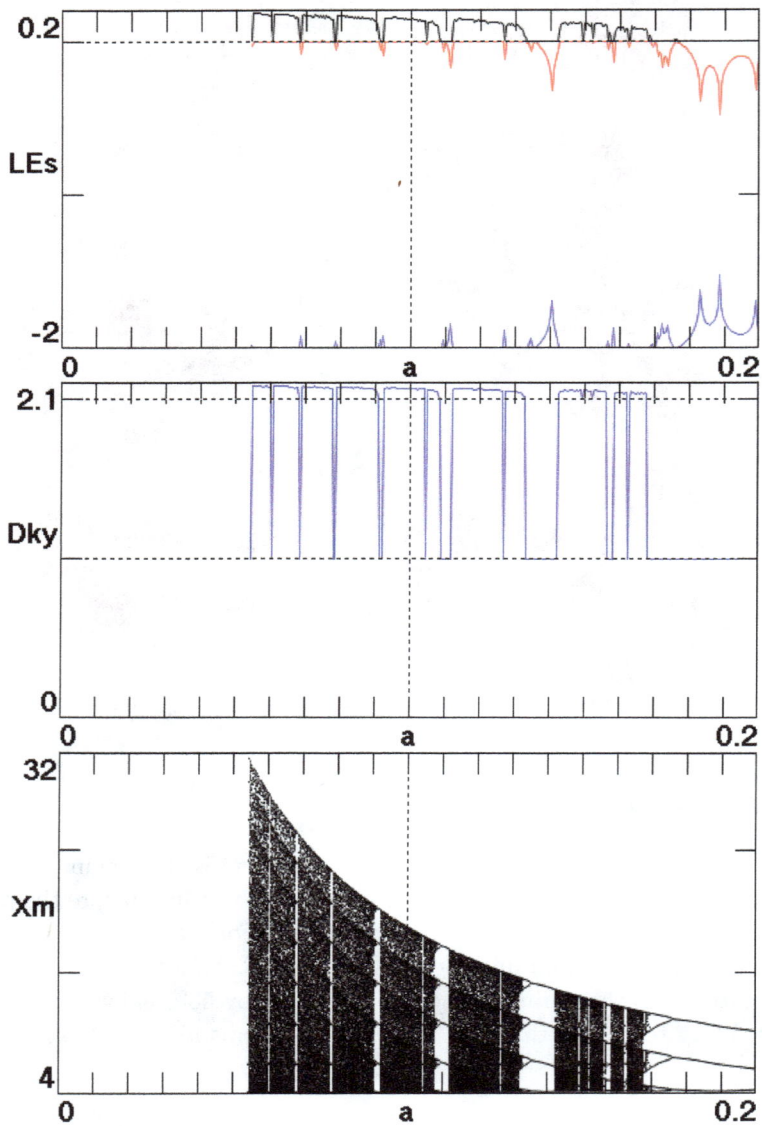

Fig. 39.4   Bifurcation diagram for Eq. (39.2) as a function of the parameter a for $b = 2$.

Chapter 40

Reflection Symmetric System

Previous chapters have included systems with inversion symmetry (three variables) and rotational symmetry (two variables). This system has reflection symmetry (one variable) taken as $x \to -x$. Since $\dot{x} = 0$ at $x = 0$, the orbit cannot cross the $x = 0$ plane, and any attractor on one side of the plane must have a mirror image on the other side.

40.1   Introduction

The simplest reflection symmetric system with quadratic nonlinearities [Sprott (2014a)] written in its most general form with an adjustable coefficient in each of the six terms is given by

$$\dot{x} = a_1 x + a_2 xy$$
$$\dot{y} = a_3 z \qquad (40.1)$$
$$\dot{z} = a_4 y + a_5 z + a_6 x^2.$$

The usual parameters are $a_1 = a_3 = a_6 = 1$, $a_2 = a_4 = -1$, $a_5 = -0.3$.

The following sections were written by the computer program that performed the optimization, carried out the analysis of the resulting system, and produced the corresponding figures, all without human intervention.

40.2   Simplified System

After about 2×10^4 trials, of which 289 were chaotic, simplified parameters for Eq. (40.1) that give chaotic solutions are $a_1 = 1$, $a_2 = 1$, $a_3 = -1$, $a_4 = 1$, $a_5 = -0.2$, $a_6 = 1$. Thus Eq. (40.1) can be written more compactly as

$$\dot{x} = yx + xy$$
$$\dot{y} = -z \qquad (40.2)$$
$$\dot{z} = y + az + x^2,$$

where $a = -0.2$.

251

Note that with six terms, Eq. (40.2) should have two independent parameters through a linear rescaling of the three variables plus time. However, one of the two parameters has a value of ± 1 and can be placed in any of the remaining five terms.

40.3   Equilibria

The system in Eq. (40.2) with $a = -0.2$ has three equilibrium points:

Equilibrium # 1 is an unstable saddle focus at $(1, -1, 0)$ with eigenvalues $(-1.0533, 0.4267 - 1.3103i, 0.4267 + 1.3103i)$ and a Poincaré index of 0 in the $z = 0$ plane.

Equilibrium # 2 is an unstable saddle focus at $(0, 0, 0)$ with eigenvalues $(1, -0.1 - 0.995i, -0.1 + 0.995i)$ and a Poincaré index of 0 in the $z = 0$ plane.

Equilibrium # 3 is an unstable saddle focus at $(-1, -1, 0)$ with eigenvalues $(-1.0533, 0.4267 - 1.3103i, 0.4267 + 1.3103i)$ and a Poincaré index of 0 in the $z = 0$ plane.

The strange attractor is self-excited, and the system is symmetric under the transformation $x \to -x$.

40.4   Attractor

Figure 40.1 shows various views of the two attractors for Eq. (40.2) with $a = -0.2$ and initial conditions $(1, -13, 32)$.

40.5   Time Series

Figure 40.2 shows the time series for the three variables along with the local value of the largest Lyapunov exponent (LL) for Eq. (40.2) with $a = -0.2$. Red color in the Lyapunov exponent indicates that the error is growing parallel to the orbit, while blue indicates growth perpendicular to the orbit. Note that the orbit passes through regions where the local Lyapunov exponent is strongly positive and other regions where it is strongly negative as is typical for a chaotic system and is also reflected by the colors in Fig. 40.1.

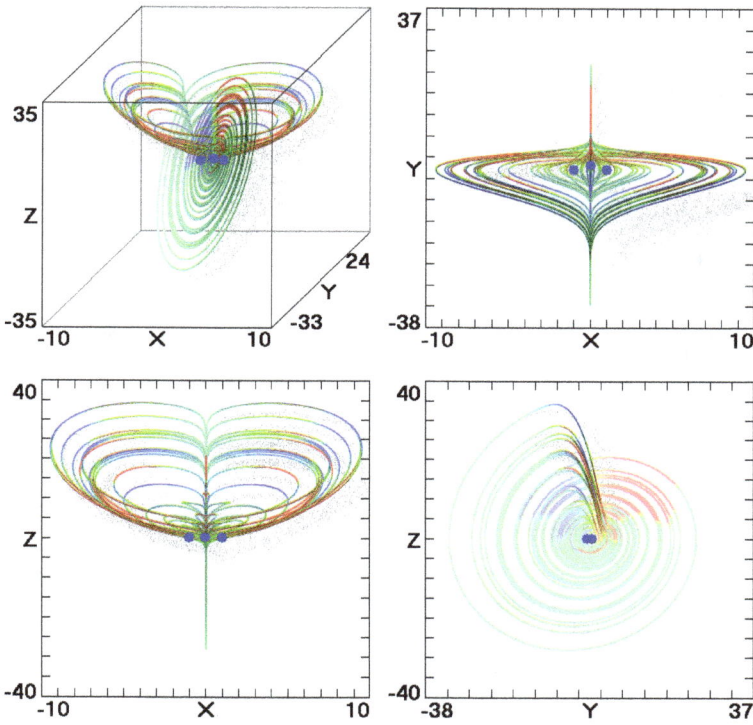

Fig. 40.1   Views of the two attractors for Eq. (40.2) with $a = -0.2$.

40.6   Lyapunov Exponents

The global Lyapunov exponents are determined by averaging the local Lyapunov exponents along the orbit. The values typically converge very slowly because of the large variation along the orbit, and an integration time of order 10^8 is required to obtain 4-digit accuracy.

The results of such a calculation for the system in Eq. (40.2) with $a = -0.2$ after a time of 1×10^7 are LE = (0.5298, 0, −0.7298) with a Kaplan–Yorke dimension of 2.7259, where the last digit in the quoted values is only an approximation. The positive value of the largest Lyapunov exponent indicates that the system is chaotic, and the negative sum of the exponents (−0.2000) indicates that the system is dissipative with a strange attractor.

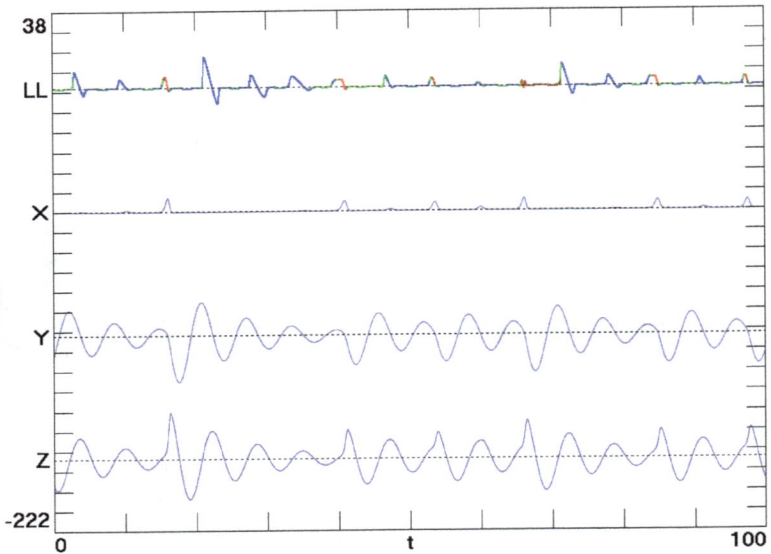

Fig. 40.2   Time series for the variables in Eq. (40.2) with $a = -0.2$ along with the local Lyapunov exponent (LL).

40.7   Basin of Attraction

Figure 40.3 shows (in red) the basin of attraction for Eq. (40.2) with $a = -0.2$ in the $z = 0$ plane. Also shown (in black) is the cross-section of the attractor in the same plane.

40.8   Bifurcations

Figure 40.4 shows the bifurcation diagram for Eq. (40.2) as a function of the parameter a from $a = -0.4$ to $a = 0$. The initial condition was taken as $(1, -13, 32)$ at $a = -0.4$ and was not changed as a slowly varied toward $a = 0$. Each of the 500 values of a was calculated for a time of about 1×10^4.

The upper plot shows the three Lyapunov exponents. The middle plot shows the Kaplan–Yorke dimension, and the lower plot shows the local maxima of x. The chaotic region is in the vicinity of $a = -0.2$, and the route to chaos is clearly shown.

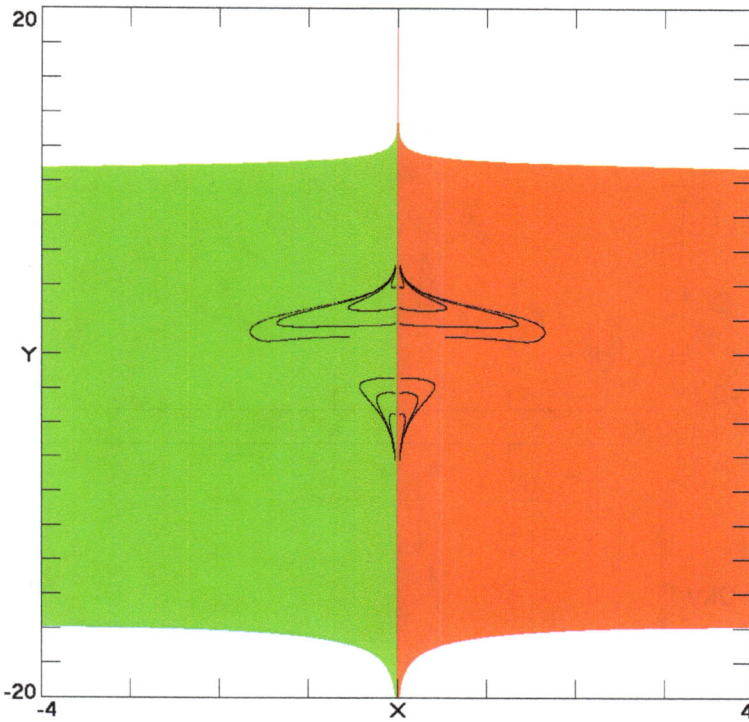

Fig. 40.3   Basin of attraction for Eq. (40.2) with $a = -0.2$ in the $z = 0$ plane.

40.9   Robustness

One measure of the robustness of a chaotic system is the amount by which the parameters can be changed from their nominal values before the probability of chaos decreases to 50% [Sprott (2022)]. For the system in Eq. (40.2) with $a = -0.2$ and initial conditions $(1, -13, 32)$, after 1057 trials, it is estimated that the parameter a can be changed by 93% before the chaos is more likely to be lost than not. Thus the system is highly robust. This result is consistent with the data in Fig. 40.4.

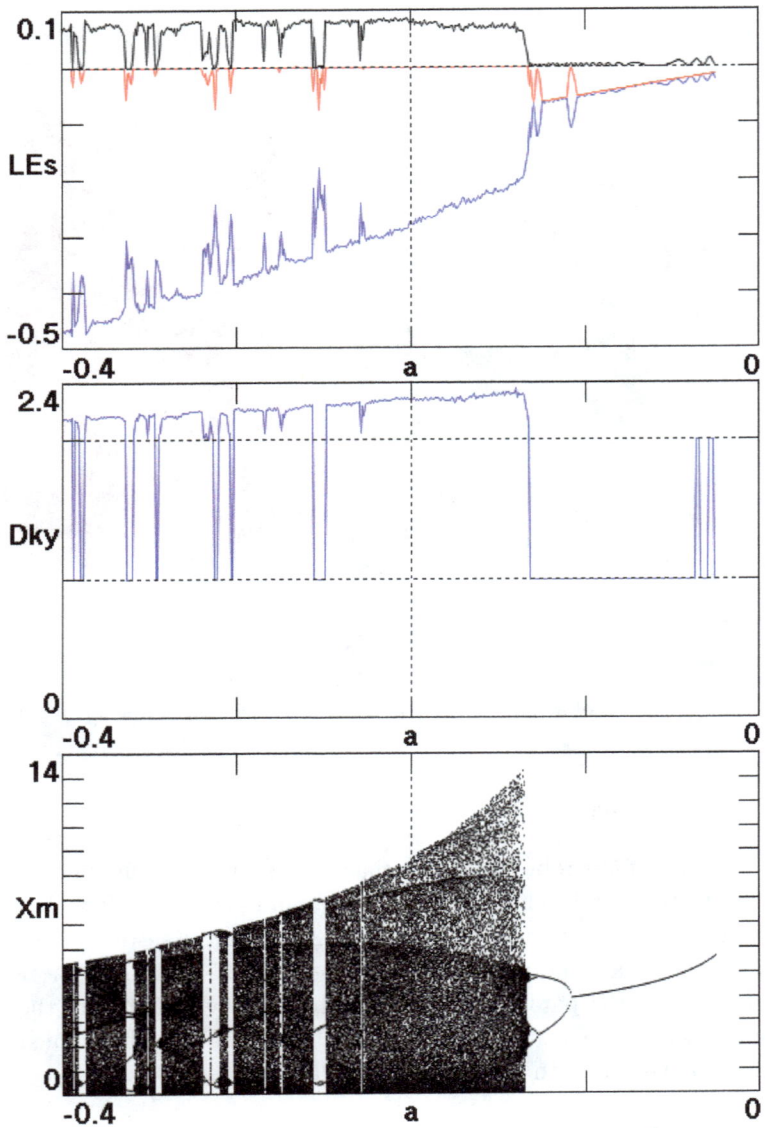

Fig. 40.4   Bifurcation diagram for Eq. (40.2) as a function of the parameter a.

Chapter 41

Butterfly System

This system has four cross product terms and admits as many as five co-existing attractors including some with a multi-wing butterfly shape for certain values of the parameters. For the chosen parameters, a symmetric strange attractor coexists with two neutrally stable equilibrium points. Like the Lorenz system, it is symmetric under a $180°$ rotation about the z axis.

41.1 Introduction

The butterfly system [Li and Sprott (2013)] written in its most general form with an adjustable coefficient in each of the seven terms is given by

$$\dot{x} = a_1 y + a_2 yz$$
$$\dot{y} = -a_3 xz + a_4 yz \qquad (41.1)$$
$$\dot{z} = -a_5 z - a_6 xy + a_7.$$

The usual parameters are $a_1 = a_2 = a_3 = a_4 = a_6 = 1$, $a_5 = 0.6$, $a_7 = 3$. The signs are preserved, and the parameters are chosen greater than zero to prevent the system from collapsing to one of the simpler systems described earlier.

The following sections were written by the computer program that performed the optimization, carried out the analysis of the resulting system, and produced the corresponding figures, all without human intervention.

41.2 Simplified System

After about 3×10^3 trials, of which 68 were chaotic, simplified parameters for Eq. (41.1) that give chaotic solutions are $a_1 = 1$, $a_2 = 1$, $a_3 = 1$, $a_4 = 1$,

$a_5 = 0.5$, $a_6 = 1$, $a_7 = 1$. Thus Eq. (41.1) can be written more compactly as

$$\dot{x} = y + yz$$
$$\dot{y} = -xz + yz \qquad\qquad (41.2)$$
$$\dot{z} = -az - xy + 1,$$

where $a = 0.5$.

Note that with seven terms, Eq. (41.2) should have three independent parameters through a linear rescaling of the three variables plus time. However, two of the three parameters have values of ± 1 and can be placed in any of the remaining six terms.

41.3 Equilibria

The system in Eq. (41.2) with $a = 0.5$ has three equilibrium points:

Equilibrium # 1 is a neutrally stable non-hyperbolic equilibrium at $(-1.2247, -1.2247, -1)$ with eigenvalues $(0 - 1.4142i, 0 + 1.4142i, -1.5)$ and a Poincaré index of 0 in the $z = -1$ plane.

Equilibrium # 2 is an unstable saddle focus at $(0, 0, 2)$ with eigenvalues $(1 - 2.2361i, 1 + 2.2361i, -0.5)$ and a Poincaré index of 1 in the $z = 2$ plane.

Equilibrium # 3 is a neutrally stable non-hyperbolic equilibrium at $(1.2247, 1.2247, -1)$ with eigenvalues $(0 - 1.4142i, 0 + 1.4142i, -1.5)$ and a Poincaré index of 0 in the $z = -1$ plane.

The strange attractor is self-excited, and the system is symmetric under the transformation $x \rightarrow -x$, $y \rightarrow -y$.

41.4 Attractor

Figure 41.1 shows various views of the attractor for Eq. (41.2) with $a = 0.5$ and initial conditions $(2, 1, 1)$.

41.5 Time Series

Figure 41.2 shows the time series for the three variables along with the local value of the largest Lyapunov exponent (LL) for Eq. (41.2) with $a = 0.5$. Red color in the Lyapunov exponent indicates that the error is growing parallel to the orbit, while blue indicates growth perpendicular to the orbit. Note that the orbit passes through regions where the local Lyapunov

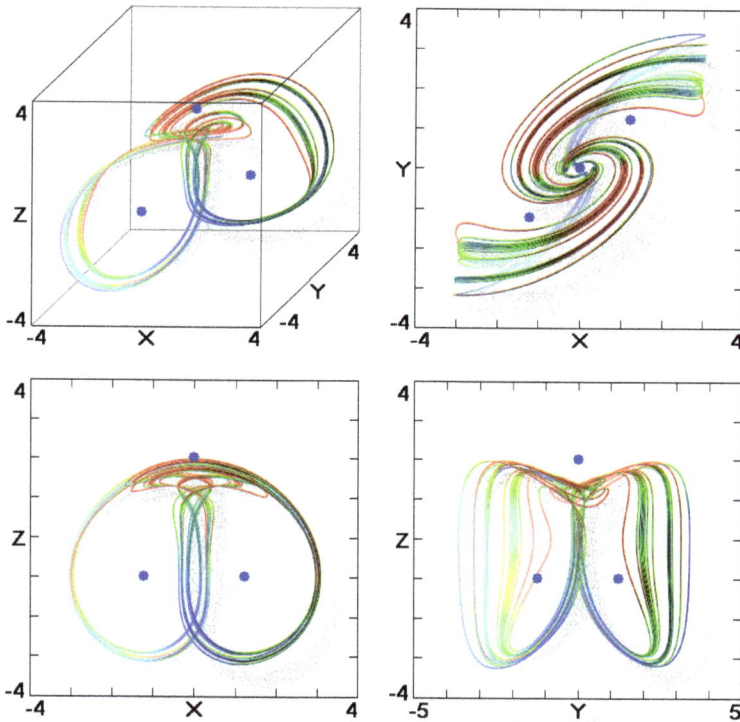

Fig. 41.1 Views of the attractor for Eq. (41.2) with $a = 0.5$ and initial conditions (2, 1, 1).

exponent is strongly positive and other regions where it is strongly negative as is typical for a chaotic system and is also reflected by the colors in Fig. 41.1.

41.6 Lyapunov Exponents

The global Lyapunov exponents are determined by averaging the local Lyapunov exponents along the orbit. The values typically converge very slowly because of the large variation along the orbit, and an integration time of order 10^8 is required to obtain 4-digit accuracy.

The results of such a calculation for the system in Eq. (41.2) with $a = 0.5$ after a time of 5×10^6 are LE = (0.0683, 0, −0.4978) with a Kaplan–Yorke dimension of 2.1371, where the last digit in the quoted values is only

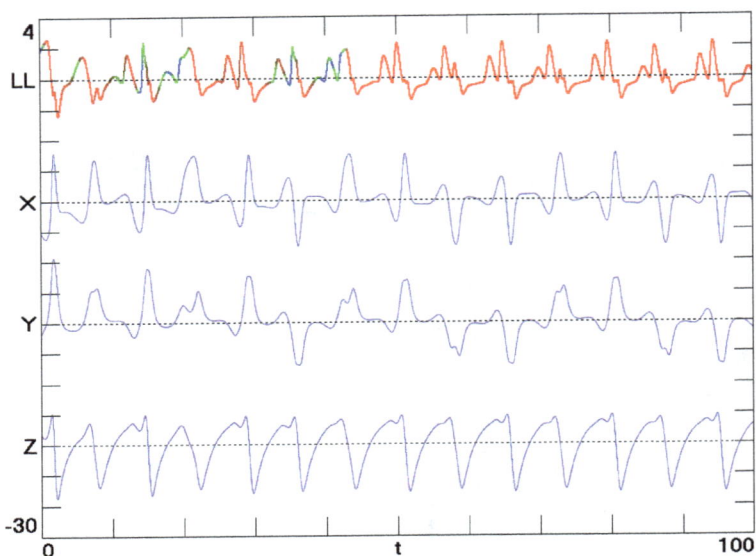

Fig. 41.2   Time series for the variables in Eq. (41.2) with $a = 0.5$ along with the local Lyapunov exponent (LL).

an approximation. The positive value of the largest Lyapunov exponent indicates that the system is chaotic, and the negative sum of the exponents (-0.4295) indicates that the system is dissipative with a strange attractor.

41.7   Basin of Attraction

Figure 41.3 shows (in red) the basin of attraction for Eq. (41.2) with $a = 0.5$ in the $z = -1$ plane. Also shown (in black) is the cross-section of the attractor in the same plane.

41.8   Bifurcations

Figure 41.4 shows the bifurcation diagram for Eq. (41.2) as a function of the parameter a from 0 to 1. The initial condition was taken as $(2, 1, 1)$ for each value of a. Each of the 500 values of a was calculated for a time of about 2×10^4.

   The upper plot shows the three Lyapunov exponents. The middle plot shows the Kaplan–Yorke dimension, and the lower plot shows the local

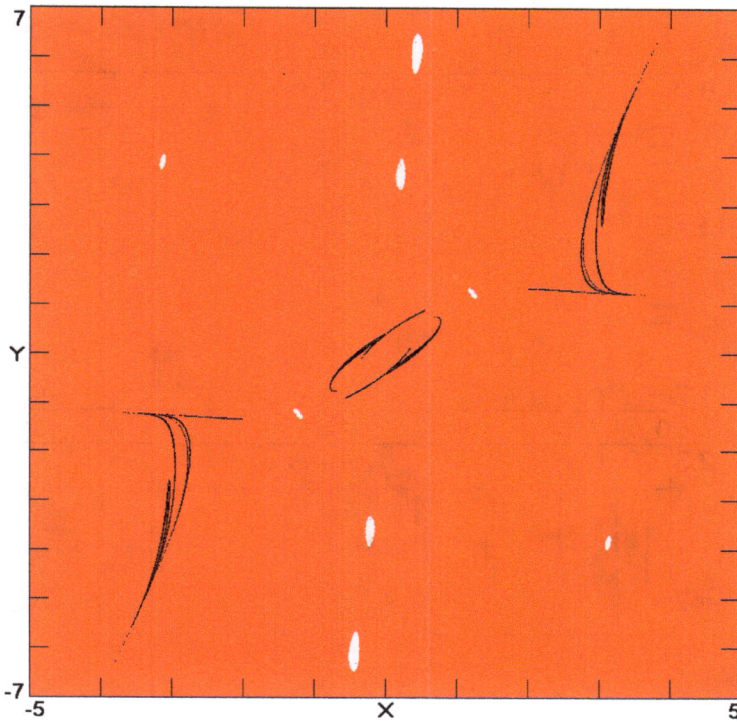

Fig. 41.3   Basin of attraction for Eq. (41.2) with $a = 0.5$ in the $z = -1$ plane.

maxima of x. The chaotic region is in the vicinity of $a = 0.5$, and the route to chaos is clearly shown.

41.9   Robustness

One measure of the robustness of a chaotic system is the amount by which the parameters can be changed from their nominal values before the probability of chaos decreases to 50% [Sprott (2022)]. For the system in Eq. (41.2) with $a = 0.5$ and initial conditions $(2, 1, 1)$, after 2181 trials, it is estimated that the parameter a can be changed by 30% before the chaos is more likely to be lost than not. Thus the system is somewhat robust. This result is consistent with the data in Fig. 41.4.

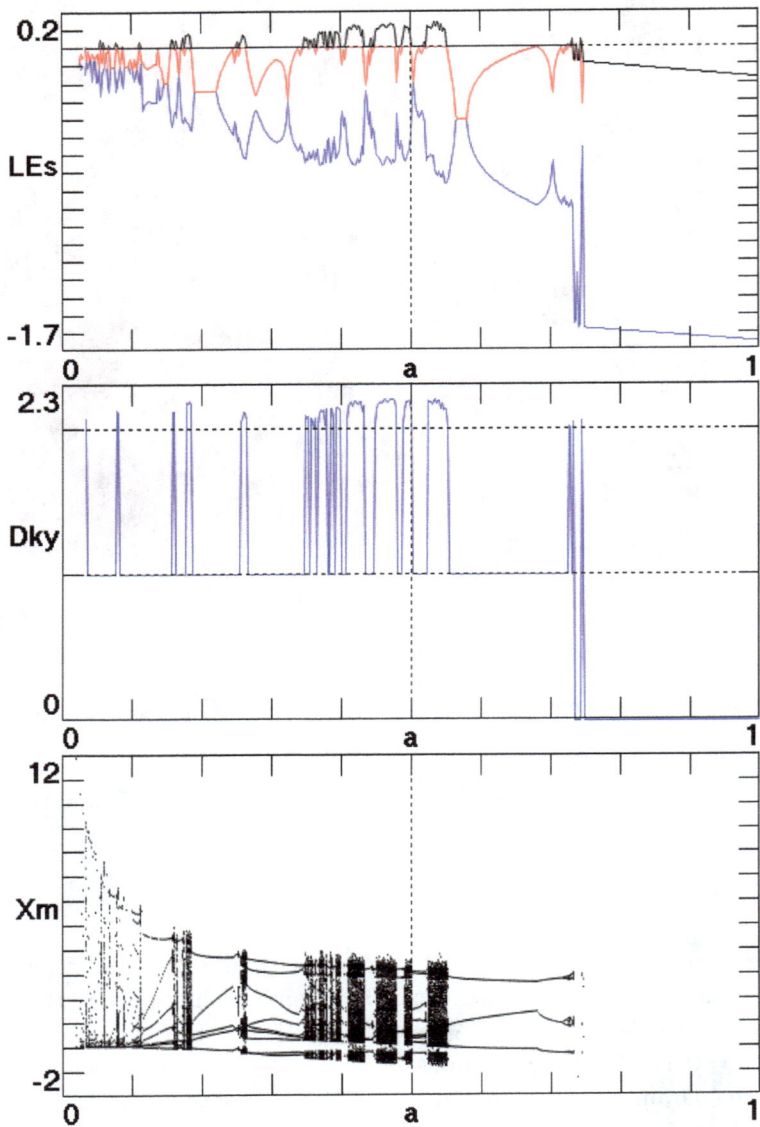

Fig. 41.4   Bifurcation diagram for Eq. (41.2) as a function of the parameter a.

Chapter 42

Line Equilibrium System

This is a simple example of a system with an infinite line of equilibria coexisting with a strange attractor that is hidden by virtue of the fact that nearly all initial conditions chosen in the vicinity of the line do not lead to the attractor. The system is a dissipative extension of the Nose–Hoover system.

42.1 Introduction

The line equilibrium (LE1) system [Jafari and Sprott (2013)] written in its most general form with an adjustable coefficient in each of the three terms is given by

$$\begin{aligned}
\dot{x} &= y \\
\dot{y} &= -x + yz \\
\dot{z} &= a_1 x + a_2 xy + a_3 xz.
\end{aligned} \qquad (42.1)$$

The usual parameters are $a_1 = a_3 = -1$, $a_2 = -15$. This system requires an especially long search since the basin of attraction is small and chaotic solutions are relatively rare. The strange attractor surrounds the line of equilibria.

The following sections were written by the computer program that performed the optimization, carried out the analysis of the resulting system, and produced the corresponding figures, all without human intervention.

42.2 Simplified System

After about 8×10^4 trials, of which 72 were chaotic, simplified parameters for Eq. (42.1) that give chaotic solutions are $a_1 = 0.3$, $a_2 = -1$, $a_3 = 0.1$.

Thus Eq. (42.1) can be written more compactly as

$$\dot{x} = y$$
$$\dot{y} = -x + yz \qquad\qquad (42.2)$$
$$\dot{z} = ax - xy + bxz,$$

where $a = 0.3$, $b = 0.1$.

Note that with six terms, Eq. (42.2) should have two independent parameters through a linear rescaling of the three variables plus time, and so the dynamics is completely captured by the given parameters, which could be put in any of the six terms, albeit with different numerical values.

42.3   Equilibria

The system in Eq. (42.2) with $a = 0.3$, $b = 0.1$ has many equilibrium points, six of which are as follows:

Equilibrium # 1 is an unstable node at $(0, 0, 3.3421)$ with eigenvalues $(0.3322, 3.0099, 0)$ and a Poincaré index of 1 in the $z = 3.3421$ plane.

Equilibrium # 2 is a neutrally stable non-hyperbolic equilibrium at $(0, 0, -5.6811)$ with eigenvalues $(-0.1818, -5.4992, 0)$ and a Poincaré index of 1 in the $z = -5.6811$ plane.

Equilibrium # 3 is a neutrally stable non-hyperbolic equilibrium at $(0, 0, -46.8279)$ with eigenvalues $(-0.0214, -46.8065, 0)$ and a Poincaré index of 1 in the $z = -46.8279$ plane.

Equilibrium # 4 is a neutrally stable non-hyperbolic equilibrium at $(0, 0, -1.8812)$ with eigenvalues $(-0.9406 - 0.3395i, -0.9406 + 0.3395i, 0)$ and a Poincaré index of 1 in the $z = -1.8812$ plane.

Equilibrium # 5 is a neutrally stable non-hyperbolic equilibrium at $(0, 0, -0.4228)$ with eigenvalues $(-0.2114 - 0.9774i, -0.2114 + 0.9774i, 0)$ and a Poincaré index of 1 in the $z = -0.4228$ plane.

Equilibrium # 6 is a neutrally stable non-hyperbolic equilibrium at $(0, 0, -14.0211)$ with eigenvalues $(-0.0717, -13.9494, 0)$ and a Poincaré index of 1 in the $z = -14.0211$ plane.

The strange attractor is self-excited, and the system has no symmetry.

42.4   Attractor

Figure 42.1 shows various views of the attractor for Eq. (42.2) with $a = 0.3$, $b = 0.1$ and initial conditions $(2, 3, -1)$.

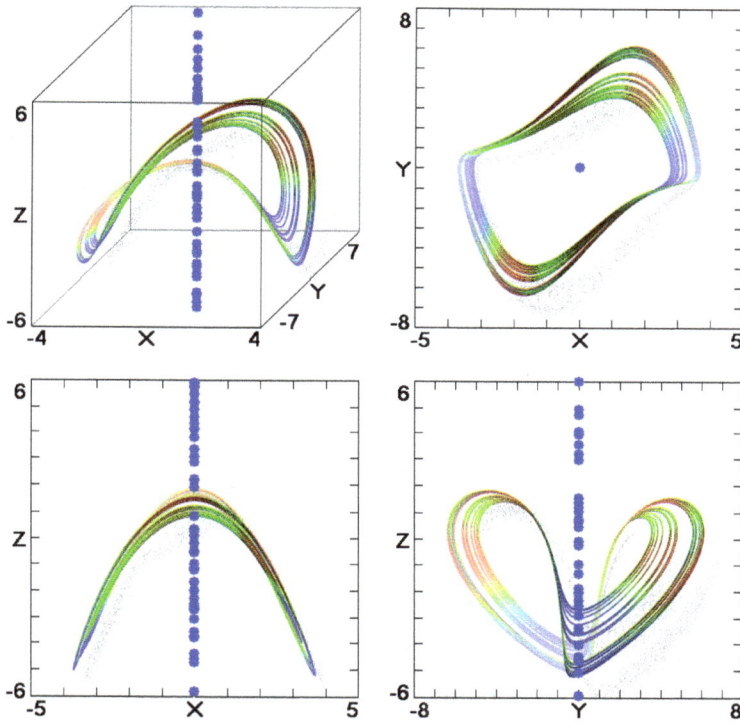

Fig. 42.1   Views of the attractor for Eq. (42.2) with $a = 0.3$, $b = 0.1$ and initial conditions $(2, 3, -1)$.

42.5   Time Series

Figure 42.2 shows the time series for the three variables along with the local value of the largest Lyapunov exponent (LL) for Eq. (42.2) with $a = 0.3$, $b = 0.1$. Red color in the Lyapunov exponent indicates that the error is growing parallel to the orbit, while blue indicates growth perpendicular to the orbit. Note that the orbit passes through regions where the local Lyapunov exponent is strongly positive and other regions where it is strongly negative as is typical for a chaotic system and is also reflected by the colors in Fig. 42.1.

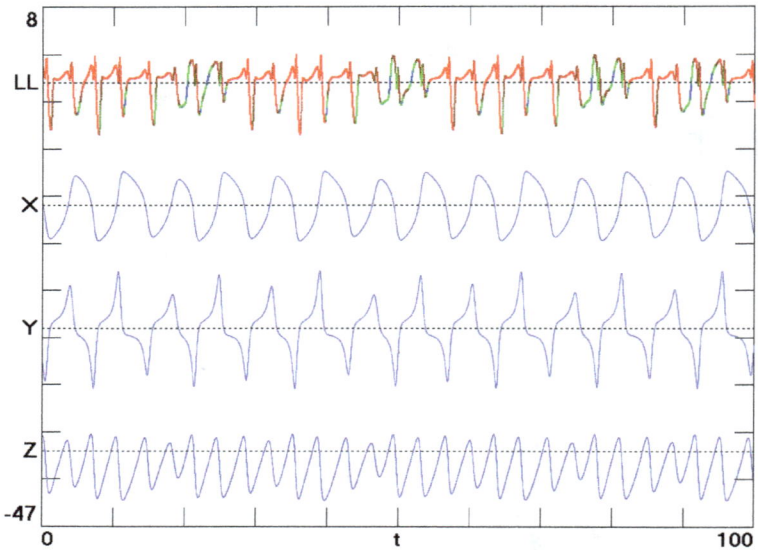

Fig. 42.2 Time series for the variables in Eq. (42.2) with $a = 0.3$, $b = 0.1$ along with the local Lyapunov exponent (LL).

42.6 Lyapunov Exponents

The global Lyapunov exponents are determined by averaging the local Lyapunov exponents along the orbit. The values typically converge very slowly because of the large variation along the orbit, and an integration time of order 10^8 is required to obtain 4-digit accuracy.

The results of such a calculation for the system in Eq. (42.2) with $a = 0.3$, $b = 0.1$ after a time of 3×10^6 are LE $= (0.0272, 0, -1.6454)$ with a Kaplan–Yorke dimension of 2.0165, where the last digit in the quoted values is only an approximation. The positive value of the largest Lyapunov exponent indicates that the system is chaotic, and the negative sum of the exponents (-1.6181) indicates that the system is dissipative with a strange attractor.

42.7 Basin of Attraction

Figure 42.3 shows (in red) the basin of attraction for Eq. (42.2) with $a = 0.3$, $b = 0.1$ in the $z = 0$ plane. Also shown (in black) is the cross-section of the attractor in the same plane.

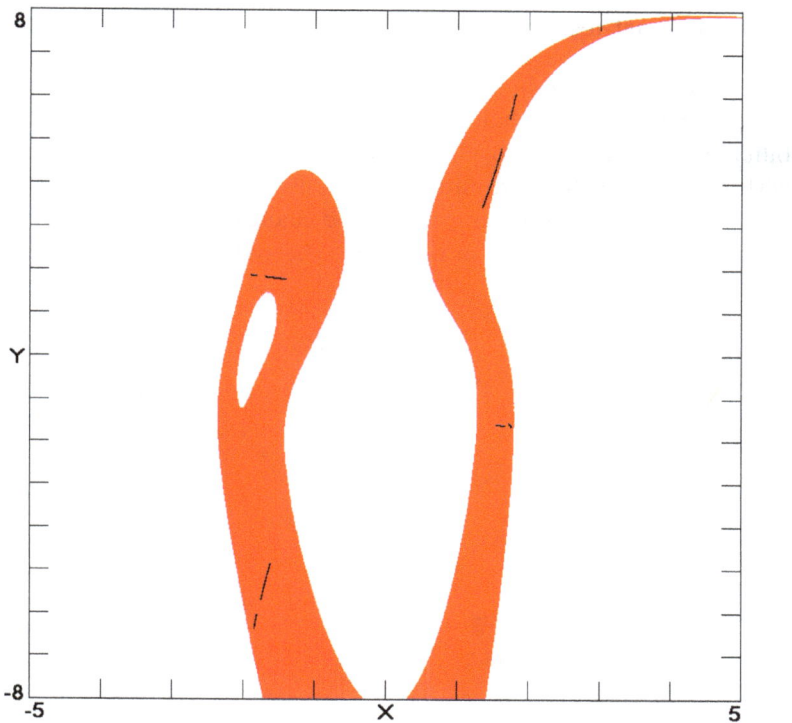

Fig. 42.3 Basin of attraction for Eq. (42.2) with $a = 0.3$, $b = 0.1$ in the $z = 0$ plane.

42.8 Bifurcations

Figure 42.4 shows the bifurcation diagram for Eq. (42.2) as a function of the parameter a from 0 to 0.6 for $b = 0.1$. The initial condition was taken as $(2, 3, -1)$ for each value of a. Each of the 500 values of a was calculated for a time of about 1×10^4.

The upper plot shows the three Lyapunov exponents. The middle plot shows the Kaplan–Yorke dimension, and the lower plot shows the local maxima of x. The chaotic region is in the vicinity of $a = 0.3$, $b = 0.1$, and the route to chaos is clearly shown.

42.9 Robustness

One measure of the robustness of a chaotic system is the amount by which the parameters can be changed from their nominal values before the probability of chaos decreases to 50% [Sprott (2022)]. For the system in Eq. (42.2) with $a = 0.3$, $b = 0.1$ and initial conditions $(2, 3, -1)$, after 5820 trials, it is estimated that the parameters can be changed by 12% before the chaos is more likely to be lost than not. Thus the system is somewhat fragile.

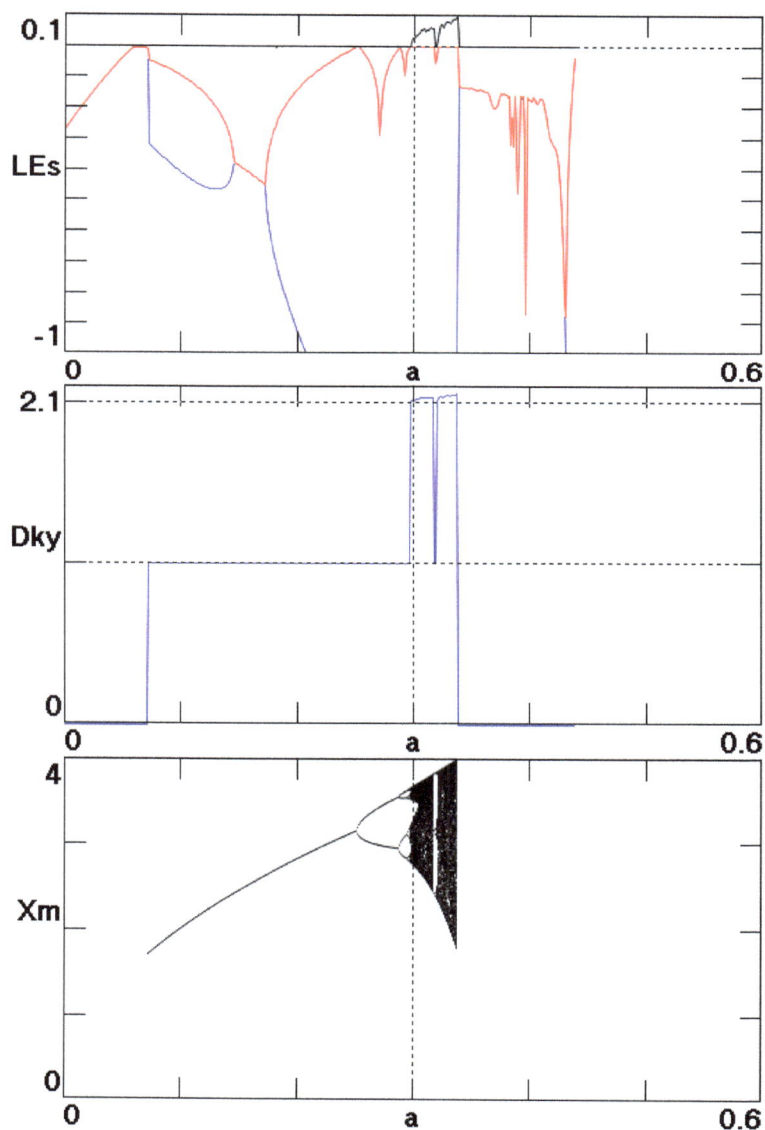

Fig. 42.4 Bifurcation diagram for Eq. (42.2) as a function of the parameter a for $b = 0.1$.

Chapter 43

Mostly Quadratic System

Most of the previous examples have more linear than nonlinear terms. This system has all quadratic terms except for a single constant term. It is apparently the simplest such system. It has inversion symmetry, time-reversal invariance, and an attractor–repellor pair.

43.1　Introduction

The mostly quadratic (Case B) system [Li and Sprott (2014)] written in its most general form with an adjustable coefficient in each of the five terms is given by

$$
\begin{aligned}
\dot{x} &= a_1 + a_2 yz \\
\dot{y} &= a_3 xz \\
\dot{z} &= a_4 y^2 + a_5 yz.
\end{aligned}
\tag{43.1}
$$

The usual parameters are $a_1 = a_2 = a_4 = 1$, $a_3 = -1$, $a_5 = 2$. No chaotic solutions are possible with only quadratic terms. The single constant term is an amplitude parameter that does not affect the dynamics but only the magnitude of the variables, which is why it cannot be set to zero since the attractor would collapse to a point.

　　The following sections were written by the computer program that performed the optimization, carried out the analysis of the resulting system, and produced the corresponding figures, all without human intervention.

43.2　Simplified System

After about 7×10^3 trials, of which 133 were chaotic, simplified parameters for Eq. (43.1) that give chaotic solutions are $a_1 = 1$, $a_2 = 1$, $a_3 = -1$,

$a_4 = 1$, $a_5 = 2$. Thus Eq. (43.1) can be written more compactly as

$$\dot{x} = 1 + yz$$
$$\dot{y} = -xz \tag{43.2}$$
$$\dot{z} = y^2 + ayz,$$

where $a = 2$.

Note that with five terms, Eq. (43.2) should have one independent parameter through a linear rescaling of the three variables plus time, and so the dynamics is completely captured by the single parameter a, which could be put in any of the five terms, albeit with a different numerical value.

43.3 Equilibria

The system in Eq. (43.2) with $a = 2$ has two equilibrium points:

Equilibrium # 1 is an unstable saddle focus at $(0, 1.4142, -0.7071)$ with eigenvalues $(-0.0754 - 0.9714i, -0.0754 + 0.9714i, 2.9793)$ and a Poincaré index of 1 in the $z = -0.7071$ plane.

Equilibrium # 2 is an unstable saddle focus at $(0, -1.4142, 0.7071)$ with eigenvalues $(0.0754 - 0.9714i, 0.0754 + 0.9714i, -2.9793)$ and a Poincaré index of 1 in the $z = 0.7071$ plane.

The strange attractor is self-excited, and the system is symmetric under the transformation $x \to -x$, $y \to -y$, $z \to -z$, $t \to -t$.

43.4 Attractor

Figure 43.1 shows various views of the attractor for Eq. (43.2) with $a = 2$ and initial conditions $(1, -1, 4)$.

43.5 Time Series

Figure 43.2 shows the time series for the three variables along with the local value of the largest Lyapunov exponent (LL) for Eq. (43.2) with $a = 2$. Red color in the Lyapunov exponent indicates that the error is growing parallel to the orbit, while blue indicates growth perpendicular to the orbit. Note that the orbit passes through regions where the local Lyapunov exponent is strongly positive and other regions where it is strongly negative as is typical for a chaotic system and is also reflected by the colors in Fig. 43.1.

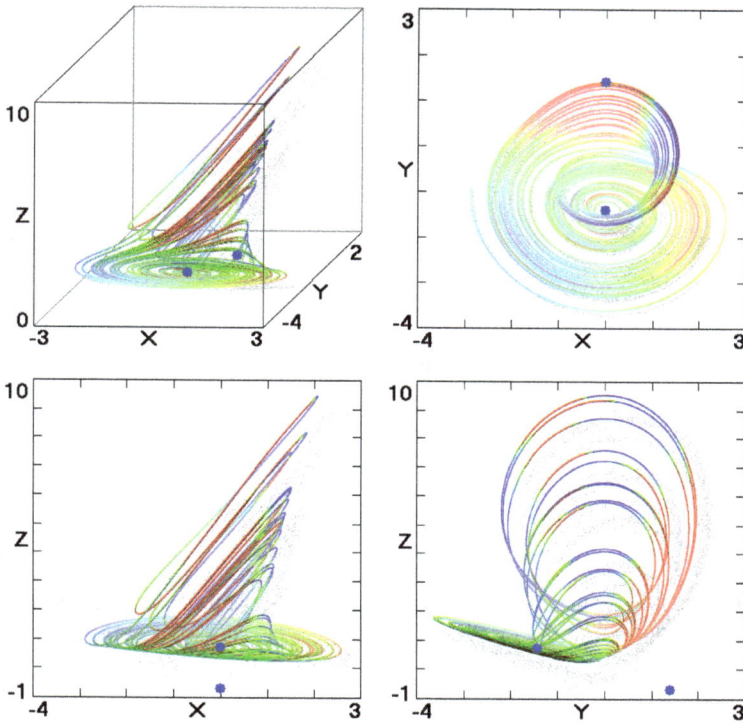

Fig. 43.1   Views of the attractor for Eq. (43.2) with $a = 2$ and initial conditions $(1, -1, 4)$.

43.6   Lyapunov Exponents

The global Lyapunov exponents are determined by averaging the local Lyapunov exponents along the orbit. The values typically converge very slowly because of the large variation along the orbit, and an integration time of order 10^8 is required to obtain 4-digit accuracy.

The results of such a calculation for the system in Eq. (43.2) with $a = 2$ after a time of 3×10^6 are LE $= (0.1520, 0, -2.3866)$ with a Kaplan–Yorke dimension of 2.0636, where the last digit in the quoted values is only an approximation. The positive value of the largest Lyapunov exponent indicates that the system is chaotic, and the negative sum of the exponents (-2.2346) indicates that the system is dissipative with a strange attractor.

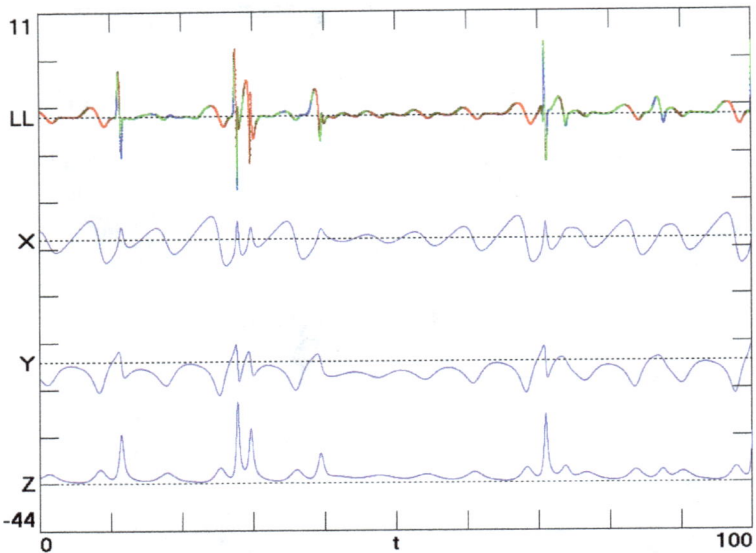

Fig. 43.2   Time series for the variables in Eq. (43.2) with $a = 2$ along with the local Lyapunov exponent (LL).

43.7   Basin of Attraction

Figure 43.3 shows (in red) the basin of attraction for Eq. (43.2) with $a = 2$ in the $z = 0.7071$ plane. Also shown (in black) is the cross-section of the attractor in the same plane.

43.8   Bifurcations

Figure 43.4 shows the bifurcation diagram for Eq. (43.2) as a function of the parameter a from 0 to 4. The initial condition was taken as $(1, -1, 4)$ at $a = 4$ and was not changed as a slowly varied toward $a = 0$. Each of the 500 values of a was calculated for a time of about 1×10^4.

The upper plot shows the three Lyapunov exponents. The middle plot shows the Kaplan–Yorke dimension, and the lower plot shows the local maxima of x. The chaotic region is in the vicinity of $a = 2$, and the route to chaos is clearly shown.

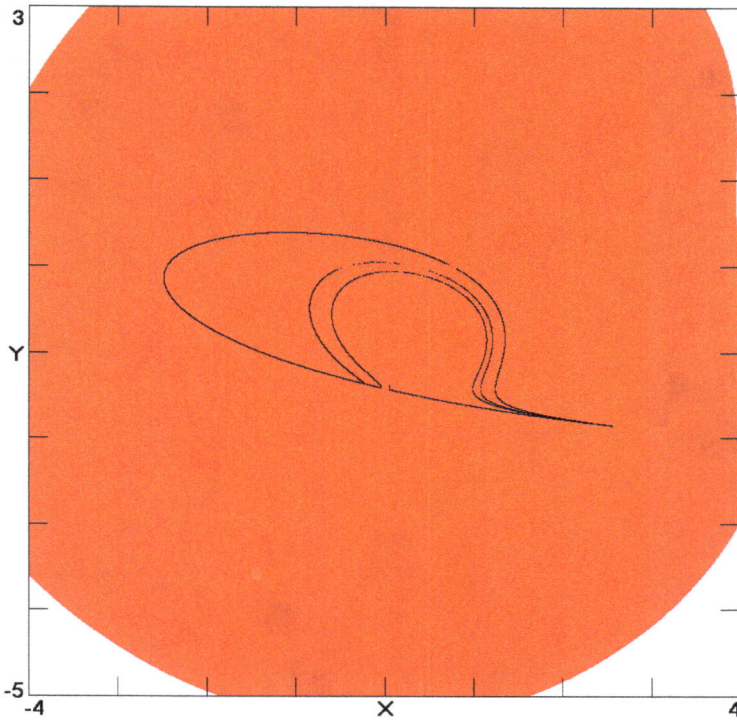

Fig. 43.3   Basin of attraction for Eq. (43.2) with $a = 2$ in the $z = 0.7071$ plane.

43.9   Robustness

One measure of the robustness of a chaotic system is the amount by which the parameters can be changed from their nominal values before the probability of chaos decreases to 50% [Sprott (2022)]. For the system in Eq. (43.2) with $a = 2$ and initial conditions $(1, -1, 4)$, after 1198 trials, it is estimated that the parameter a can be changed by 88% before the chaos is more likely to be lost than not. Thus the system is highly robust. This result is consistent with the data in Fig. 43.4.

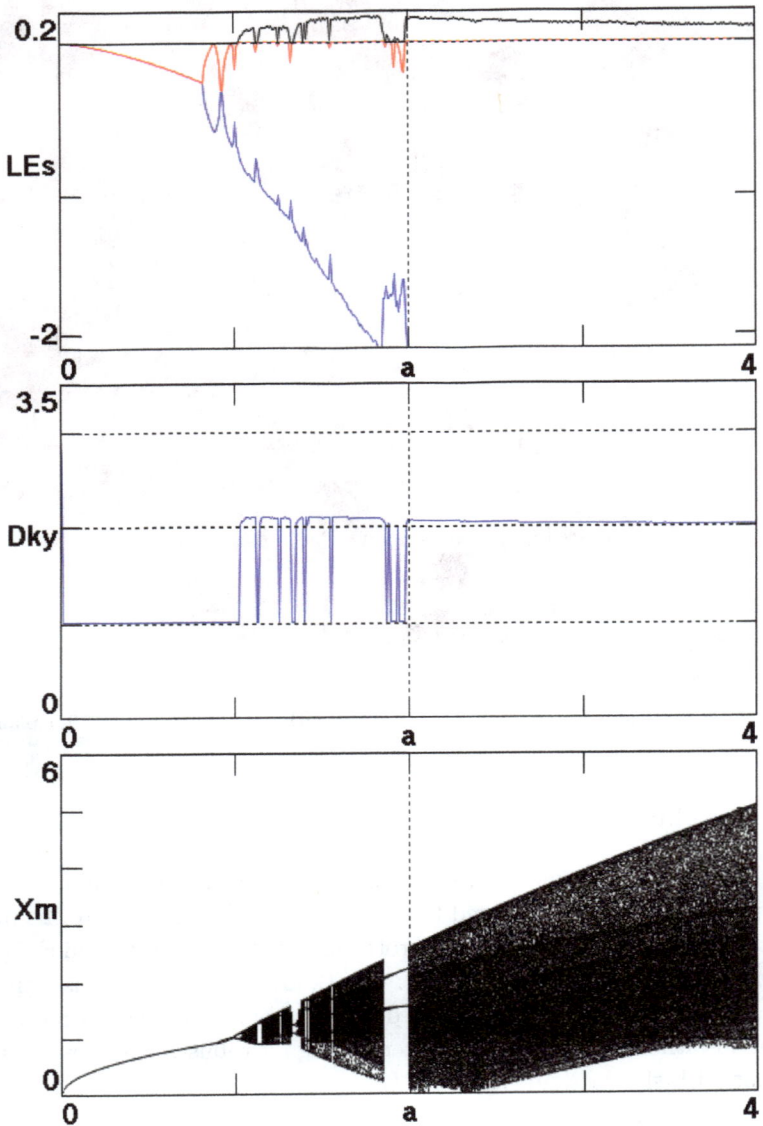

Fig. 43.4 Bifurcation diagram for Eq. (43.2) as a function of the parameter a.

Chapter 44

Dissipative–Conservative System

This is an unusual system with a dissipative region for some initial conditions and a conservative region for other initial conditions for the same value of the parameters. There are no equilibrium points, and thus the strange attractor is hidden. The system is time-reversal invariant with a symmetric attractor–repellor pair.

44.1 Introduction

The dissipative–conservative system [Sprott (2014b)] written in its most general form with an adjustable coefficient in each of the nine terms is given by

$$
\begin{aligned}
\dot{x} &= a_1 y + a_2 xy + a_3 xz \\
\dot{y} &= a_4 - a_5 x^2 + a_6 yz \\
\dot{z} &= a_7 x - a_8 x^2 - a_9 y^2.
\end{aligned}
\tag{44.1}
$$

The usual parameters are $a_1 = a_3 = a_4 = a_6 = a_7 = a_8 = a_9 = 1$, $a_2 = a_5 = 2$. The number of terms and their signs are preserved.

The following sections were written by the computer program that performed the optimization, carried out the analysis of the resulting system, and produced the corresponding figures, all without human intervention.

44.2 Simplified System

After about 2×10^4 trials, of which 30 were chaotic, simplified parameters for Eq. (44.1) that give chaotic solutions are $a_1 = 1$, $a_2 = 1$, $a_3 = 1$, $a_4 = 1$, $a_5 = 1$, $a_6 = 1$, $a_7 = 1$, $a_8 = 0.61$, $a_9 = 1$. Thus Eq. (44.1) can be written

277

more compactly as

$$\dot{x} = y + xy + xz$$
$$\dot{y} = 1 - x^2 + yz$$
$$\dot{z} = x - ax^2 - y^2,$$

$$(44.2)$$

where $a = 0.61$.

Note that with nine terms, Eq. (44.2) should have five independent parameters through a linear rescaling of the three variables plus time. However, four of the five parameters have values of ± 1 and can be placed in any of the remaining eight terms.

44.3 Equilibria

The system in Eq. (44.2) with $a = 0.61$ has no equilibrium points. The strange attractor is apparently hidden, and the system is symmetric under the transformation $y \to -y$, $z \to -z$, $t \to -t$.

44.4 Attractor

Figure 44.1 shows various views of the attractor for Eq. (44.2) with $a = 0.61$ and initial conditions $(0, -1, 0)$.

44.5 Time Series

Figure 44.2 shows the time series for the three variables along with the local value of the largest Lyapunov exponent (LL) for Eq. (44.2) with $a = 0.61$. Red color in the Lyapunov exponent indicates that the error is growing parallel to the orbit, while blue indicates growth perpendicular to the orbit. Note that the orbit passes through regions where the local Lyapunov exponent is strongly positive and other regions where it is strongly negative as is typical for a chaotic system and is also reflected by the colors in Fig. 44.1.

44.6 Lyapunov Exponents

The global Lyapunov exponents are determined by averaging the local Lyapunov exponents along the orbit. The values typically converge very slowly because of the large variation along the orbit, and an integration time of order 10^8 is required to obtain 4-digit accuracy.

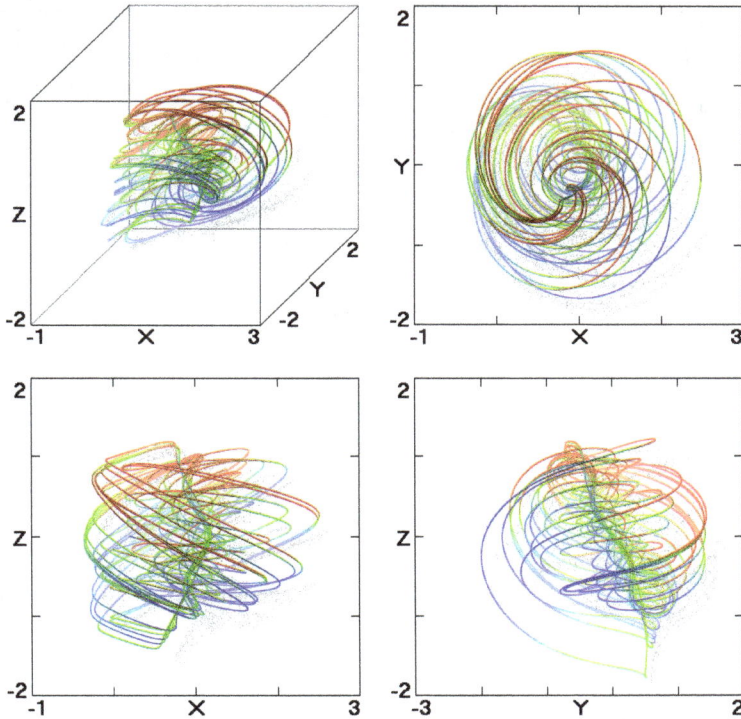

Fig. 44.1   Views of the attractor for Eq. (44.2) with $a = 0.61$ and initial conditions $(0, -1, 0)$.

The results of such a calculation for the system in Eq. (44.2) with $a = 0.61$ after a time of 4×10^6 are LE = $(0.0689, 0, -0.2306)$ with a Kaplan–Yorke dimension of 2.2987, where the last digit in the quoted values is only an approximation. The positive value of the largest Lyapunov exponent indicates that the system is chaotic, and the negative sum of the exponents (-0.1616) indicates that the system is dissipative with a strange attractor.

44.7   Basin of Attraction

Figure 44.3 shows (in red) the basin of attraction for Eq. (44.2) with $a = 0.61$ in the $z = 0$ plane. Also shown (in black) is the cross-section of the attractor in the same plane.

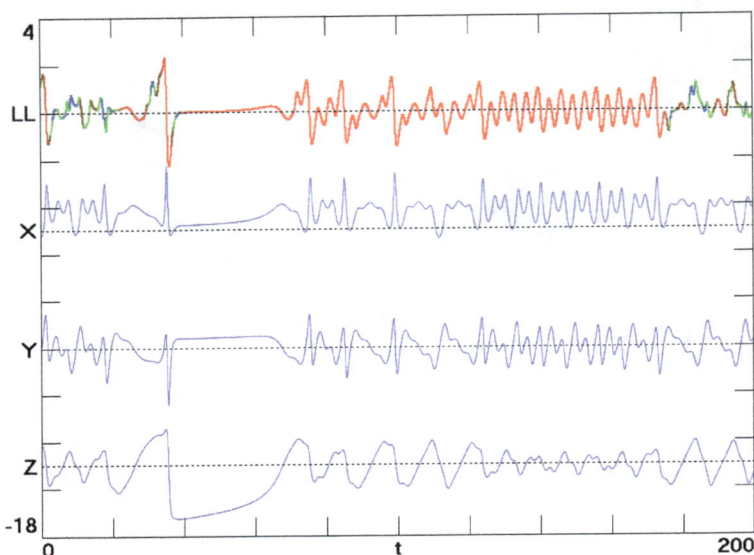

Fig. 44.2   Time series for the variables in Eq. (44.2) with $a = 0.61$ along with the local Lyapunov exponent (LL).

44.8   Bifurcations

Figure 44.4 shows the bifurcation diagram for Eq. (44.2) as a function of the parameter a from 0 to 1.22. The initial condition was taken as $(0, -1, 0)$ for each value of a. Each of the 500 values of a was calculated for a time of about 2×10^4.

The upper plot shows the three Lyapunov exponents. The middle plot shows the Kaplan–Yorke dimension, and the lower plot shows the local maxima of x. The chaotic region is in the vicinity of $a = 0.61$, and the route to chaos is clearly shown.

44.9   Robustness

One measure of the robustness of a chaotic system is the amount by which the parameters can be changed from their nominal values before the probability of chaos decreases to 50% [Sprott (2022)]. For the system in Eq. (44.2) with $a = 0.61$ and initial conditions $(0, -1, 0)$, after 2691 trials, it is estimated that the parameter a can be changed by 9.4% before the chaos is

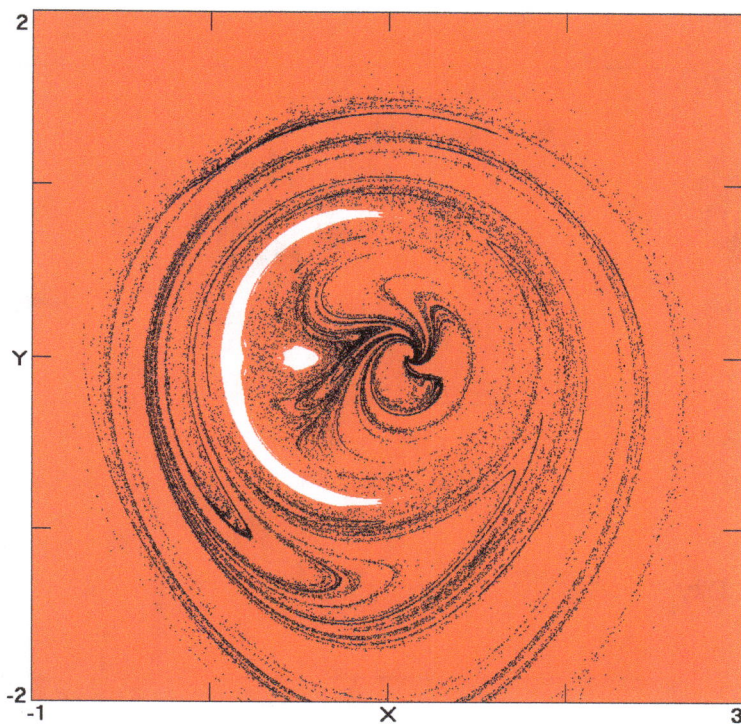

Fig. 44.3   Basin of attraction for Eq. (44.2) with $a = 0.61$ in the $z = 0$ plane.

more likely to be lost than not. Thus the system is somewhat fragile. This result is consistent with the data in Fig. 44.4.

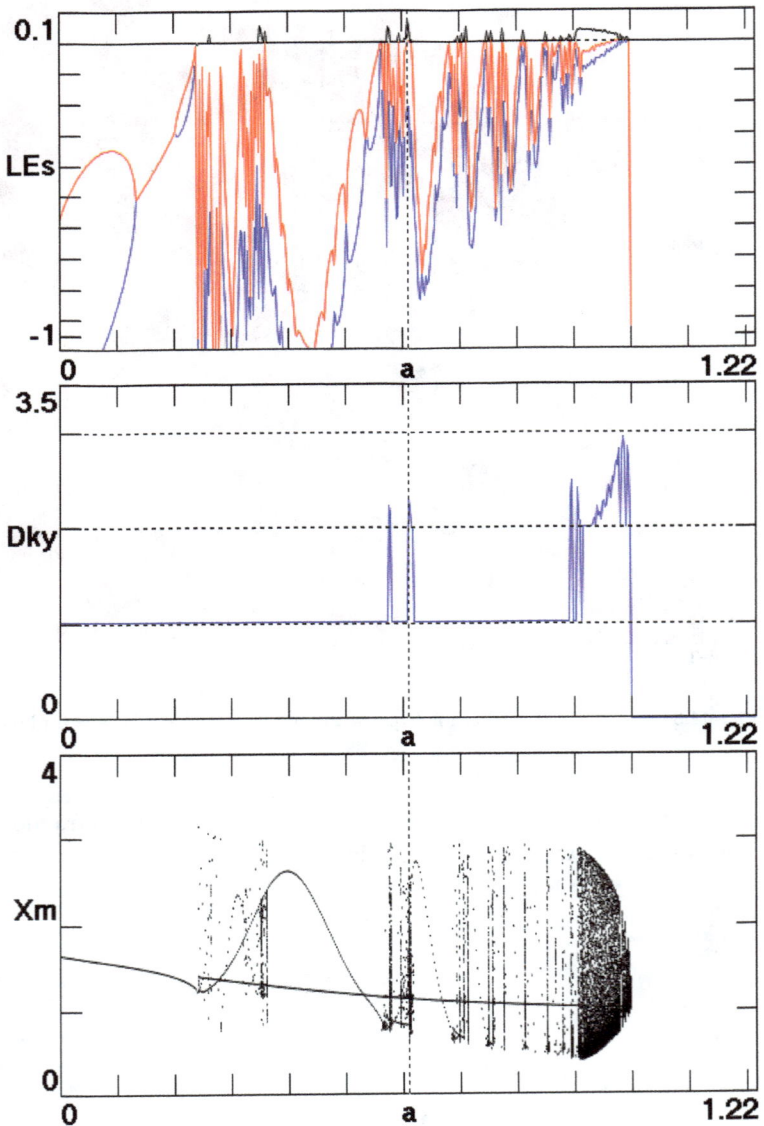

Fig. 44.4   Bifurcation diagram for Eq. (44.2) as a function of the parameter *a*.

Chapter 45

Time-Reversible Reflection-Invariant System

Previous examples have included dissipative systems that are time-reversible with inversion and rotational symmetry. The case here completes the picture with a time-reversible system that is reflection invariant with a symmetric attractor–repellor pair. It is also unusual in that it has three lines of equilibria at $(0, 0, z)$ and $(\pm 1, y, 0)$ and conservative periodic orbits coexisting with the strange attractor.

45.1 Introduction

The time-reversible reflection-invariant system [Sprott (2015)] written in its most general form with an adjustable coefficient in each of the six terms is given by

$$
\begin{aligned}
\dot{x} &= -a_1 yz \\
\dot{y} &= a_2 xz + a_3 yz + a_4 z^3 \\
\dot{z} &= a_5 x - a_6 x^3.
\end{aligned}
\tag{45.1}
$$

The usual parameters are $a_1 = a_3 = a_4 = a_5 = a_6 = 1$, $a_2 = 2$. It is written here with the signs constrained and positive parameters to facilitate the search.

The following sections were written by the computer program that performed the optimization, carried out the analysis of the resulting system, and produced the corresponding figures, all without human intervention.

45.2 Simplified System

After about 1×10^4 trials, of which eleven were chaotic, simplified parameters for Eq. (45.1) that give chaotic solutions are $a_1 = 1$, $a_2 = 2$, $a_3 = 1$,

$a_4 = 1$, $a_5 = 1$, $a_6 = 1$. Thus Eq. (45.1) can be written more compactly as

$$\dot{x} = -yz$$
$$\dot{y} = axz + yz + z^3 \qquad (45.2)$$
$$\dot{z} = x - x^3,$$

where $a = 2$.

Note that with six terms, Eq. (45.2) should have two independent parameters through a linear rescaling of the three variables plus time. However, one of the two parameters has a value of ± 1 and can be placed in any of the remaining five terms.

45.3 Equilibria

The system in Eq. (45.2) with $a = 2$ has many equilibrium points, six of which are as follows:

Equilibrium # 1 is an unstable saddle focus at $(-1, 0, -1.4142)$ with eigenvalues $(0.338 - 2.3019i, 0.338 + 2.3019i, -2.0902)$ and a Poincaré index of 1 in the $z = -1.4142$ plane.

Equilibrium # 2 is an unstable saddle focus at $(-1, 0, 1.4142)$ with eigenvalues $(-0.338 - 2.3019i, -0.338 + 2.3019i, 2.0902)$ and a Poincaré index of 1 in the $z = 1.4142$ plane.

Equilibrium # 3 is a neutrally stable non-hyperbolic equilibrium at $(1, 0.0097, 0)$ with eigenvalues $(0, 0, 0)$ and a Poincaré index of 0 in the $z = 0$ plane. Despite the neutral linear stability, the equilibrium is nonlinearly unstable.

Equilibrium # 4 is a neutrally stable non-hyperbolic equilibrium at $(1, 1.7323, 0)$ with eigenvalues $(0, 0, 0)$ and a Poincaré index of 0 in the $z = 0$ plane. Despite the neutral linear stability, the equilibrium is nonlinearly unstable.

Equilibrium # 5 is a neutrally stable non-hyperbolic equilibrium at $(0, 0, -0)$ with eigenvalues $(0, 0, 0)$ and a Poincaré index of 1 in the $z = 0$ plane. Despite the neutral linear stability, the equilibrium is nonlinearly unstable.

Equilibrium # 6 is a neutrally stable non-hyperbolic equilibrium at $(1, 0.0341, 0)$ with eigenvalues $(0, 0, 0)$ and a Poincaré index of 0 in the $z = 0$ plane. Despite the neutral linear stability, the equilibrium is nonlinearly unstable.

The strange attractor is self-excited, and the system is symmetric under the transformation $z \to -z$, $t \to -t$.

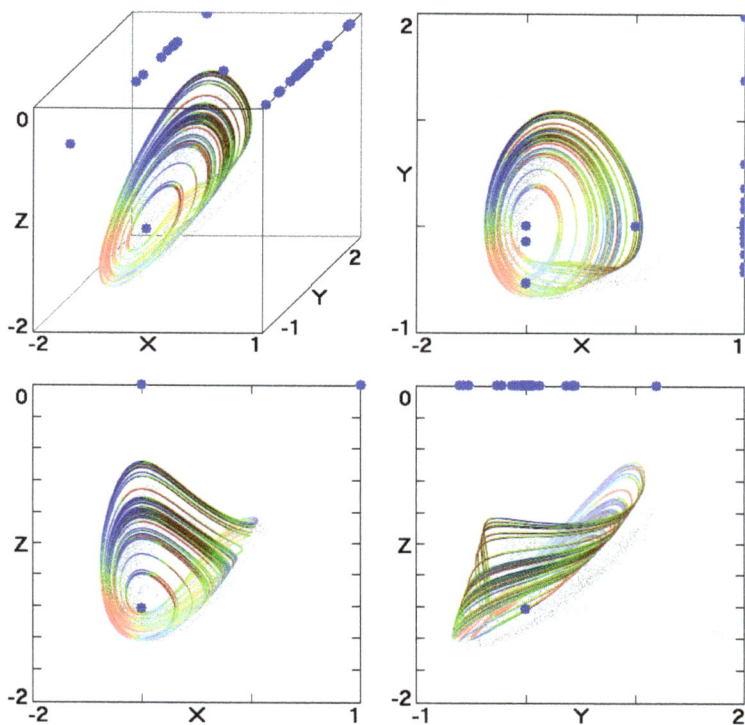

Fig. 45.1 Views of the attractor for Eq. (45.2) with $a = 2$ and initial conditions $(-1, 1, -1)$.

45.4 Attractor

Figure 45.1 shows various views of the attractor for Eq. (45.2) with $a = 2$ and initial conditions $(-1, 1, -1)$.

45.5 Time Series

Figure 45.2 shows the time series for the three variables along with the local value of the largest Lyapunov exponent (LL) for Eq. (45.2) with $a = 2$. Red color in the Lyapunov exponent indicates that the error is growing parallel to the orbit, while blue indicates growth perpendicular to the orbit. Note that the orbit passes through regions where the local Lyapunov exponent is strongly positive and other regions where it is strongly negative as is typical for a chaotic system and is also reflected by the colors in Fig. 45.1.

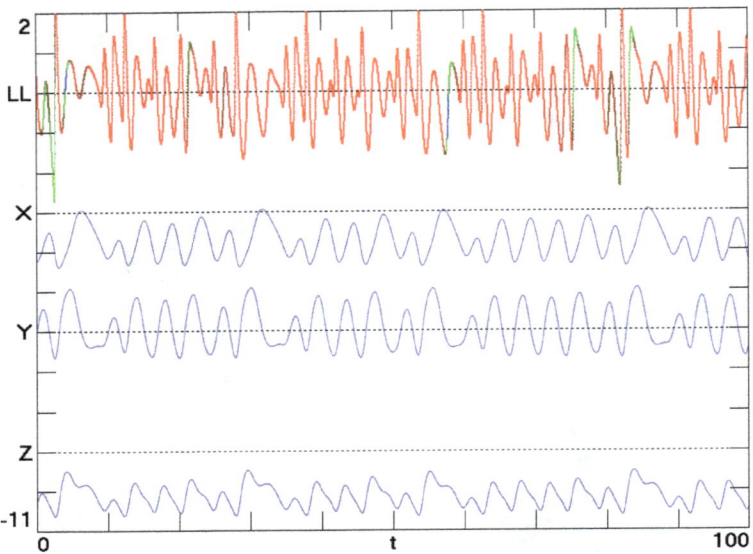

Fig. 45.2   Time series for the variables in Eq. (45.2) with $a = 2$ along with the local
Lyapunov exponent (LL).

45.6   Lyapunov Exponents

The global Lyapunov exponents are determined by averaging the local Lyapunov exponents along the orbit. The values typically converge very slowly
because of the large variation along the orbit, and an integration time of
order 10^8 is required to obtain 4-digit accuracy.

The results of such a calculation for the system in Eq. (45.2) with $a = 2$
after a time of 3×10^6 are LE $= (0.0893,\ 0,\ -1.2271)$ with a Kaplan–
Yorke dimension of 2.0727, where the last digit in the quoted values is only
an approximation. The positive value of the largest Lyapunov exponent
indicates that the system is chaotic, and the negative sum of the exponents
(-1.1378) indicates that the system is dissipative with a strange attractor.

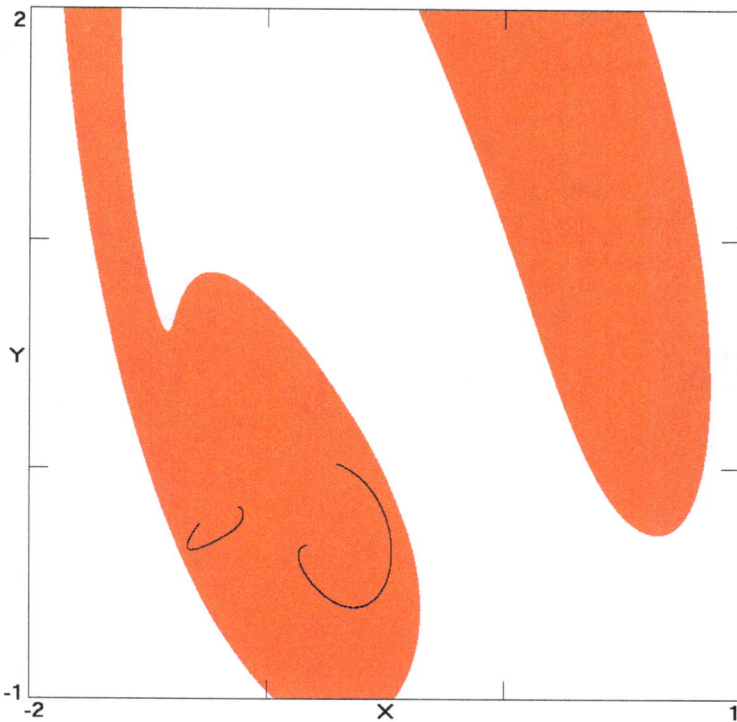

Fig. 45.3 Basin of attraction for Eq. (45.2) with $a = 2$ in the $z = -1.4142$ plane.

45.7 Basin of Attraction

Figure 45.3 shows (in red) the basin of attraction for Eq. (45.2) with $a = 2$ in the $z = -1.4142$ plane. Also shown (in black) is the cross-section of the attractor in the same plane.

45.8 Bifurcations

Figure 45.4 shows the bifurcation diagram for Eq. (45.2) as a function of the parameter a from 0 to 4. The initial condition was taken as $(-1, 1, -1)$ at $a = 4$ and was not changed as a slowly varied toward $a = 0$. Each of the 500 values of a was calculated for a time of about 1×10^4.

The upper plot shows the three Lyapunov exponents. The middle plot shows the Kaplan–Yorke dimension, and the lower plot shows the local maxima of x. The chaotic region is in the vicinity of $a = 2$, and the route to chaos is clearly shown.

45.9   Robustness

One measure of the robustness of a chaotic system is the amount by which the parameters can be changed from their nominal values before the probability of chaos decreases to 50% [Sprott (2022)]. For the system in Eq. (45.2) with $a = 2$ and initial conditions $(-1, 1, -1)$, after 4972 trials, it is estimated that the parameter a can be changed by 8.2% before the chaos is more likely to be lost than not. Thus the system is somewhat fragile. This result is consistent with the data in Fig. 45.4.

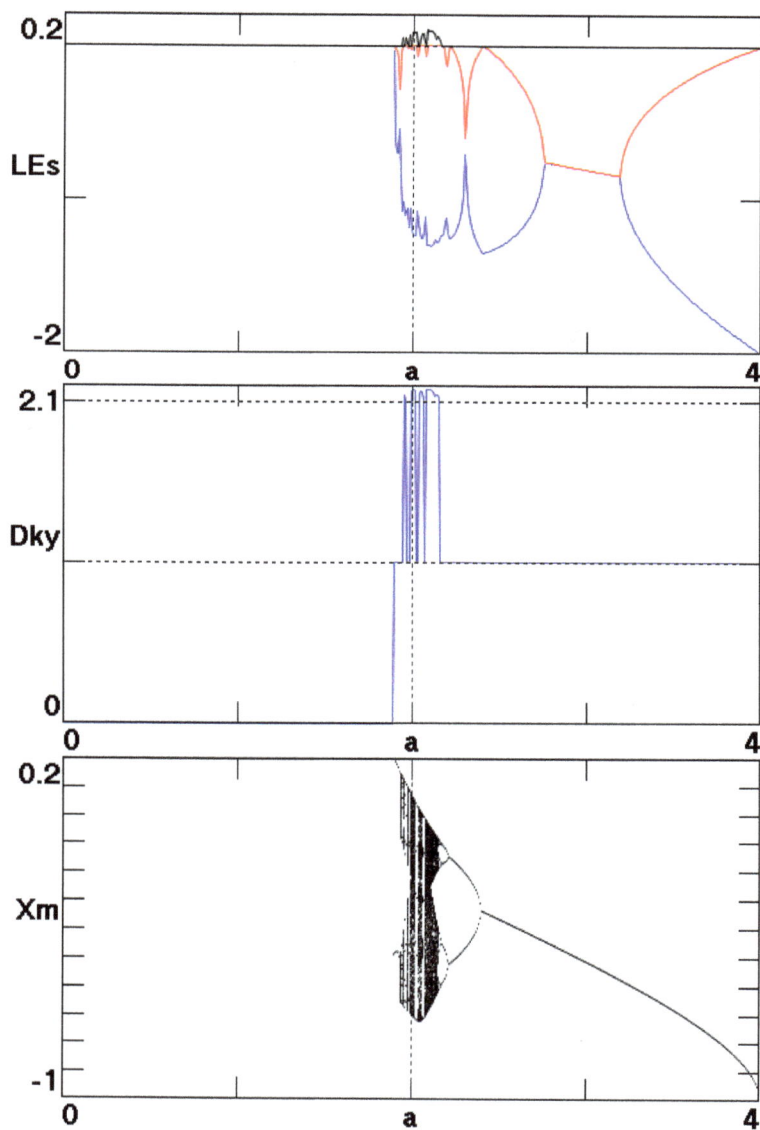

Fig. 45.4   Bifurcation diagram for Eq. (45.2) as a function of the parameter *a*.

Chapter 46

Plane Equilibrium System

Chaotic systems can be constructed with surfaces of equilibria, one simple example of which is shown here. If the surface is closed or extends to infinity in a three-dimensional state space, it divides the space into two regions that the orbit cannot cross since the flow vector is zero on the surface. More complicated or more numerous surfaces can divide the space into even more segments, some of which may contain coexisting attractors. This system is invariant under a 180° rotation about the z axis with a symmetric pair of strange attractors.

46.1   Introduction

A simple system (ES2) with a plane of equilibria [Jafari *et al.* (2016)] written in its most general form with an adjustable coefficient in each of the five terms is given by

$$\dot{x} = a_1 xz$$
$$\dot{y} = a_2 xz^2 + a_3 x^2 y \qquad (46.1)$$
$$\dot{z} = a_4 x^2 + a_5 xy.$$

The usual parameters are $a_1 = a_5 = 1$, $a_2 = 3$, $a_3 = a_4 = -1$. The y and z variables have been exchanged for consistency with previous cases. Since each term contains x, the entire $x = 0$ plane is an equilibrium, and this modification can be applied to any chaotic system often without destroying the chaos.

The following sections were written by the computer program that performed the optimization, carried out the analysis of the resulting system, and produced the corresponding figures, all without human intervention.

46.2 Simplified System

After about 2×10^4 trials, of which 86 were chaotic, simplified parameters for Eq. (46.1) that give chaotic solutions are $a_1 = 1$, $a_2 = 3$, $a_3 = -1$, $a_4 = -1$, $a_5 = 1$. Thus Eq. (46.1) can be written more compactly as

$$\dot{x} = xz$$
$$\dot{y} = axz^2 - x^2y \qquad (46.2)$$
$$\dot{z} = -x^2 + xy,$$

where $a = 3$.

Note that with five terms, Eq. (46.2) should have one independent parameter through a linear rescaling of the three variables plus time, and so the dynamics is completely captured by the single parameter a, which could be put in any of the five terms, albeit with a different numerical value.

46.3 Equilibria

The system in Eq. (46.2) with $a = 3$ has many equilibrium points, six of which are as follows:

Equilibrium # 1 is an unstable node at $(0, -6.3810, 3.2265)$ with eigenvalues $(3.2265, 0, 0)$ and a Poincaré index of 0 in the $z = 3.2265$ plane.

Equilibrium # 2 is an unstable node at $(0, 27.1650, 12.8190)$ with eigenvalues $(12.819, 0, 0)$ and a Poincaré index of 0 in the $z = 12.819$ plane.

Equilibrium # 3 is a neutrally stable non-hyperbolic equilibrium at $(0, 6.1437, -3.2398)$ with eigenvalues $(-3.2398, 0, 0)$ and a Poincaré index of 0 in the $z = -3.2398$ plane.

Equilibrium # 4 is a neutrally stable non-hyperbolic equilibrium at $(0, 0, 0)$ with eigenvalues $(0, 0, 0)$ and a Poincaré index of 0 in the $z = 0$ plane.

Equilibrium # 5 is an unstable node at $(0, -27.1394, 8.4077)$ with eigenvalues $(8.4077, 0, 0)$ and a Poincaré index of 0 in the $z = 8.4077$ plane.

Equilibrium # 6 is an unstable node at $(0, 10.0292, 20.4351)$ with eigenvalues $(20.4351, 0, 0)$ and a Poincaré index of 0 in the $z = 20.4351$ plane.

The strange attractor is self-excited, and the system is symmetric under the transformation $x \rightarrow -x$, $y \rightarrow -y$.

46.4 Attractor

Figure 46.1 shows various views of the two attractors for Eq. (46.2) with $a = 3$ and initial conditions $(1, 2, -1)$.

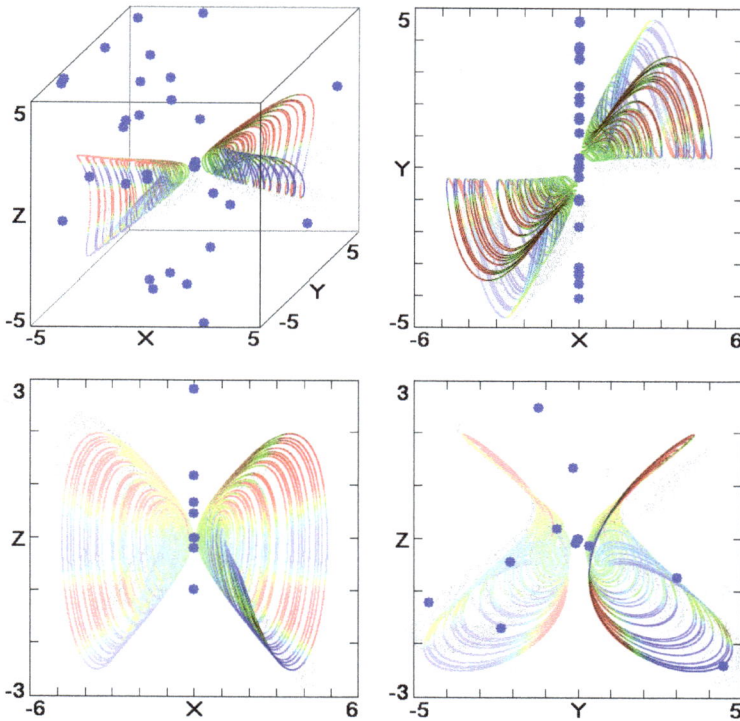

Fig. 46.1   Views of the two attractors for Eq. (46.2) with $a = 3$.

46.5   Time Series

Figure 46.2 shows the time series for the three variables along with the local value of the largest Lyapunov exponent (LL) for Eq. (46.2) with $a = 3$. Red color in the Lyapunov exponent indicates that the error is growing parallel to the orbit, while blue indicates growth perpendicular to the orbit. Note that the orbit passes through regions where the local Lyapunov exponent is strongly positive and other regions where it is strongly negative as is typical for a chaotic system and is also reflected by the colors in Fig. 46.1.

46.6   Lyapunov Exponents

The global Lyapunov exponents are determined by averaging the local Lyapunov exponents along the orbit. The values typically converge very slowly

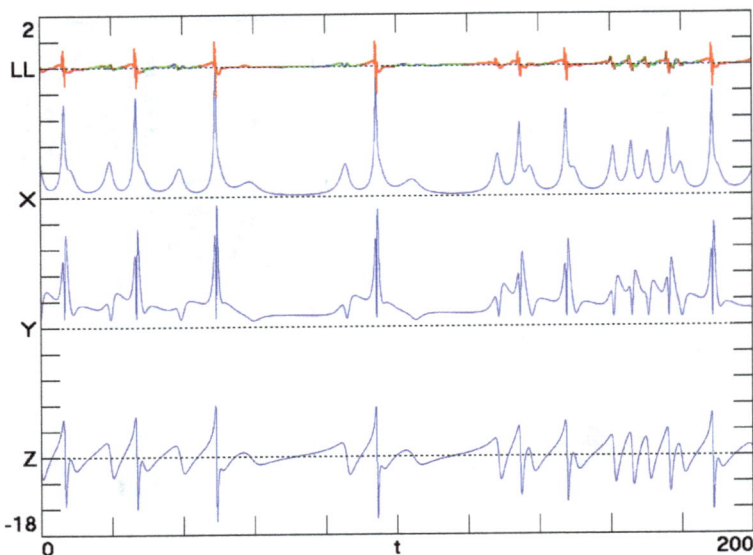

Fig. 46.2   Time series for the variables in Eq. (46.2) with $a = 3$ along with the local Lyapunov exponent (LL).

because of the large variation along the orbit, and an integration time of order 10^8 is required to obtain 4-digit accuracy.

The results of such a calculation for the system in Eq. (46.2) with $a = 3$ after a time of 4×10^6 are LE $= (0.0643, 0, -0.8273)$ with a Kaplan–Yorke dimension of 2.0777, where the last digit in the quoted values is only an approximation. The positive value of the largest Lyapunov exponent indicates that the system is chaotic, and the negative sum of the exponents (-0.7629) indicates that the system is dissipative with a strange attractor.

46.7   Basin of Attraction

Figure 46.3 shows (in red) the basin of attraction for Eq. (46.2) with $a = 3$ in the $z = 0$ plane. Also shown (in black) is the cross-section of the attractor in the same plane.

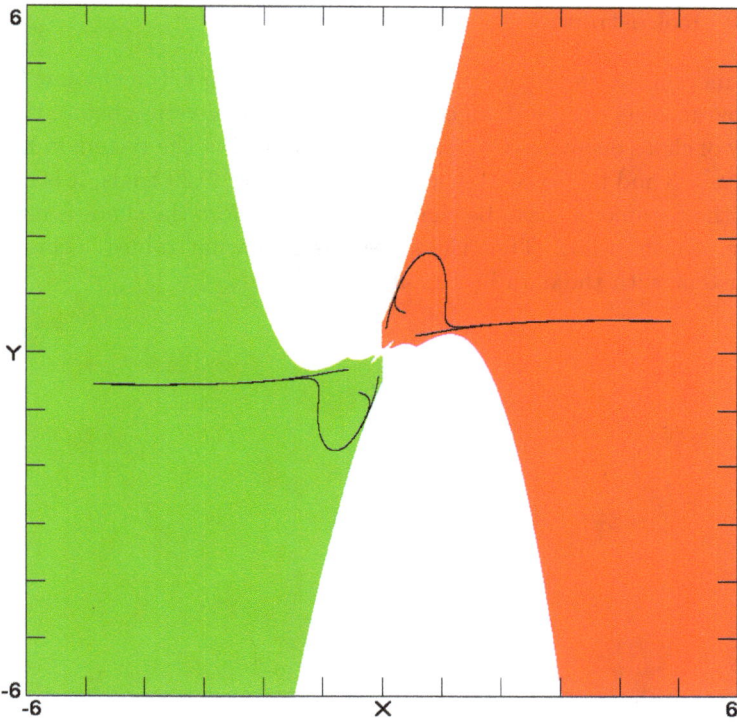

Fig. 46.3   Basin of attraction for Eq. (46.2) with $a = 3$ in the $z = 0$ plane.

46.8   Bifurcations

Figure 46.4 shows the bifurcation diagram for Eq. (46.2) as a function of the parameter a from 0 to 6. The initial condition was taken as $(1, 2, -1)$ at $a = 6$ and was not changed as a slowly varied toward $a = 0$. Each of the 500 values of a was calculated for a time of about 2×10^4.

The upper plot shows the three Lyapunov exponents. The middle plot shows the Kaplan–Yorke dimension, and the lower plot shows the local maxima of x. The chaotic region is in the vicinity of $a = 3$, and the route to chaos is clearly shown.

46.9   Robustness

One measure of the robustness of a chaotic system is the amount by which the parameters can be changed from their nominal values before the probability of chaos decreases to 50% [Sprott (2022)]. For the system in Eq. (46.2) with $a = 3$ and initial conditions $(1,\ 2,\ -1)$, after 2139 trials, it is estimated that the parameter a can be changed by 43% before the chaos is more likely to be lost than not. Thus the system is somewhat robust. This result is consistent with the data in Fig. 46.4.

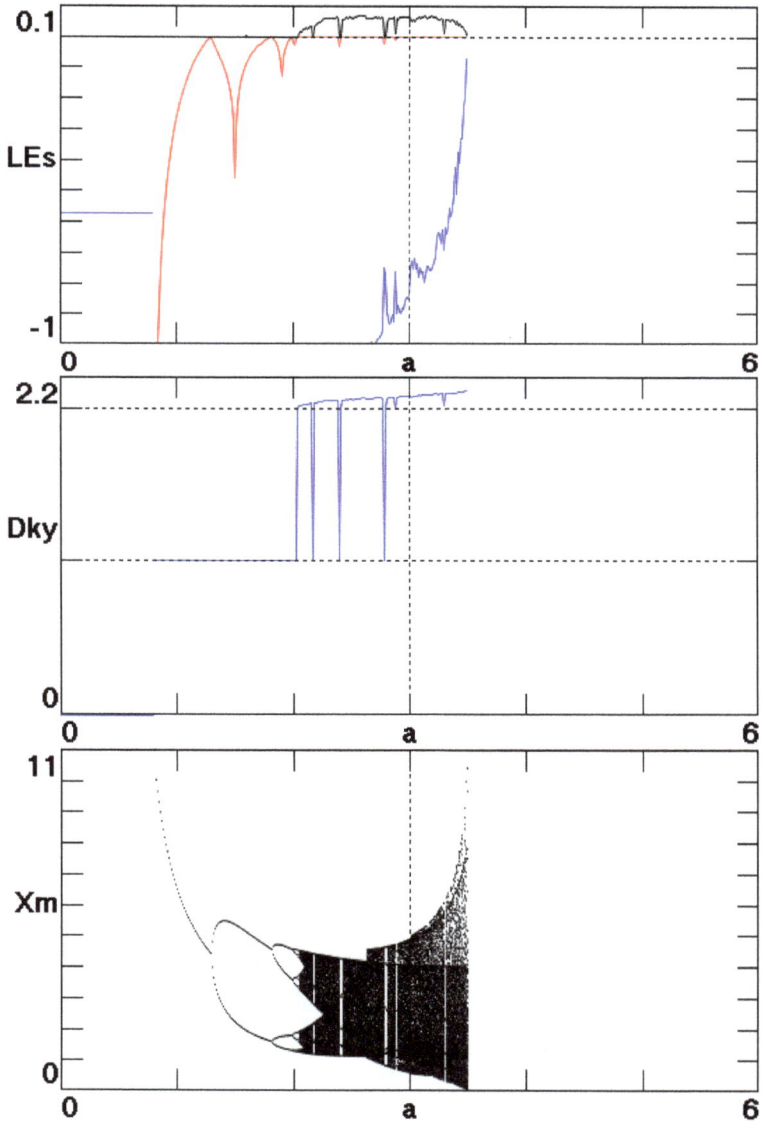

Fig. 46.4   Bifurcation diagram for Eq. (46.2) as a function of the parameter a.

Chapter 47

Forced Ueda System

There are many examples of two-dimensional periodically forced nonlinear oscillators that are chaotic. They can all be written in a three-dimensional autonomous form. The case shown here is one of the oldest and simplest such example. Periodically forced systems do not have equilibrium points, and so the attractor is hidden despite being globally attracting.

47.1 Introduction

The periodically forced Ueda (1979) system written in its most general form with an adjustable coefficient in each of the five terms is given by

$$
\begin{aligned}
\dot{x} &= a_1 y \\
\dot{y} &= -a_2 x^3 - a_3 y + a_4 \sin z \\
\dot{z} &= a_5.
\end{aligned}
\tag{47.1}
$$

The usual parameters are $a_1 = a_2 = a_5 = 1$, $a_3 = 0.05$, $a_4 = 7.5$. The system has chaotic conservative solutions for $a_3 = 0$, but we consider here only the dissipative case with $a_3 > 0$. Since z appears only as the argument of $\sin z$, nothing is lost by restricting z to the range $-\pi < z < \pi$ except for purposes of calculating the Lyapunov exponents.

    The following sections were written by the computer program that performed the optimization, carried out the analysis of the resulting system, and produced the corresponding figures, all without human intervention.

47.2 Simplified System

After about 1×10^4 trials, of which 57 were chaotic, simplified parameters for Eq. (47.1) that give chaotic solutions are $a_1 = 5$, $a_2 = 1$, $a_3 = 0.1$,

$a_4 = 1$, $a_5 = 1$. Thus Eq. (47.1) can be written more compactly as

$$\dot{x} = ay$$
$$\dot{y} = -x + by \cos x + \sin z \qquad (47.2)$$
$$\dot{z} = 1,$$

where $a = 5$, $b = 0.1$.

Note that with five terms, Eq. (47.2) should have two independent parameters through a linear rescaling of the three variables plus time, and so the dynamics is completely captured by the given parameters, which could be put in any of the five terms, albeit with different numerical values.

47.3 Equilibria

The system in Eq. (47.2) with $a = 5$, $b = 0.1$ has no equilibrium points. The strange attractor is apparently hidden, and the system has no symmetry.

47.4 Attractor

Figure 47.1 shows various views of the attractor for Eq. (47.2) with $a = 5$, $b = 0.1$ and initial conditions $(0, 1, -3)$.

47.5 Time Series

Figure 47.2 shows the time series for the three variables along with the local value of the largest Lyapunov exponent (LL) for Eq. (47.2) with $a = 5$, $b = 0.1$. Red color in the Lyapunov exponent indicates that the error is growing parallel to the orbit, while blue indicates growth perpendicular to the orbit. Note that the orbit passes through regions where the local Lyapunov exponent is strongly positive and other regions where it is strongly negative as is typical for a chaotic system and is also reflected by the colors in Fig. 47.1.

47.6 Lyapunov Exponents

The global Lyapunov exponents are determined by averaging the local Lyapunov exponents along the orbit. The values typically converge very slowly because of the large variation along the orbit, and an integration time of order 10^8 is required to obtain 4-digit accuracy.

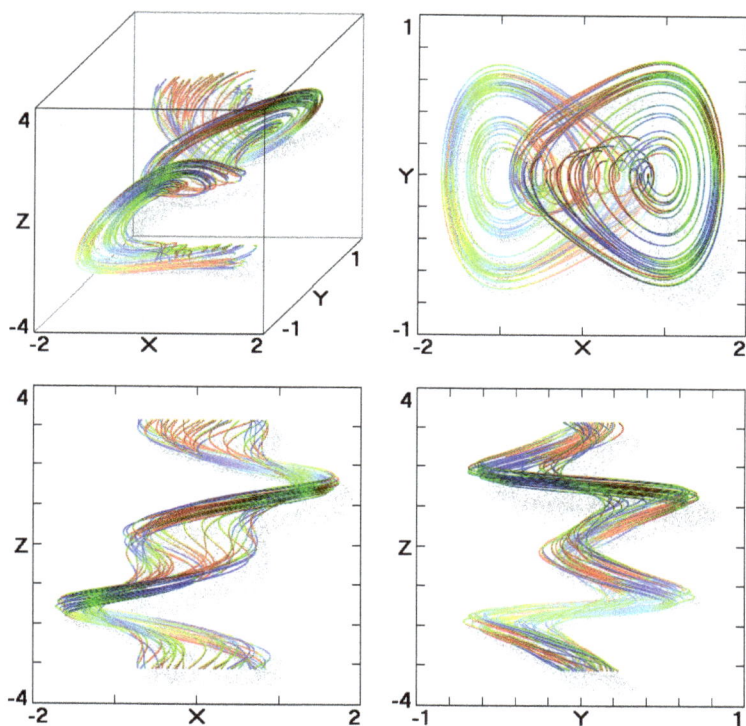

Fig. 47.1   Views of the attractor for Eq. (47.2) with $a = 5$, $b = 0.1$ and initial conditions $(0, 1, -3)$.

The results of such a calculation for the system in Eq. (47.2) with $a = 5$, $b = 0.1$ after a time of 3×10^6 are LE $= (0.1173, 0, -0.2173)$ with a Kaplan–Yorke dimension of 2.5397, where the last digit in the quoted values is only an approximation. The positive value of the largest Lyapunov exponent indicates that the system is chaotic, and the negative sum of the exponents (-0.1000) indicates that the system is dissipative with a strange attractor.

47.7   Basin of Attraction

Figure 47.3 shows (in red) the basin of attraction for Eq. (47.2) with $a = 5$, $b = 0.1$ in the $z = 0$ plane. Also shown (in black) is the cross-section of the attractor in the same plane.

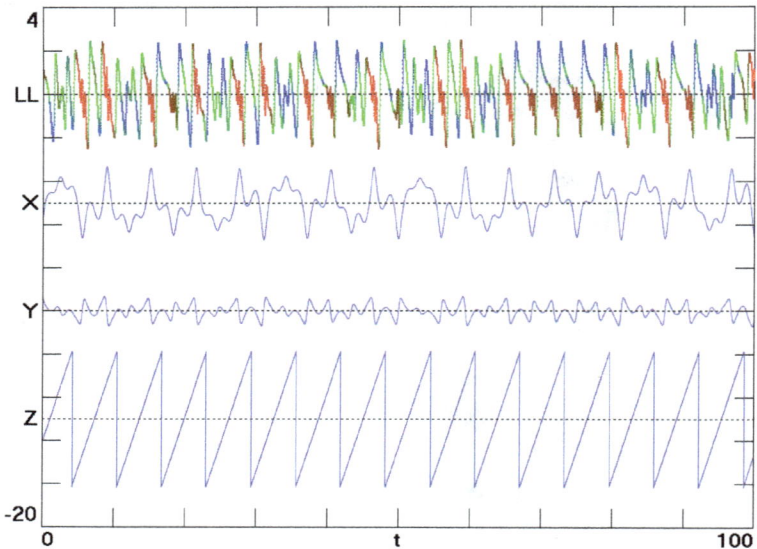

Fig. 47.2   Time series for the variables in Eq. (47.2) with $a = 5$, $b = 0.1$ along with the local Lyapunov exponent (LL).

47.8   Bifurcations

Figure 47.4 shows the bifurcation diagram for Eq. (47.2) as a function of the parameter a from 0 to 10 for $b = 0.1$. The initial condition was taken as $(0, 1, -3)$ at $a = 10$ and was not changed as a slowly varied toward $a = 0$. Each of the 500 values of a was calculated for a time of about 1×10^4.

The upper plot shows the three Lyapunov exponents. The middle plot shows the Kaplan–Yorke dimension, and the lower plot shows the local maxima of x. The chaotic region is in the vicinity of $a = 5$, $b = 0.1$, and the route to chaos is clearly shown.

47.9   Robustness

One measure of the robustness of a chaotic system is the amount by which the parameters can be changed from their nominal values before the probability of chaos decreases to 50% [Sprott (2022)]. For the system in Eq. (47.2) with $a = 5$, $b = 0.1$ and initial conditions $(0, 1, -3)$, after 2459 trials, it is

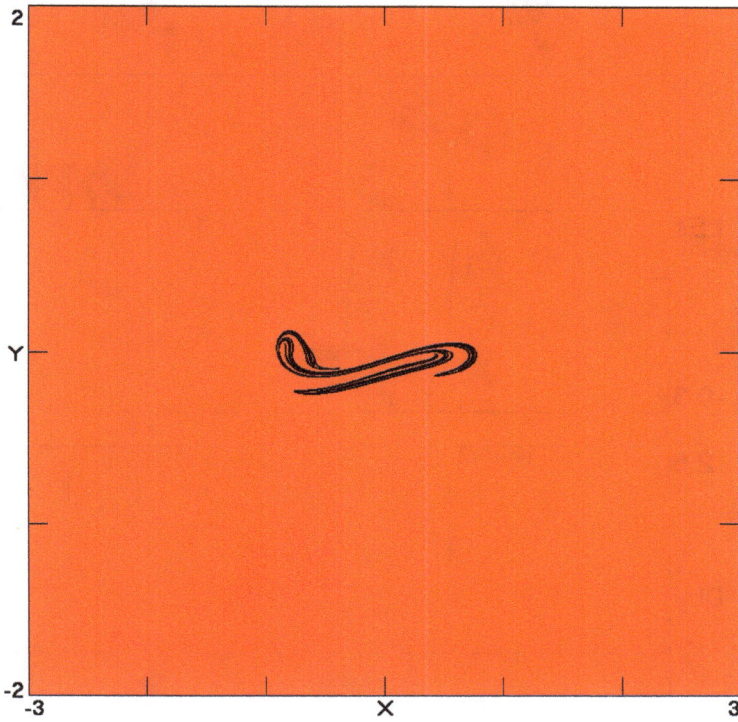

Fig. 47.3   Basin of attraction for Eq. (47.2) with $a = 5$, $b = 0.1$ in the $z = 0$ plane.

estimated that the parameters can be changed by 24% before the chaos is more likely to be lost than not. Thus the system is somewhat robust.

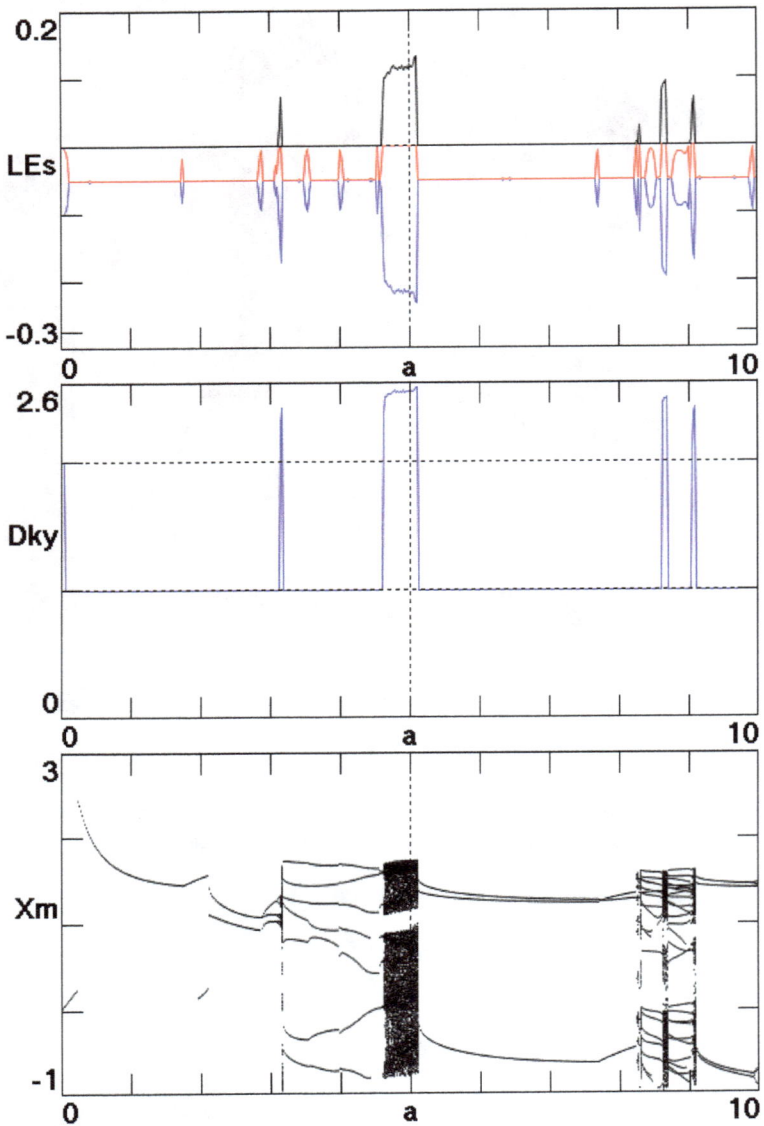

Fig. 47.4   Bifurcation diagram for Eq. (47.2) as a function of the parameter a for $b = 0.1$.

Chapter 48

Megastable System

The previous chapter showed a periodically forced system with the non-linearity in the restoring force. The system here is a linear oscillator with a spatially periodic nonlinearity in the damping, resulting in a countable infinity of nested attractors, including limit cycles, tori, and strange attractors. Such systems are said to be 'megastable'.

48.1 Introduction

The periodically forced megastable system [Sprott *et al.* (2017)] written in its most general form with an adjustable coefficient in each of the five terms is given by

$$
\begin{aligned}
\dot{x} &= a_1 y \\
\dot{y} &= -a_2 x + a_3 y \cos x + a_4 \sin z \\
\dot{z} &= a_5 .
\end{aligned}
\tag{48.1}
$$

The usual parameters are $a_1 = a_2 = a_3 = 1$, $a_4 = 4.3$, $a_5 = 0.85$. Since z appears only as the argument of $\sin z$, nothing is lost by restricting z to the range $-\pi < z < \pi$ except for purposes of calculating the Lyapunov exponents.

The following sections were written by the computer program that performed the optimization, carried out the analysis of the resulting system, and produced the corresponding figures, all without human intervention.

48.2 Simplified System

After about 4×10^4 trials, of which 56 were chaotic, simplified parameters for Eq. (48.1) that give chaotic solutions are $a_1 = 1$, $a_2 = 0.45$, $a_3 = 1$,

$a_4 = 1$, $a_5 = 0.5$. Thus Eq. (48.1) can be written more compactly as

$$\dot{x} = y$$
$$\dot{y} = -ax + y\cos x + \sin z \qquad (48.2)$$
$$\dot{z} = b,$$

where $a = 0.45$, $b = 0.5$.

Note that with five terms, Eq. (48.2) should have three independent parameters through a linear rescaling of the three variables plus time. However, one of the three parameters has a value of ± 1 and can be placed in any of the remaining three terms.

48.3 Equilibria

The system in Eq. (48.2) with $a = 0.45$, $b = 0.5$ has no equilibrium points. The strange attractor is apparently hidden, and the system has no symmetry.

48.4 Attractor

Figure 48.1 shows various views of one of the infinitely many attractors for Eq. (48.2) with $a = 0.45$, $b = 0.5$ and initial conditions (5, 3, 0).

48.5 Time Series

Figure 48.2 shows the time series for the three variables along with the local value of the largest Lyapunov exponent (LL) for Eq. (48.2) with $a = 0.45$, $b = 0.5$. Red color in the Lyapunov exponent indicates that the error is growing parallel to the orbit, while blue indicates growth perpendicular to the orbit. Note that the orbit passes through regions where the local Lyapunov exponent is strongly positive and other regions where it is strongly negative as is typical for a chaotic system and is also reflected by the colors in Fig. 48.1.

48.6 Lyapunov Exponents

The global Lyapunov exponents are determined by averaging the local Lyapunov exponents along the orbit. The values typically converge very slowly because of the large variation along the orbit, and an integration time of order 10^8 is required to obtain 4-digit accuracy.

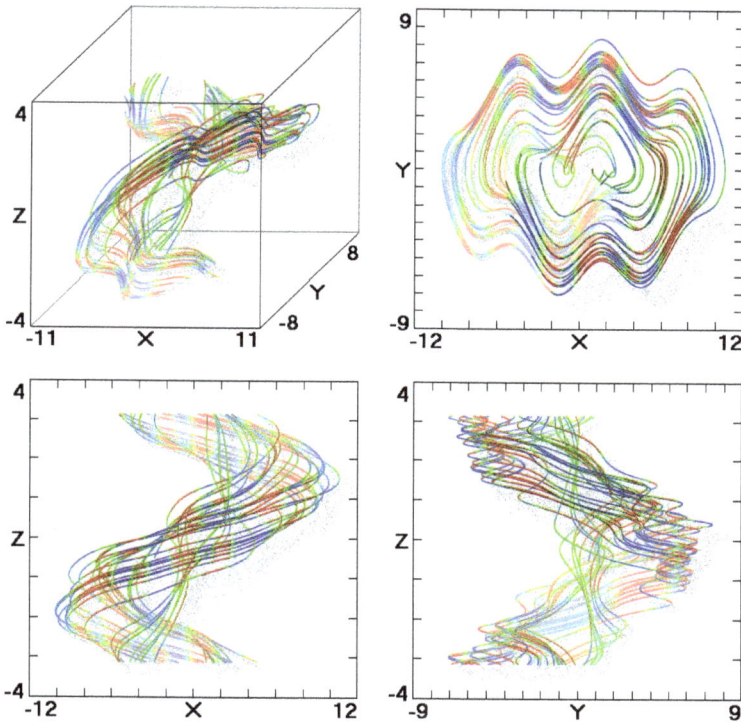

Fig. 48.1   Views of one of the infinitely many attractors for Eq. (48.2) with $a = 0.45$, $b = 0.5$ and initial conditions (5, 3, 0).

The results of such a calculation for the system in Eq. (48.2) with $a = 0.45$, $b = 0.5$ after a time of 2×10^6 are LE = (0.0199, 0, −0.0442) with a Kaplan–Yorke dimension of 2.4491, where the last digit in the quoted values is only an approximation. The positive value of the largest Lyapunov exponent indicates that the system is chaotic, and the negative sum of the exponents (−0.0243) indicates that the system is dissipative with a strange attractor.

48.7   Basin of Attraction

Figure 48.3 shows (in red) the basin of attraction for Eq. (48.2) with $a = 0.45$, $b = 0.5$ in the $z = 0$ plane. Also shown (in black) is the cross-section of the attractors in the same plane.

308   *Elegant Automation*

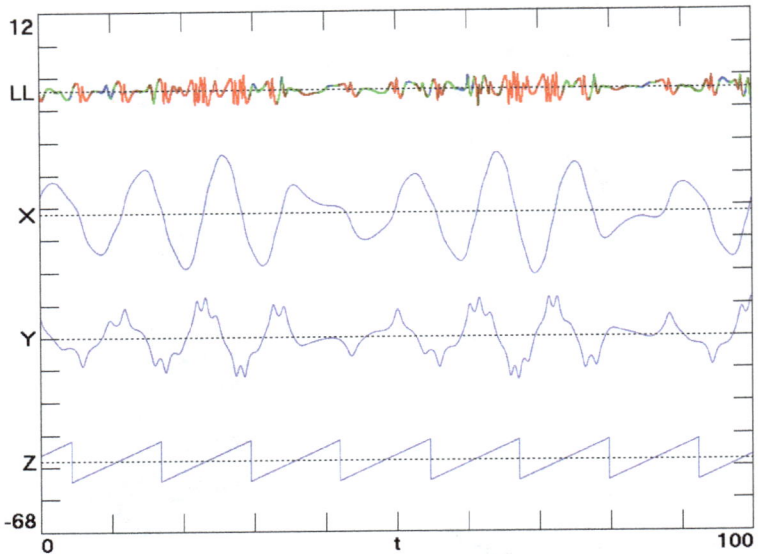

Fig. 48.2   Time series for the variables in Eq. (48.2) with $a = 0.45$, $b = 0.5$ along with the local Lyapunov exponent (LL).

48.8   Bifurcations

Figure 48.4 shows the bifurcation diagram for Eq. (48.2) as a function of the parameter a from 0 to 0.9 for $b = 0.5$. The initial condition was taken as (5, 3, 0) for each value of a. Each of the 500 values of a was calculated for a time of about 1×10^4.

The upper plot shows the three Lyapunov exponents. The middle plot shows the Kaplan–Yorke dimension, and the lower plot shows the local maxima of x. The chaotic region is in the vicinity of $a = 0.45$, $b = 0.5$, and the route to chaos is clearly shown.

48.9   Robustness

One measure of the robustness of a chaotic system is the amount by which the parameters can be changed from their nominal values before the probability of chaos decreases to 50% [Sprott (2022)]. For the system in Eq. (48.2) with $a = 0.45$, $b = 0.5$ and initial conditions (5, 3, 0), after 10246 trials, it

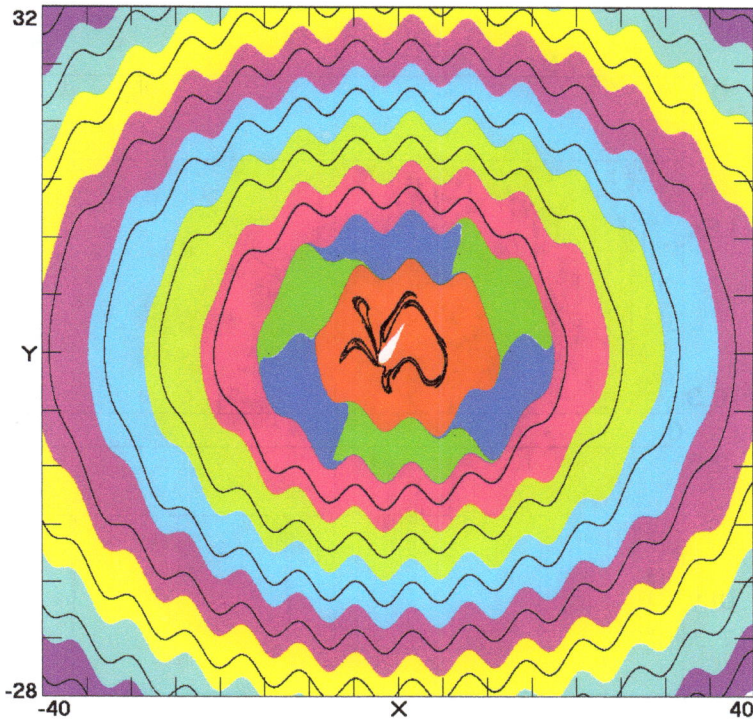

Fig. 48.3   Basin of attraction for Eq. (48.2) with $a = 0.45$, $b = 0.5$ in the $z = 0$ plane.

is estimated that the parameters can be changed by 2.1% before the chaos is more likely to be lost than not. Thus the system is somewhat fragile.

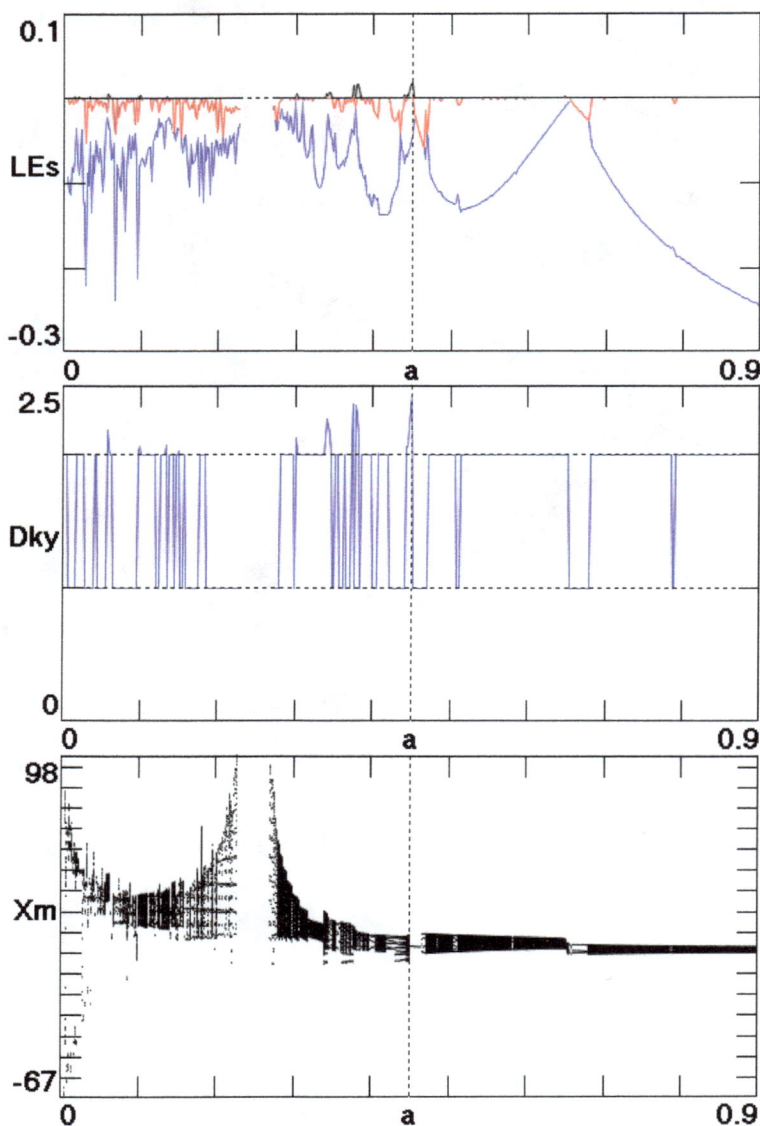

Fig. 48.4   Bifurcation diagram for Eq. (48.2) as a function of the parameter a for $b = 0.5$.

Chapter 49

Attracting Torus System

Tori are common in three-dimensional conservative systems and in two-dimensional periodically driven systems, but attracting tori are rare in three-dimensional autonomous systems. The case here, which is a variant of the Nosé–Hoover system, is perhaps the simplest example where such a torus exists along with a strange attractor or by itself.

49.1 Introduction

The attracting torus system [Mehrabbeik *et al.* (2022)] written in its most general form with an adjustable coefficient in each of the six terms is given by

$$
\begin{aligned}
\dot{x} &= a_1 y \\
\dot{y} &= -a_2 x - a_3 z y \\
\dot{z} &= a_4 y^2 - a_5 + a_6 z.
\end{aligned}
\tag{49.1}
$$

The usual parameters are $a_1 = a_2 = a_3 = a_4 = 1$, $a_5 = 4$, $a_6 = 0.1$. The number of terms and their signs are preserved to keep the system from collapsing to a simple Nosé–Hoover system. The $a_6 z$ term is antidamping and provides positive feedback to the thermostat.

The following sections were written by the computer program that performed the optimization, carried out the analysis of the resulting system, and produced the corresponding figures, all without human intervention.

49.2 Simplified System

After about 3×10^5 trials, of which 57 were chaotic, simplified parameters for Eq. (49.1) that give chaotic solutions are $a_1 = 1$, $a_2 = 1$, $a_3 = 1$, $a_4 = 1$,

$a_5 = 4$, $a_6 = 0.1$. Thus Eq. (49.1) can be written more compactly as

$$
\begin{aligned}
\dot{x} &= y \\
\dot{y} &= -x - zy \\
\dot{z} &= y^2 - a + bz,
\end{aligned}
\tag{49.2}
$$

where $a = 4$, $b = 0.1$.

Note that with six terms, Eq. (49.2) should have two independent parameters through a linear rescaling of the three variables plus time, and so the dynamics is completely captured by the given parameters, which could be put in any of the six terms, albeit with different numerical values.

49.3   Equilibria

The system in Eq. (49.2) with $a = 4$, $b = 0.1$ has an unstable saddle node at $(0, 0, 40)$ with eigenvalues $(-0.025, -39.975, 0.1)$ and a Poincaré index of 1 in the $z = 40$ plane.

The strange attractor is self-excited, and the system is symmetric under the transformation $x \to -x$, $y \to -y$.

49.4   Attractor

Figure 49.1 shows various views of the attractor for Eq. (49.2) with $a = 4$, $b = 0.1$ and initial conditions $(0, 2, 1)$.

49.5   Time Series

Figure 49.2 shows the time series for the three variables along with the local value of the largest Lyapunov exponent (LL) for Eq. (49.2) with $a = 4$, $b = 0.1$. Red color in the Lyapunov exponent indicates that the error is growing parallel to the orbit, while blue indicates growth perpendicular to the orbit. Note that the orbit passes through regions where the local Lyapunov exponent is strongly positive and other regions where it is strongly negative as is typical for a chaotic system and is also reflected by the colors in Fig. 49.1.

49.6   Lyapunov Exponents

The global Lyapunov exponents are determined by averaging the local Lyapunov exponents along the orbit. The values typically converge very slowly

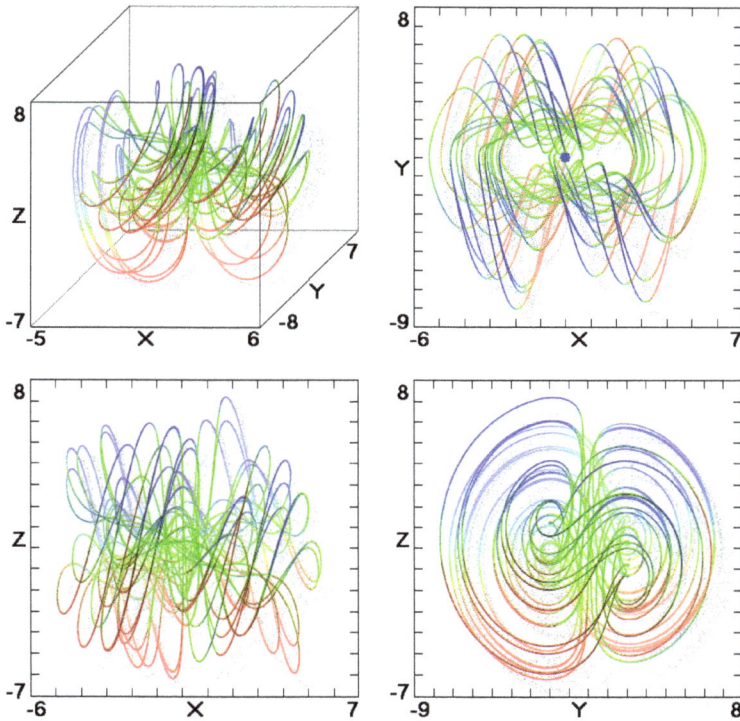

Fig. 49.1   Views of the attractor for Eq. (49.2) with $a = 4$, $b = 0.1$ and initial conditions (0, 2, 1).

because of the large variation along the orbit, and an integration time of order 10^8 is required to obtain 4-digit accuracy.

The results of such a calculation for the system in Eq. (49.2) with $a = 4$, $b = 0.1$ after a time of 3×10^6 are LE = $(0.0661, 0, -0.0981)$ with a Kaplan–Yorke dimension of 2.6734, where the last digit in the quoted values is only an approximation. The positive value of the largest Lyapunov exponent indicates that the system is chaotic, and the negative sum of the exponents (-0.0320) indicates that the system is dissipative with a strange attractor.

Elegant Automation

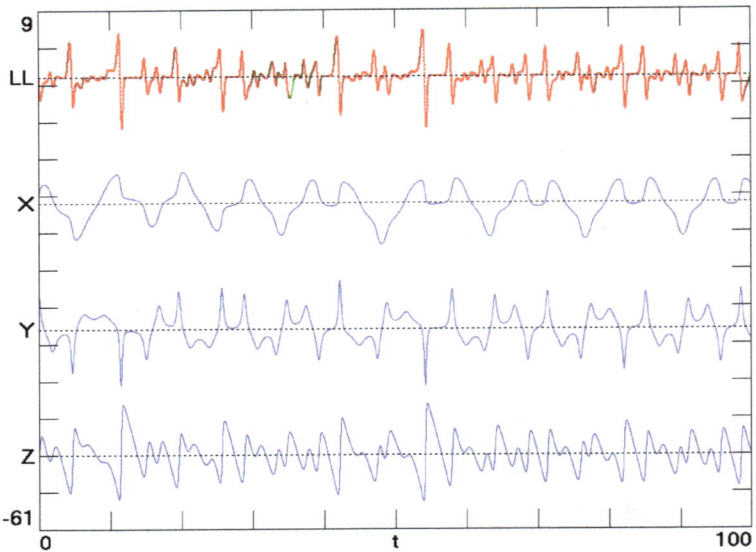

Fig. 49.2   Time series for the variables in Eq. (49.2) with $a = 4$, $b = 0.1$ along with the local Lyapunov exponent (LL).

49.7   Basin of Attraction

Figure 49.3 shows (in red) the basin of attraction for Eq. (49.2) with $a = 4$, $b = 0.1$ in the $z = 0$ plane. Also shown (in black) is the cross-section of the attractor in the same plane.

49.8   Bifurcations

Figure 49.4 shows the bifurcation diagram for Eq. (49.2) as a function of the parameter a from 0 to 8 for $b = 0.1$. The initial condition was taken as $(0, 2, 1)$ at $a = 8$ and was not changed as a slowly varied toward $a = 0$. Each of the 500 values of a was calculated for a time of about 1×10^4.

The upper plot shows the three Lyapunov exponents. The middle plot shows the Kaplan–Yorke dimension, and the lower plot shows the local maxima of x. The chaotic region is in the vicinity of $a = 4$, $b = 0.1$, and the route to chaos is clearly shown.

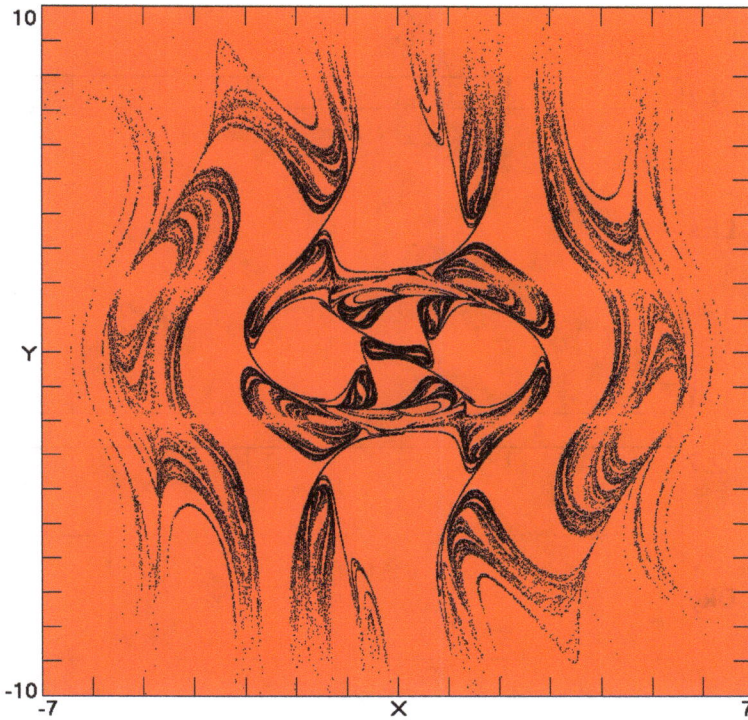

Fig. 49.3 Basin of attraction for Eq. (49.2) with $a = 4$, $b = 0.1$ in the $z = 0$ plane.

49.9 Robustness

One measure of the robustness of a chaotic system is the amount by which the parameters can be changed from their nominal values before the probability of chaos decreases to 50% [Sprott (2022)]. For the system in Eq. (49.2) with $a = 4$, $b = 0.1$ and initial conditions $(0, 2, 1)$, after 2185 trials, it is estimated that the parameters can be changed by 35% before the chaos is more likely to be lost than not. Thus the system is somewhat robust.

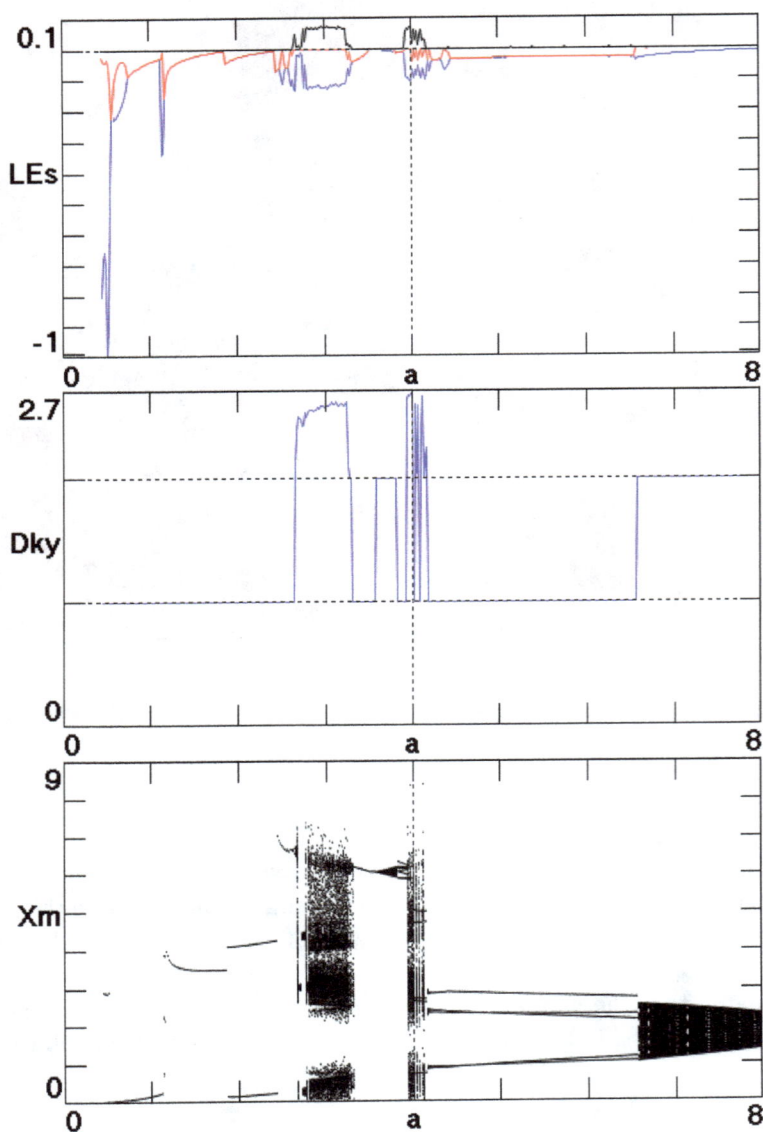

Fig. 49.4   Bifurcation diagram for Eq. (49.2) as a function of the parameter a for $b = 0.1$.

Chapter 50

Buncha System

Buncha Munmuangsaen studied a variant of the Nosé–Hoover system in which the thermostat controls the average speed rather than the average energy of the linear oscillator. The resulting system has strange attractors, limit cycles and invariant tori. The system is time-reversible with an attractor–repellor pair and is apparently fully ergodic for some choices of the parameters despite lacking equilibrium points and thus being hidden.

50.1 Introduction

The Buncha system [Munmuangsaen *et al.* (2015)] written in its most general form with an adjustable coefficient in each of the five terms is given by

$$
\begin{aligned}
\dot{x} &= a_1 y \\
\dot{y} &= -a_2 x - a_3 z y \\
\dot{z} &= a_4 |y| - a_5.
\end{aligned}
\tag{50.1}
$$

The usual parameters are $a_1 = a_2 = a_3 = a_4 = 1$, $a_5 = 5$. The parameters are constrained to be positive, and only dissipative solutions are considered.

The following sections were written by the computer program that performed the optimization, carried out the analysis of the resulting system, and produced the corresponding figures, all without human intervention.

50.2 Simplified System

After about 2×10^4 trials, of which 164 were chaotic, simplified parameters for Eq. (50.1) that give chaotic solutions are $a_1 = 0.3$, $a_2 = 1$, $a_3 = 1$,

$a_4 = 1$, $a_5 = 1$. Thus Eq. (50.1) can be written more compactly as

$$\dot{x} = ay$$
$$\dot{y} = -x - zy \qquad (50.2)$$
$$\dot{z} = |y| - 1,$$

where $a = 0.3$.

Note that with five terms, Eq. (50.2) should have one independent parameter through a linear rescaling of the three variables plus time, and so the dynamics is completely captured by the single parameter a, which could be put in any of the five terms, albeit with a different numerical value.

50.3   Equilibria

The system in Eq. (50.2) with $a = 0.3$ has no equilibrium points. The strange attractor is apparently hidden, and the system is symmetric under the transformation $y \to -y$, $z \to -z$, $t \to -t$.

50.4   Attractor

Figure 50.1 shows various views of the attractor for Eq. (50.2) with $a = 0.3$ and initial conditions $(-1, 5, 0)$.

50.5   Time Series

Figure 50.2 shows the time series for the three variables along with the local value of the largest Lyapunov exponent (LL) for Eq. (50.2) with $a = 0.3$. Red color in the Lyapunov exponent indicates that the error is growing parallel to the orbit, while blue indicates growth perpendicular to the orbit. Note that the orbit passes through regions where the local Lyapunov exponent is strongly positive and other regions where it is strongly negative as is typical for a chaotic system and is also reflected by the colors in Fig. 50.1.

50.6   Lyapunov Exponents

The global Lyapunov exponents are determined by averaging the local Lyapunov exponents along the orbit. The values typically converge very slowly because of the large variation along the orbit, and an integration time of order 10^8 is required to obtain 4-digit accuracy.

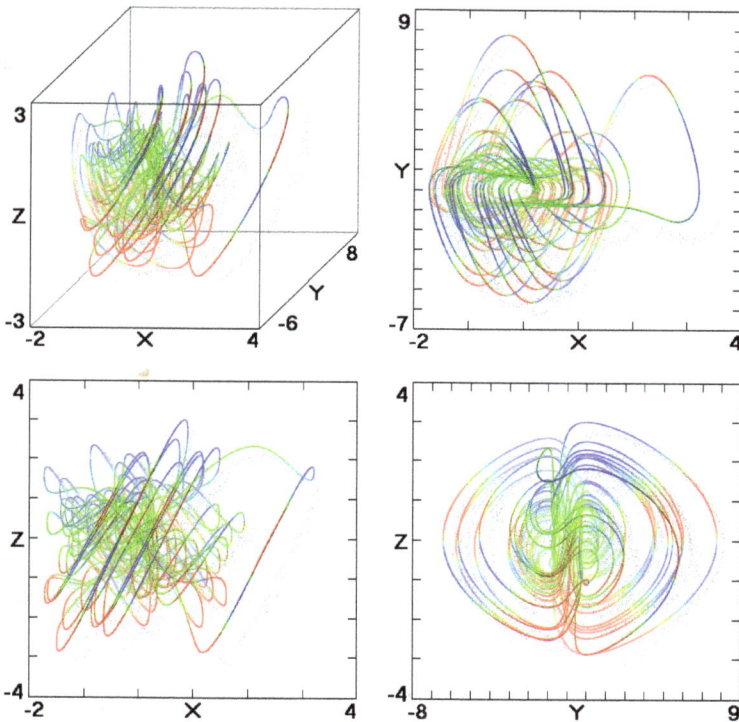

Fig. 50.1 Views of the attractor for Eq. (50.2) with $a = 0.3$ and initial conditions $(-1, 5, 0)$.

The results of such a calculation for the system in Eq. (50.2) with $a = 0.3$ after a time of 3×10^6 are LE $= (0.0715, 0, -0.0765)$ with a Kaplan–Yorke dimension of 2.9344, where the last digit in the quoted values is only an approximation. The positive value of the largest Lyapunov exponent indicates that the system is chaotic, and the negative sum of the exponents (-0.0050) indicates that the system is dissipative with a strange attractor.

50.7 Basin of Attraction

Figure 50.3 shows (in red) the basin of attraction for Eq. (50.2) with $a = 0.3$ in the $z = 0$ plane. Also shown (in black) is the cross-section of the attractor in the same plane.

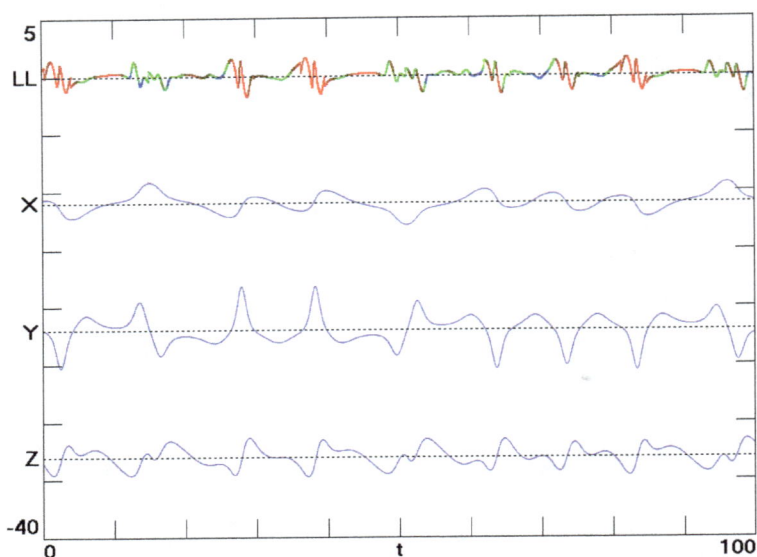

Fig. 50.2   Time series for the variables in Eq. (50.2) with $a = 0.3$ along with the local Lyapunov exponent (LL).

50.8   Bifurcations

Figure 50.4 shows the bifurcation diagram for Eq. (50.2) as a function of the parameter a from 0 to 0.6. The initial condition was taken as $(-1, 5, 0)$ at $a = 0.6$ and was not changed as a slowly varied toward $a = 0$. Each of the 500 values of a was calculated for a time of about 2×10^4.

The upper plot shows the three Lyapunov exponents. The middle plot shows the Kaplan–Yorke dimension, and the lower plot shows the local maxima of x. The chaotic region is in the vicinity of $a = 0.3$, and the route to chaos is clearly shown.

50.9   Robustness

One measure of the robustness of a chaotic system is the amount by which the parameters can be changed from their nominal values before the probability of chaos decreases to 50% [Sprott (2022)]. For the system in Eq. (50.2) with $a = 0.3$ and initial conditions $(-1, 5, 0)$, after 1138 trials, it is estimated that the parameter a can be changed by 98% before the chaos is

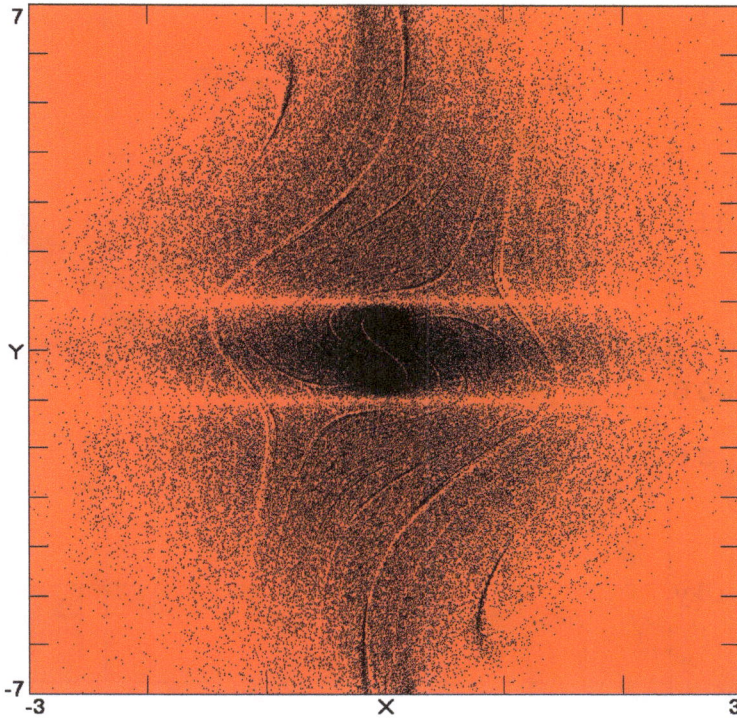

Fig. 50.3   Basin of attraction for Eq. (50.2) with $a = 0.3$ in the $z = 0$ plane.

more likely to be lost than not. Thus the system is highly robust. This result is consistent with the data in Fig. 50.4.

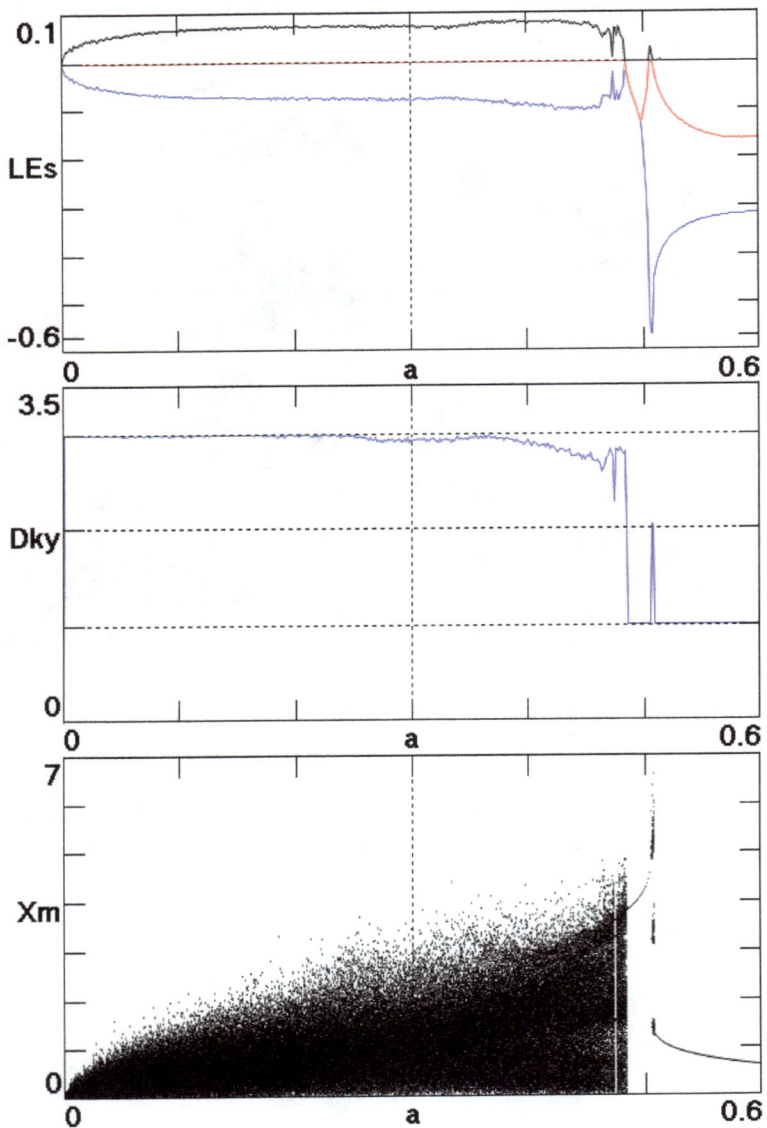

Fig. 50.4   Bifurcation diagram for Eq. (50.2) as a function of the parameter a.

Chapter 51

Signum Thermostat System

This chapter concludes the book with an even fifty systems by considering a simple conservative linear oscillator with a signum thermostat that makes it fully ergodic with a chaotic sea that has a precisely Gaussian distribution of the x and y variables. It is a variant of the Nosé–Hoover system in which the 'bang-bang' thermostat turns on and off abruptly rather than linearly. The system is time-reversible and rotationally invariant with no equilibrium points.

51.1 Introduction

The signum thermostat system [Sprott (2018)] written in its most general form with an adjustable coefficient in each of the five terms is given by

$$
\begin{aligned}
\dot{x} &= a_1 y \\
\dot{y} &= -a_2 x - a_3 \operatorname{sgn}(z) y \\
\dot{z} &= a_4 y^2 - a_5.
\end{aligned}
\tag{51.1}
$$

The usual parameters are $a_1 = a_2 = a_4 = a_5 = 1$, $a_3 = 2$. The parameters are constrained to be positive, and only conservative solutions exist. While there are chaotic solutions for $a_3 = 1$, they are not fully ergodic.

The following sections were written by the computer program that performed the optimization, carried out the analysis of the resulting system, and produced the corresponding figures, all without human intervention.

51.2 Simplified System

After about 2×10^4 trials, of which 164 were chaotic, simplified parameters for Eq. (51.1) that give chaotic solutions are $a_1 = 1$, $a_2 = 1$, $a_3 = 2$, $a_4 = 1$,

$a_5 = 1$. Thus Eq. (51.1) can be written more compactly as

$$\dot{x} = y$$
$$\dot{y} = -x - a\,\mathrm{sgn}(z)y \qquad\qquad (51.2)$$
$$\dot{z} = y^2 - 1,$$

where $a = 2$.

Note that with five terms, Eq. (51.2) should have one independent parameter through a linear rescaling of the three variables plus time, and so the dynamics is completely captured by the single parameter a, which could be put in any of the five terms, albeit with a different numerical value.

51.3   Equilibria

The system in Eq. (51.2) with $a = 2$ has no equilibrium points. The chaotic sea is apparently hidden, and the system is symmetric under the transformation $y \to -y$, $z \to -z$, $t \to -t$.

51.4   Chaotic Sea

Figure 51.1 shows various views of the chaotic sea for Eq. (51.2) with $a = 2$ and initial conditions $(1, -2, 0)$.

51.5   Time Series

Figure 51.2 shows the time series for the three variables along with the local value of the largest Lyapunov exponent (LL) for Eq. (51.2) with $a = 2$. Red color in the Lyapunov exponent indicates that the error is growing parallel to the orbit, while blue indicates growth perpendicular to the orbit. Note that the orbit passes through regions where the local Lyapunov exponent is strongly positive and other regions where it is strongly negative as is typical for a chaotic system and is also reflected by the colors in Fig. 51.1.

51.6   Lyapunov Exponents

The global Lyapunov exponents are determined by averaging the local Lyapunov exponents along the orbit. The values typically converge very slowly because of the large variation along the orbit, and an integration time of order 10^8 is required to obtain 4-digit accuracy.

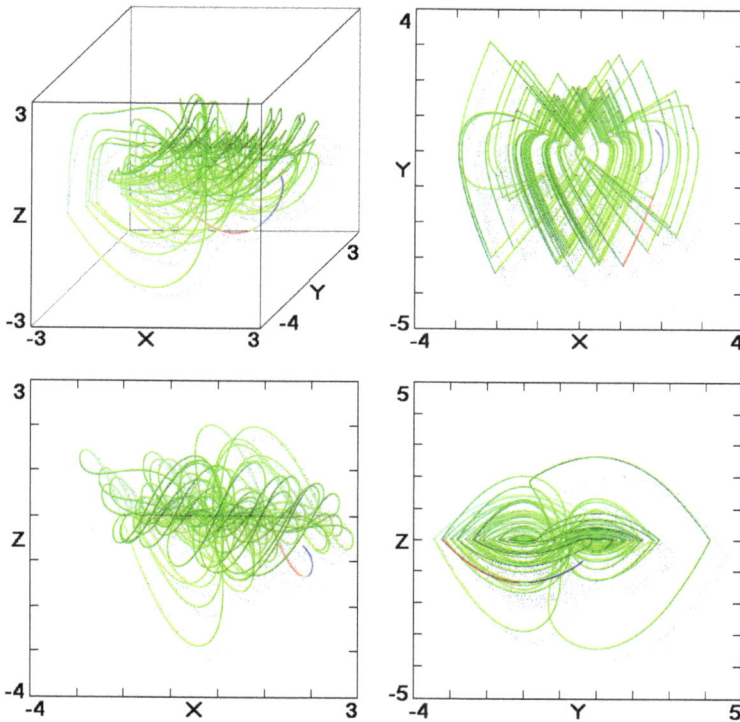

Fig. 51.1   Views of the chaotic sea for Eq. (51.2) with $a = 2$ and initial conditions (1, -2, 0).

The results of such a calculation for the system in Eq. (51.2) with $a = 2$ after a time of 4×10^6 are LE $= (0.3032, 0, -0.3032)$ with a Kaplan–Yorke dimension of 3.0, where the last digit in the quoted values is only an approximation. The positive value of the largest Lyapunov exponent indicates that the system is chaotic, and the zero sum of the exponents indicates that the system is conservative with a chaotic sea.

51.7   Extent of the Chaotic Sea

Figure 51.3 shows (in red) the extent of the chaotic sea for Eq. (51.2) with $a = 2$ in the $z = 0$ plane. Also shown (in black) is the cross-section of the chaotic sea in the same plane.

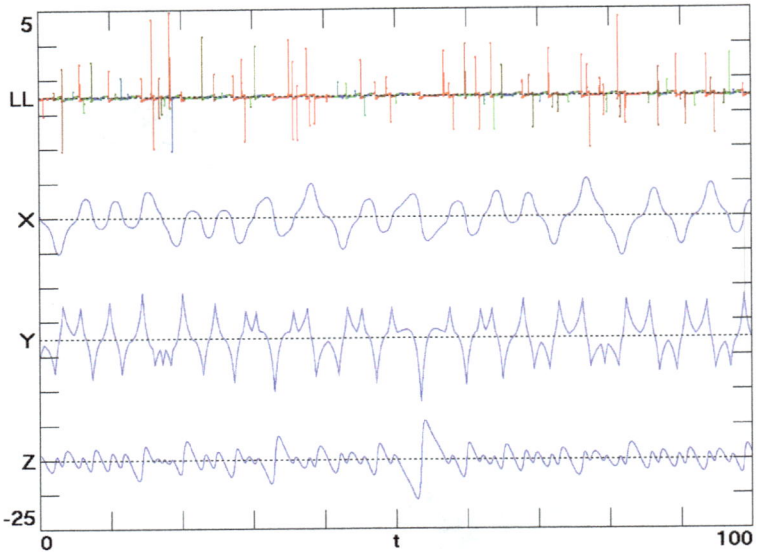

Fig. 51.2   Time series for the variables in Eq. (51.2) with $a = 2$ along with the local Lyapunov exponent (LL).

51.8   Bifurcations

Figure 51.4 shows the bifurcation diagram for Eq. (51.2) as a function of the parameter a from 0 to 4. The initial condition was taken as $(1, -2, 0)$ at $a = 4$ and was not changed as a slowly varied toward $a = 0$. Each of the 500 values of a was calculated for a time of about 2×10^4.

The upper plot shows the three Lyapunov exponents. The middle plot shows the Kaplan–Yorke dimension, and the lower plot shows the local maxima of x. The chaotic region is in the vicinity of $a = 2$, and the route to chaos is clearly shown.

51.9   Robustness

One measure of the robustness of a chaotic system is the amount by which the parameters can be changed from their nominal values before the probability of chaos decreases to 50% [Sprott (2022)]. For the system in Eq. (51.2) with $a = 2$ and initial conditions $(1, -2, 0)$, after 1046 trials, it is estimated that the parameter a can be changed by 93% before the chaos is more likely

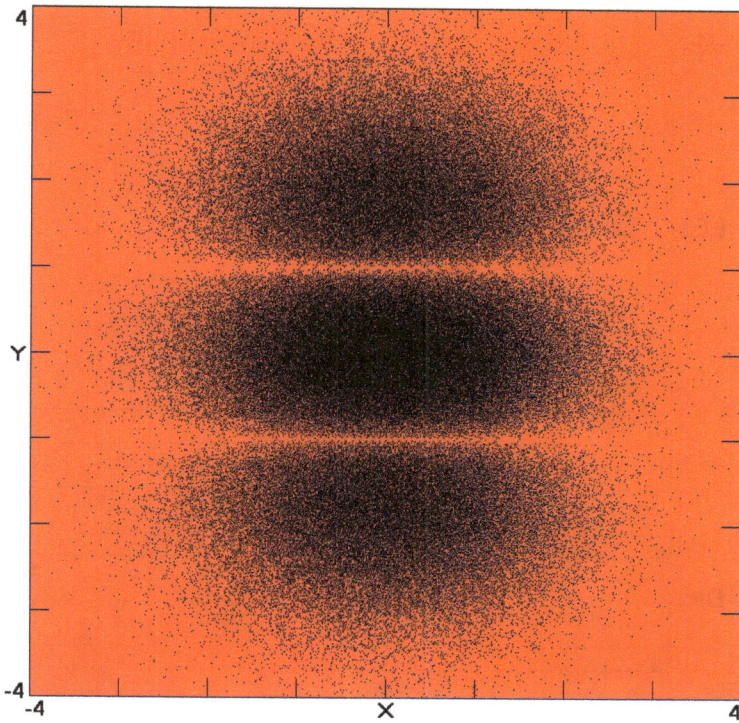

Fig. 51.3   Extent of the chaotic sea for Eq. (51.2) with $a = 2$ in the $z = 0$ plane.

to be lost than not. Thus the system is highly robust. This result is consistent with the data in Fig. 51.4.

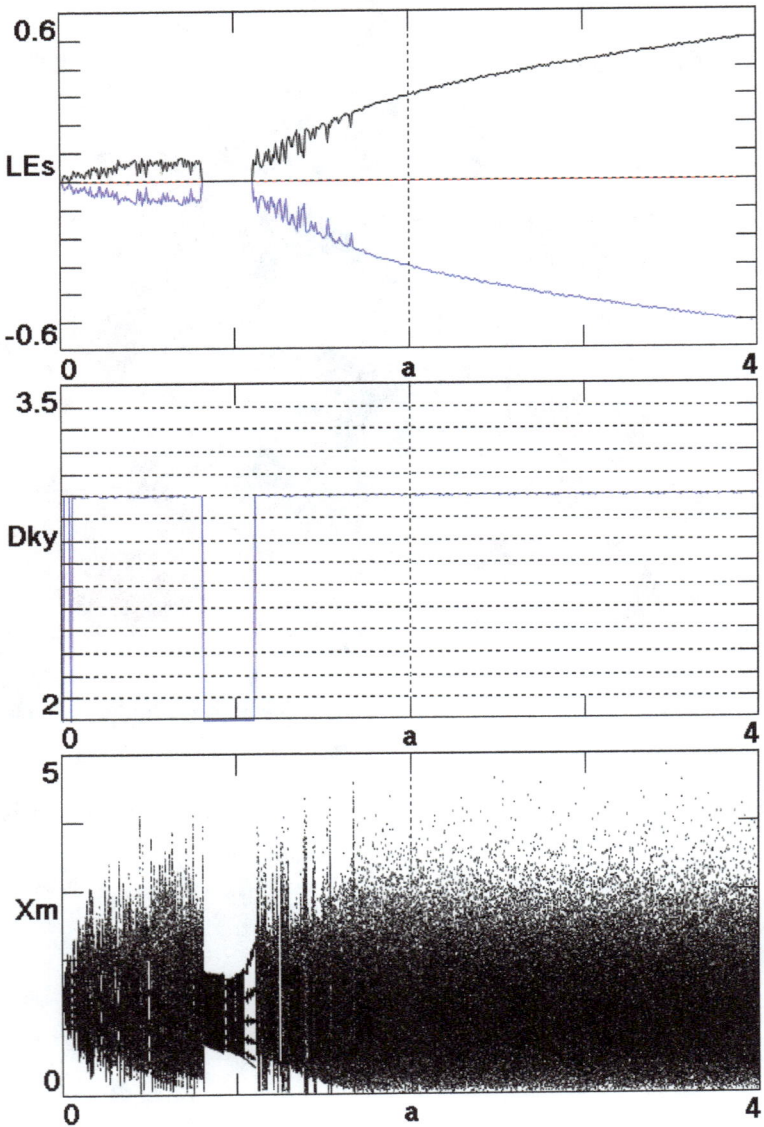

Fig. 51.4   Bifurcation diagram for Eq. (51.2) as a function of the parameter *a*.

Bibliography

Aleksandrov, A. G. (2022). The Poincaré index and its applications, *Universe* **8**, pp. 223–238.

Algaba, A., Fernández-Sánchez, F., Merino, M., and Rodríguez-Luis, A. J. (2013). Chen's attractor exists if Lorenz repulsor exists: The Chen system is a special case of the Lorenz system, *Chaos* **23**, pp. 033108-1–6.

Arnéodo, A., Coullet, P., and Tresser, C. (1981). A possible new mechanism for the onset of turbulence, *Phys. Lett. A* **81**, pp. 197–201.

Chen, G. and Ueta, T. (1999). Yet another chaotic attractor, *Int. J. Bifurcat. Chaos* **9**, pp. 1465–1466.

Grebogi, C., Ott, E., Pelikan, S., and Yorke, J. A. (1985). Exterior dimension of fat fractals, *Phys. Lett. A* **110**, pp. 1–4.

Elwakil, A. S. and Kennedy, M. P. (2001). Construction of classes of circuit-independent chaotic oscillators using passive-only nonlinear devices, *IEEE T. Circuits Syst.-I* **CS48**, pp. 289–307.

Hoover, W. G. (1985). Canonical dynamics: Equilibrium phase-space distributions, *Phys. Rev. A* **31**, pp. 1695–1697.

Jafari, S. and Sprott, J. C. (2013). Simple chaotic flows with a line equilibrium, *Chaos Soliton Fract.* **57**, pp. 79–84.

Jafari, S., Sprott, J. C., and Nazarimehr, F. (2015). Recent new examples of hidden attractors, *Eur. Phys. J. Special Topics* **224**, pp. 1469–1476.

Jafari, S., Sprott, J. C., Pham, V.-T., Volos, C., and Li, C. (2016). Simple chaotic flows with surfaces of equilibria, *Nonlin. Dyn.* **86**, pp. 1349–1358.

Kaplan, J. and Yorke, J. (1979). Chaotic behavior of multidimensional difference equations. In *Functional Differential Equations and Approximation of Fixed Points, Lecture Notes in Mathematics* **730** (ed. H.-O. Peitgen and H.-O. Walther), pp. 228–237 (Springer, Berlin).

Leipnik, R. B. and Newton, T. A. (1981). Double strange attractors in rigid body motion with linear feedback control, *Phys. Lett. A* **86**, pp. 63–67.

Leonov, G. A. and Kuznetsov, N. V. (2013). Hidden attractors in dynamical systems. From hidden oscillations in Hilbert–Kolomogorov, Aizerman, and Kalman problems to hidden chaotic attractor in Chua cirucits, *Int. J. Bifurcat. Chaos* **23**, pp. 1330002-1–69.

Li, C. and Sprott, J. C. (2013). Multistability in a butterfly flow, *Int. J. Bifurcat. Chaos* **23**, pp. 1350199-1-10.

Li, C. and Sprott, J. C. (2014). Chaotic flows with a single nonquadratic term, *Phys. Lett. A* **378**, pp. 178–183.

Linz, S. J. and Sprott, J. C. (1999). Elementary chaotic flow, *Phys. Lett. A* **259**, pp. 240–245.

Liouville, J. (1838). Note on the theory of the variation of arbitrary constants (in French), *J. Math. Pure Appl.* **3**, pp. 342–349.

Lorenz, E. N. (1963). Deterministic nonperiodic flow, *J. Atmos. Sci.* **20**, pp. 130–141.

Lorenz, E. N. (1984). Irregularity: A fundamental property of the atmosphere, *Tellus* **36A**, pp. 98–110.

Malasoma, J.-M. (2000). What is the simplest dissipative chaotic jerk equation which is parity invariant?, *Phys. Lett. A* **264**, pp. 383–389.

Matsumoto, T., Chua, L. O., and Komuro, M. (1985). The double scroll, *IEEE Trans. Circ. Syst.* **33**, pp. 797–818.

Mehrabbeik, M., Jafari, S., and Sprott, J. C. (2022). A simple three-dimensional quadratic flow with an attracting torus, *Phys. Lett. A* **451**, pp. 128427-1-8.

Moor, J. (2006). The Dartmouth College Artificial Intelligence Conference: The next fifty years, *AI Mag.* **27**(4), pp. 87–91.

Moore, D. W. and Spiegel, E. A. (1966). A thermally excited non-linear oscillator, *Astrophys. J.* **143**, pp. 871–887.

Munmuangsaen, B. and Srisuchinwong, B. (2009). A new five-term simple chaotic attractor, *Phys. Lett. A* **373**, pp. 4038–4043.

Munmuangsaen, B., Sprott, J. C., Thio, W. J., Buscarino, A., and Fortuna, L. (2015). A simple chaotic flow with a continuously adjustable attractor dimension, *Int. J. Bifurcat. Chaos* **25**, pp. 1530036-1-12.

Nosé, S. (1984). A molecular dynamics method for simulations in the canonical ensemble, *Mol. Phys.* **52**, pp. 255–268.

Press, W. H., Teukolsky, S. A., Vetterling, W. T., and Flannery, B. P. (2007). *Numerical Recipes: The Art of Scientific Computing* (3rd edn) (Cambridge University Press, Cambridge).

Rabinovich, M. I. and Fabrikant, A. L. (1979). Stochastic self-modulation of waves in nonequilibrium media, *Sov. Phys. JETP* **50**, pp. 311–317.

Rikitake, T. (1958). Oscillations of a system of disk dynamos, *Math. Proc. Camb. Phil. Soc.* **54**, pp. 89–105.

Rössler, O. E. (1976). An equation for continuous chaos, *Phys. Lett. A* **71**, pp. 155–157.

Rössler, O. E. (1979). Continuous chaos – four protype equations, *Ann. New York Acad. Sci.* **316**, pp. 376–392.

Sprott, J. C. (1993a). Automatic generation of strange attractors, *Computers & Graphics* **17**, pp. 223–232.

Sprott, J. C. (1993b). *Strange Attractors: Creating Patterns in Chaos* (M & T Books, New York).

Sprott, J. C. (1994). Some simple chaotic flows, *Phys. Rev. E* **50**, pp. R647–650.

Sprott, J. C. (1997). Simplest dissipative chaotic flow, *Phys. Lett. A* **228**, pp. 271–274.

Sprott, J. C. (2003). *Chaos and Time-Series Analysis* (Oxford University Press, Oxford).

Sprott, J. C. (2009). Simplifications of the Lorenz attractor, *Nonlin. Dynam. Psychol.* **13**, pp. 271–278.

Sprott, J. C. (2010). *Elegant Chaos: Algebraically Simple Chaotic Flows* (World Scientific, Singapore).

Sprott, J. C. (2011). A proposed standard for the publication of new chaotic systems, *Int. J. Bifurcat. Chaos* **21**, pp. 2391–2394.

Sprott, J. C. (2014a). Simplest chaotic flows with involutional symmetries, *Int. J. Bifurcat. Chaos* **14**, pp. 1450009-1-9.

Sprott, J. C. (2014b). A dynamical system with a strange attractor and invariant tori, *Phys. Lett. A* **378**, pp. 1361–1363.

Sprott, J. C. (2015). Symmetric time-reversible flows with a strange attractor, *Int. J. Bifurcat. Chaos* **25**, pp. 1550078-1-7.

Sprott, J. C. (2018). Ergodicity of one-dimensional oscillators with a signum thermostat, *Comp. Meth. Sci. Tech.* **24**, pp. 169–176.

Sprott, J. C. (2019). *Elegant Fractals: Automated Generation of Computer Art* (World Scientific, Singapore).

Sprott, J. C. (2022). Quantifying the robustness of a chaotic system, *Chaos* **32**, pp. 0331244-1-6.

Sprott, J. C. (2022b). Artificial intelligence study of the system JCS-08-13-2022, *Int. J. Bifurcat. Chaos* **32**, pp. 2230028-1-4.

Sprott, J. C. and Thio, W. J. (2022). *Elegant Circuits: Simple Chaotic Oscillators* (World Scientific, Singapore).

Sprott, J. C. and Xiong, A. (2015). Classifying and quantifying basins of attraction, *Chaos* **25**, pp. 083101-1-7.

Sprott, J. C., Wang, X., and Chen, G. (2013). Coexistence of point, periodic and strange attractors, *Int. J. Bifurcat. Chaos* **23**, pp. 1350093-1-5.

Sprott, J. C., Jafari, S., Khalaf, A. J. M., and Kapitaniak, T. (2017). Megastability: Coexistence of a countable infinity of nested attractors in a periodically-forced oscillator with spatially-periodic damping, *Eur. Phys. J. Special Topics* **226**, pp. 1979–1985.

Sprott, J. C., Hoover, W. G., and Hoover, C. G. (2023). *Elegant Simulations: From Simple Oscillators to Many-Body Systems* (World Scientific, Singapore).

Thomas, R. (1999). Deterministic chaos seen in terms of feeback circuits: Analysis, synthesis, 'labyrinth chaos', *Int. J. Bifurcat. Chaos* **4**, pp. 1889–1905.

Ueda, Y. (1979). Randomly transitional phenomena in the system governed by Duffing's equation, *J. Stat. Phys.* **20**, pp. 181–196.

Umberger, D. K. and Farmer, J. D. (1985). Fat fractals on the energy surface, *Phys. Rev. Lett.* **55**, pp. 661–664.

van der Schrier, G. and Maas, L. R. M. (2000). The diffusionless Lorenz equations; Shil'nikov bifurcations and reduction to an explicit map, *Phys. Nonlinear Phenom.* **141**, pp. 19–36.

Wang, X. and Chen, G. (2012). A chaotic system with only one stable equilibrium, *Commun. Nonlin. Sci. Numer. Simul.* **17**, pp. 1264–1272.

Wang, X., Kuznetsov, N. V., and Chen, G., editors (2021). *Chaotic Systems with Multistability and Hidden Attractors* (Springer, Cham, Switzerland).

Wei, Z. (2011). Dynamical behaviors of a chaotic system with no equilibria, *Phys. Lett. A* **376**, pp. 102–108.

Wolf, A., Swift, J. B., Swinney, H. L., and Vastano, J. A. (1985). Determining Lyapunov exponents from a time series, *Phys. Nonlinear Phenom.* **16**, pp. 285–317.

Zhang, F. and Heidel, J. (1997). Non-chaotic behaviour in three-dimensional quadratic systems, *Nonlinearity* **10**, pp. 1289–1303.

Zhou, W., Xu, Y., Lu, H., and Pan, L. (2008). On dynamics analysis of a new chaotic attractor, *Phys. Lett. A* **372**, pp. 5773–5777.

Index

About the Author

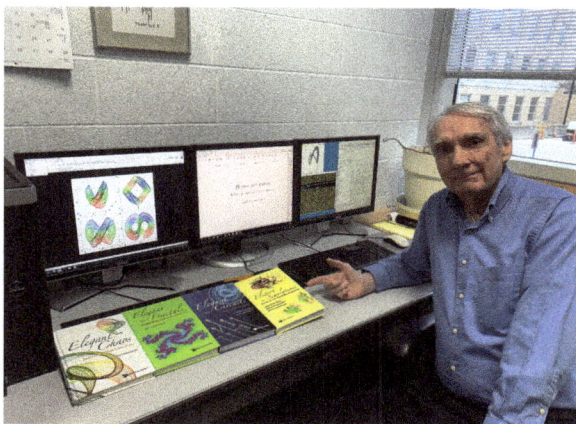

JULIEN CLINTON SPROTT is Emeritus Professor of Physics at the University of Wisconsin–Madison. After graduating from MIT in 1964 and receiving a PhD in Physics from the University of Wisconsin in 1969, he spent two years at Oak Ridge National Laboratory before returning to Wisconsin. For twenty five years he worked in plasma physics and controlled nuclear fusion but became interested in chaos in 1988. He is author of over five hundred technical papers and fourteen books including *Chaos and Time-Series Analysis* [Sprott (2003)], *Elegant Chaos* [Sprott (2010)], *Elegant Fractals* [Sprott (2019)], *Elegant Circuits* [Sprott and Thio (2022)], and *Elegant Simulations* [Sprott *et al.* (2023)]. He has produced forty hour-long videos of his popular public presentations of *The Wonders of Physics*. He retired in 2008 and resides in Madison, Wisconsin where he continues research and writing. His award-winning web site is at `https://sprott.physics.wisc.edu/`.

www.ingramcontent.com/pod-product-compliance
Lightning Source LLC
Chambersburg PA
CBHW050538190326
41458CB00007B/1833